Cloud Dynamics

Advances in Earth and Planetary Sciences

Cloud Dynamics

Proceedings of a Symposium
held at the Third General Assembly of IAMAP,
Hamburg, West Germany, 17-28 August, 1981

Edited by

E. M. Agee
Department of Geosciences, Purdue University, U.S.A.

and

T. Asai
Ocean Research Institute, University of Tokyo, Japan

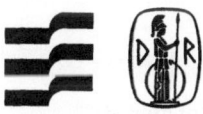

Terra Scientific Publishing Company/Tokyo
D. Reidel Publishing Company/Dordrecht, Boston, London

Library of Congress Cataloging in Publication Data
Main entry under title:

Cloud dynamics

(Advances in earth and planetary sciences)
Includes index.
1.   Convection (Meteorology)–Congresses.   2.   Cloud physics–Congresses.
I.   Agee, E. M. (Ernest M.), 1942–   .   II.   Asai, T. (Tomio)   III.   International
Association of Meteorology and Atmospheric Physics. General Assembly (3rd : 1981 :
Hamburg, Germany)   IV.   Series.
QC921.6.C65C56   1982   551.57'6   82–12320
ISBN-13:978-94-009-7892-8   e-ISBN-13:978-94-009-7890-4
DOI: 10.1007/978-94-009-7890-4

Published by D. Reidel Publishing Company,
P.O. Box 17, 3300 AA Dordrecht, Holland,
in co-publication with Terra Scientific Publishing Company, Tokyo, Japan

Sold and distributed in the U.S.A. and Canada
by Kluwer Boston Inc.,
190 Old Derby Street, Hingham, MA 02043, U.S.A.,
in Japan by Terra Scientific Publishing Company,
307 Shibuyadai-haim, 4-17 Sakuragaoka-cho, Shibuya-ku, Tokyo 150, Japan

In all other countries, sold and distributed
by Kluwer Academic Publishers Group,
P.O. Box 322, 3300 AH Dordrecht, Holland

D. Reidel Publishing Company is a member of the Kluwer Group

TABLE OF CONTENTS

PREFACE

The contents of this volume are based largely on a symposium on cloud dynamics held in Hamburg, Germany, as a part of the Third General Assembly of the International Association of Meteorology and Atmospheric Physics (IAMAP), 17-28 August 1981. Plans for this symposium were initiated by the Cloud Dynamics Group, appointed through the International Commission for Dynamic Meteorology (ICDM), IAMAP. Members of this group are Bruce Morton - Chairman, Australia; Tomio Asai - Secretary, Japan; E. M. Agee, USA; V. Andreyev, Bulgaria; K. Fradrich, FRG; M. W. Moncrieff, UK; R. S. Pastushkov, USSR; J. P. Chao, PR China; Bh. V. Ramanamurti, India; W. Roach, UK; J. Simpson, USA; J. T. Steiner, New Zealand; M. Yanai, USA. E. M. Agee served as convener of the symposium in Hamburg.

Dynamic meteorology has grown rapidly and covered a great variety of fields since the foundation of ICDM, making it increasingly difficult for the Commission to manage all activities. In order to improve this situation, some ad-hoc Working Groups within ICDM were organized at the 1977 IAMAP meeting in Seattle. Each of these working groups, such as the one in Cloud Dynamics, deals with a particular sub-field of dynamic meteorology.

Cloud dynamics has become of increasing interest to meteorologists since the 1960s, and has established itself as an important subject area in modern meteorology. Recent technological developments have provided a large amount of observational data on various aspects of clouds, which have aided the numerical modeling of complex cloud systems. A gap between observational and theoretical studies is being filled and both are being undertaken cooperatively and complementary to the advancement of each.

The outline of this volume corresponds to the four sessions of the symposium: 1) Shallow Convective Systems - Observations, 2) Shallow Convective Systems - Models, 3) Deep Convective Systems - Observations, and 4) Deep Convective Systems - Models. Twenty three conference papers and two invited papers, are included, which represent contributions from eight different countries. Introductions to the respective sessions on shallow and deep convection have been prepared by the convener.

Main topics in the present volume are confined primarily to convective clouds. The other important subject fields, such as stratified clouds and interaction of cloud ensembles with larger-scale disturbances, remain to be considered in more detail in the future.

E. M. Agee and T. Asai (eds.), Cloud Dynamics, vii.
Copyright © 1982 by D. Reidel Publishing Company.

SHALLOW CONVECTIVE SYSTEMS

OBSERVATIONS AND MODELS

OBSERVATIONS AND MODEL

AN INTRODUCTION TO SHALLOW CONVECTIVE SYSTEMS

Ernest M. Agee

Department of Geosciences
Purdue University, West Lafayette, Indiana 47907 USA

ABSTRACT

A brief review of classical Benard-Rayleigh convective theory and observations is made, with particular emphasis on its relevance to shallow convective cloud systems in the Earth's atmosphere. The importance of satellite observations in helping establish this relevance is also considered, especially in the identification and study of mesoscale cellular convection (MCC), as well as observations of other convective phenomena occurring in the planetary boundary layer.

1. INTRODUCTION

A cloud can be defined as a visible ensemble or aggregrate of hydrometeors, consisting of minute water droplets and/or ice crystals. Three conditions often viewed as necessary for cloud formation are 1) a proper distribution of hygroscopic muclei, 2) a sufficient amount of water vapor, and 3) proper dynamics or convective motion that raises parcels or layers to their lifting condensation level. Much of the study of clouds has focused on their microphysical processes, evidenced by the voluminous material on cloud physics, with lesser attention paid to the dynamics of cloud formation (except possibly for the more notorious clouds like the cumulonimbus). Clouds are generally classified according to their formation and/or appearance. The ten basic cloud *genera* are cirrus, cirrocumulus, cirrostratus, altocumulus, altostratus, nimbostratus, stratocumulus, stratus, cumulus and cumulonimbus. In addition there are 14 other species based on pecularities in shape or structure, as well as a number of other varieties, supplementary forms and accessory clouds based on peculiar arrangement and form. This volume addresses

3

E. M. Agee and T. Asai (eds.), Cloud Dynamics, 3–30.

largely, the dynamics of cloud formation and their behavior, with particular emphasis on a) the motion field responsible for the formation and structure of cumulus and stratocumulus in the planetary boundary layer, and b) the development of the cumulonimbus with its frequent manifestation of high winds, heavy rain, hail and sometimes tornadoes.

In general one might say that some form of convection is responsible for all cloud formation. In fluid mechanics, *convection* is defined as mass motion within a fluid resulting in the transport of fluid properties such as heat and momentum. Meteorologists, however, usually think of convection as vertical air currents and transport, with the term *advection* introduced to indicate the horizontal transport. By including both free and forced convection, all dynamical mechanisms or processes leading to cloud formation would be considered. *Free* convective motion, sometimes referred to as gravitational convection, is caused by thermally-induced density differences in a fluid that produce a buoyant force resulting in hydrostatic instability. Thermal differences in the atmosphere may apply to either ambient dry-bulb temperatures or virtual temperatures (when moisture buoyancy is significant).

Free or buoyantly driven convection in the atmosphere arises from sufficiently heating the bottom and/or cooling the top of the fluid layer. This can occur in a rather uniform manner or non-uniformly due to local heating or cooling, as discussed below. Two common physical processes in the atmosphere that yield quasi-uniform heating from below and cooling from above are: 1) *sensible heating* at the bottom of the atmosphere due to cold air advection over a warmer water surface, as shown in Figure 1, and 2) *infrared radiative cooling* at the top of a cloudy layer in the presence of overlying drier air. In principle, free convection in the atmosphere arising from these thermodynamic processes approaches the case of ideal cellular convection (as seen best in laboratory demonstrations), which is illustrated in the satellite cloud photography of mesoscale cellular convection seen in Figure 1. This cloud pattern is contained within the planetary boundary layer ($\sim$ 1.5 km deep) and consists of cumulus and stratocumulus genera.

As previously implied, convective overturning is not restricted to uniform heating (or cooling) of atmospheric layers, but can occur due to non-uniform local heating, as might be associated with induced thermals over parking lots or plowed fields, along highway systems or in city heat island circulations, as shown in Figure 2. Implicit in examples such as these are free convective phenomena that occur in response to non-uniform heating at the earth's surface (and subsequent sensible heating of the lower atmosphere), rather than uniform heating from below. A

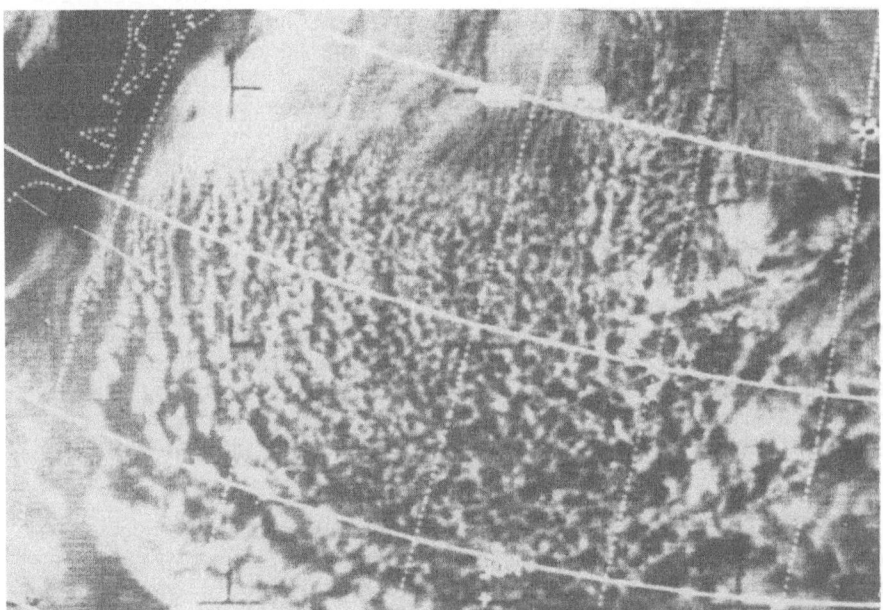

Figure 1. ESSA 7 imagery of mesoscale cellular convection off the East Coast of the United States at 1842 GMT, 16 January 1969. The convective cloud pattern formed in response to cold air advected from the North American continent out over the warmer Gulf Stream.

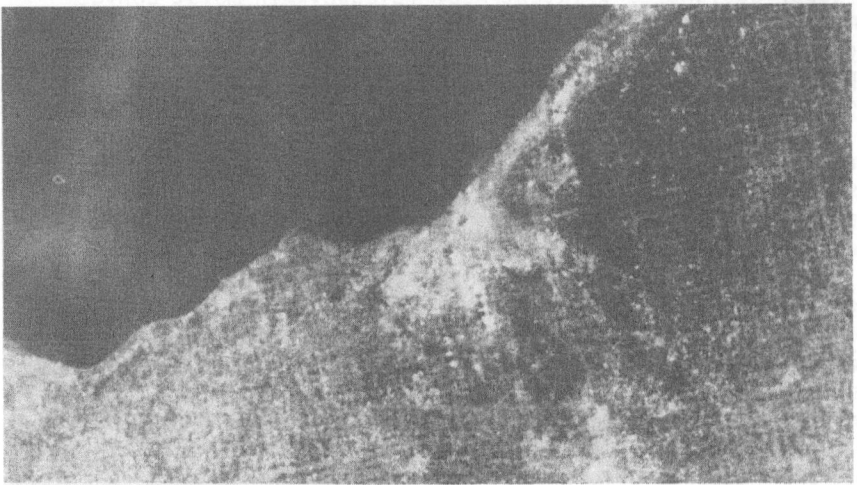

Figure 2. ERTS satellite imagery of cumulus clouds over Cleveland, Ohio (USA) at 1408 GMT on 4 September 1973. A city heat island circulation may have helped induce this convective cloud.

convection cell (or cells) formed in this manner is characterized
by upward motion at the center, away from the heat source and toward the heat sink, with downward motion in the cell's outer regions. In contrast, arrays of convection cells due to uniform
heating from below and/or cooling from above can form with either
ascending or descending motion at the center of the cell with
compensating motion in the opposite direction in the peripheral
regions.

According to general concepts already mentioned, other convective motions occur in the atmosphere that do not fit the definition of free convection. Such motions will be labeled *forced*
convection since they arise from external or mechanical lifting
forces other than a thermally driven buoyant force. Common atmospheric examples are orographic lifting, frontal lifting, gravity
wave displacements and vertical motion induced by horizontal convergence of the wind field. All of these mechanisms can produce
an array of cloud types and species.

Not only can convection be classified as either free or
forced, but it can also be labeled as dry or moist. *Dry* convection, generally, represents convective overturning in the absence
of clouds or the presence of non-precipitating clouds. *Moist*
convection, therefore, always implies the manifestation of precipitating cloud systems. These definitions of dry and moist
convection are given in terms of *observed* features of convective
systems. It is optional in dry convection models as to whether
or not the momentum and energy equations contain the effects of
water vapor buoyancy and latent heat release. In moist convection models the latent heat release must always be included.
Accordingly, possible combinations of convective systems in the
atmosphere are free-dry, free-moist, forced-dry and forced-moist.
Figure 1, for example, represents a free-dry convective system
since the mesoscale cellular convection is buoyantly-driven and
non-precipitating. An example of free-dry convection in clear
air is given in Figure 3, similar to that studied by Hardy and
Ottersten [1]. While convective phenomena observed in the atmosphere can be classified as free or forced and as dry or moist,
it is important to realize that convective systems may pass
through various stages in this simple classification scheme. For
instance, a cold front motion may force the formation of a convective cloud line that grows to the precipitation stage, resulting
in the release of large amounts of latent heat during the condensation process. This in turn allows the cloud to develop to even
greater heights as a free convective system and to persist in time
without any required external forcing mechanism. In this example
the convective stages are forced-dry, forced-moist, and free-moist.
Such a system propagating with the front or with a gravity wave
may actually exist as a forced-free-moist convective system, implying that both external forcing and buoyancy are driving mechanisms.

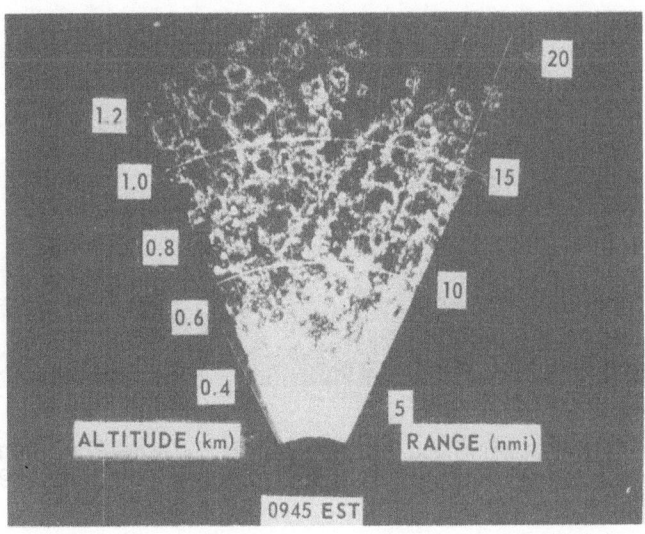

Figure 3. An example of convective cells in clear air as seen by radar at Wallops Island, Virginia (USA). PPI photograph is taken at S-band on 5 June 1968 at 2⁰ elevation angle. Representative cell diameter and convective depth are, respectively, 1.4 km and 1.2 km. Photo provided by Konrad [2].

Finally, the terms shallow and deep convection are now introduced, as these two classifications for convection are used to divide this volume (as was done in the cloud dynamics symposium). Clouds that occur within the Earth's atmosphere are principally confined to the troposphere (the first 10 to 15 kilometers). Thermals and cloud formations restricted to the first 1 to 2 kilometers (essentially the planetary boundary layer) represent *shallow* convection, since they are confined to a vertical depth one order of magnitude less than the depth of the troposphere. A good example of shallow convection is the mesoscale convective cloud pattern shown in Figure 1, as well as the thermal pattern in clear air shown in Figure 3. Another well-known example of shallow convection is the pattern of cloud streets shown in Figure 4. As already indicated by definition, convection does not necessarily require the formation of a cloud, but is precursive to such. *Deep* convection can now be defined as that which extends through a large depth of the troposphere. This type of convection will virtually always produce clouds of large vertical extent that precipitate. The best example of a deep convective system is the thunderstorm, which will receive the principle focus of attention in Part II of this volume.

Figure 4.   Convective cloud streets in the planetary boundary
layer.   Photo taken over the Florida penninsula in May 1970
by E. Agee.

2.   EARLY STUDIES

The historical aspects of early recognition and study of
convective phenomena are almost as fascinating as the scientific
results from the studies themselves.   A good historical summary
has been given by Chandrasekhar [3], who recognizes Count Rumford
[4] as the discoverer of the phenomenon of thermal convection in
fluids as it is understood today.   William Prout [5] was the
first to introduce the term convection into the field of science,
as expressed below in his own words: "There is at present no
single term in our language employed to denote this mode of propagation of heat; but we venture to propose for that purpose the
term *convection* (*convectio*, a carrying or converging), which not
only expresses the leading facts, but also accords very well with
the two other terms (conduction and radiation)."   A brief account
of early laboratory study has been given by Sir David Brunt [6]
in the Compendium of Meteorology.

The single most improtant study that has promoted the investigation of convection and the development of a body of knowledge
must be attributed to the dissertation research by *Henri Bénard*
at the University of Paris at the end of the nineteenth century.
His work included the first quantitative experiments on the onset
of thermal instability and the recognition of the role of viscosity in the phenomenon.   Benard's [7] experiment consisted of a
thin layer of spermaceti (whale oil) about 1 mm thick that was
heated from below by a horizontal metallic plate maintained at
uniform temperature.   The initially motionless fluid layer upon
achieving instability rapidly resolved itself into a number of

convective cells, with ascending motion at the cell centers and descending motion at the common boundary between each cell and its neighbors. Brunt [8] was the first to apply the name "Bénard cell" to imply the typical convection cell which occurs in unstable fluid layers initially at rest.

2.1 Rayleigh's model

The experimental results by Bénard stimulated Lord Rayleigh [9] to explain the convection theoretically. Even though the slow transition into a steady pattern of regular hexagons was of interest, Lord Rayleigh's model was formulated primarily to examine the conditions leading to the onset of convective motions. The physical restraints and assumptions incorporated in the model were: a non-rotating gravitating fluid with constant viscosity ν and conductivity κ initially at rest between two free surfaces; and the *assumptions of Boussinesq* [10] which are: (1) The fluctuations in density which appear with the advent of motion result principally from thermal (as opposed to pressure) effects, and (2) In the equations for the rate of change of momentum and mass, density variations may be neglected except when coupled to the gravitational acceleration in the buoyancy force. Applying the concept of small disturbances (frequently referred to as first-order perturbation theory) and the Boussinesq approximations yields the following form of the momentum, continuity, state and energy equations in the Rayleigh model:

$$\frac{\partial u_i}{\partial t} = \frac{-1}{\rho_m} \frac{\partial p}{\partial x_i} + \nu \nabla^2 u_i - \frac{g\rho}{\rho_m} \delta_{i3} \tag{1}$$

$$\frac{\partial u_i}{\partial x_i} = 0 \tag{2}$$

$$\rho = -\rho_m \alpha(T) \tag{3}$$

$$\frac{\partial T}{\partial t} + u_3 \frac{d\overline{T}}{dx_3} = \kappa \nabla^2 T \tag{4}$$

where $\overline{T}$ is the undisturbed temperature, α is the coefficient of thermal expansion and ρ_m is the mean density of the fluid layer. The above *linearized* form of the Boussinesq equations was solved using the hypothesis that the velocities, and the departures of pressure and temperature from their equilibrium values were proportional to $e^{ilx} e^{imy} e^{\sigma t}$, where l and m are components of the horizontal wavenumber and σ represents a complex stability function.

Rayleigh's model yielded two important results. *Firstly*, a criterion was found which represents a *sufficient* condition for the onset of free convective motion:

$$\frac{\rho_T - \rho_B}{\rho_m} > \frac{27\pi^4}{4gd^3} \kappa\nu \tag{5}$$

where d is the convective depth and ρ_T and ρ_B are the densities of the top and bottom of the layer, respectively. Equation (5) implies that a given fluid of fixed depth must exceed a critical value of density difference between the top and bottom of the layer in order for motion to occur. Also, it follows that the shallower the layer the greater the density difference required, or convection occurs "easier" in deeper layers. In addition, for a given density difference there is a lower limit to the depth in which motion can occur. Substituting (3) into (5) allows one to express the instability condition alternatively as

$$\frac{g\alpha\beta d^4}{\kappa\nu} > \frac{27\pi^4}{4}, \tag{6}$$

where β is the temperature lapse rate (-T/d) through the fluid layer. The left-hand side of (6) is called the 'characteristic number' or *Rayleigh number* (R), a term adopted by Chandrasekhar [11]. The right-hand side of (6) is the *critical value* of the Rayleigh number (R_c) required for the onset of convection. *Secondly*, the model shows that the most unstable disturbance has a unique wavenumber (where $\lambda \equiv \sqrt{1^2 + m^2}$), corresponding to R_c. For the free-free boundary conditions it was found that $R_c = 27\pi^4/4$ and $\lambda_c = \pi/\sqrt{2}$, as shown in the stability diagram presented in Figure 5. The four cases of Rayleigh's convection model (with free-free boundary conditions) have been conveniently summarized by Platzman [12].

Since Lord Rayleigh solved a linearized set of equations, the model did not yield a prescribed geometry of the convective pattern. In fact, Lord Rayleigh assumed a horizontal plan form that corresponded to an array of square convection cells. The degree of cell flatness can be defined as the *aspect ratio* (A), which is equal to the cell diameter (L) divided by the convective depth (d). In Rayleigh's square cells the aspect ratio for unit depth is $2\sqrt{2}$ or 2.828. Since Bénard's experiments on spermaceti gave A values of 3.27 to 3.34, the proper degree of cell flatness was also not obtained in Rayleigh's results.

In addition to *free-free* boundary conditions, *rigid-free* and *rigid-rigid* represent the other possibilities for the bounding surfaces. Jeffreys [13] and [14] reexamined Rayleigh's problem for the cases of *rigid-free* and *rigid-rigid* boundary conditions

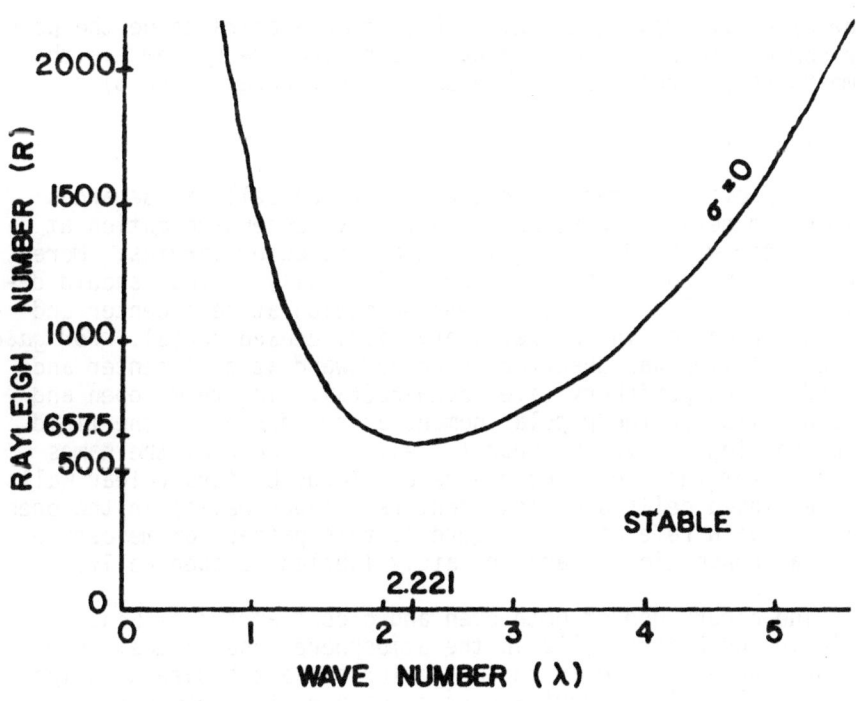

Figure 5. Stability diagram of convective disturbances in a motionless fluid as a function of Rayleigh number R and wavenumber
λ. The marginal stability curve ($\sigma=0$) shown corresponds to Lord
Rayleigh's [9] model results for *free-free* boundaries. Minimum
R_c = 657.511 and λ_c = 2.2214.

and found successively larger values for R_c and λ_c, which were
later confirmed by Pellew and Southwell [15]. Chandrasekhar [3]
has summarized these parameters characterizing the marginal state
of convection for the three cases of bounding surfaces.

    Further comparisons of Lord Rayleigh's findings and Bénard's
experiment resulted in an even more serious discrepancy, namely,
the critical value of the Rayleigh number required to produce
convection. The value of the Rayleigh number in Bénard's experiments was always less than the predicted minimum value required
at the onset of motion, which led Bénard to later suspect that
the role played by surface tension forces might be responsible.
In fact, Low and Brunt [16] were the first to notice that the
temperature gradients in Bénard's experiments were at least ten-
fold less than that required by Rayleigh's theory. Bénard [17]
and [18] recognized the discrepancy and later estimated the ratio
of the different gradients to be 10^{-4} to 10^{-5}. Perhaps

appropriately, *Bénard-Rayleigh convection* appears to be the popular nomenclature for describing free convective systems in the
atmosphere (as depicted, for example, in Figures 1 and 3).

2.2 Circulation direction

As previously mentioned the convection cells in Bénard's
original experiment were observed to have ascending motion at
cell center with descending motion at the outer margins. More
specifically, convection in *liquids* initially at rest should develop hexagonal cells with ascending motion at cell center and
descending motion in the periphery (i.e. *closed cells*). For *gases*,
the circulation was observed to be downward at cell center and
upward in the periphery (i.e. *open cells*). The terms open and
closed cells are the popular nomenclature adopted in the field
of meteorology, given by Hubert [19]. The logic of the names
follows from the usual occurrence of cloudy centers (clear walls)
in the closed cells and clear centers (cloudy walls) in the open
cells. Again referring to Figure 1, this pattern of mesoscale
cellular convection is appropriately labeled as open cells.

The occurrence of both open and closed arrays of Bénard-
Rayleigh convection cells in the atmosphere poses a most interesting problem, since in the molecular sense the gaseous atmosphere should only support arrays of open cells. However, as
illustrated in Figure 6, convection cells also occur in the atmosphere as closed circulation systems and even sometimes as coexisting patterns of open and closed cells. The closed cells seen
in Figure 6 have a remarkably similar appearance to the Bénard-
Rayleigh cells created in the laboratory by Chandra [20]. The
validity of such laboratory demonstrations as *simulations* of convective patterns seen in the atmosphere is not a trite question,
and one that deserves more attention than will be given here. As
previously pointed out, the solution of the linearized Boussinesq
equations does not yield the geometry or mode of convection. Rayleigh and successors assumed a square (or rectangular) plan-form
of cells, even though Bénard observed hexagons. Christopherson's
[21] mathematical shape function for a hexagonal array allowed
Pellew and Southwell [15] to re-examine Rayleigh's model for the
case of hexagonal cells, as well as for a wider range of boundary
conditions. The classical linear solution of the relative vertical
velocity field in the hexagonal cell, as obtained by Pellew and
Southwell, is presented in Figure 7. The nonlinear version of
this pattern has been given by Krishnamurti [22]. The zero-value
vertical velocity isopleth is circular shaped, surrounding a core
of ascending (or descending) motion with compensating reversed
flow in the periphery. As shown in Figure 7, extrema are at the
cell center and the six corners of the hexagon. Also, the magnitude of the vertical motion at cell center is twice that at the vertex. Similarity between this pattern and atmospheric convection

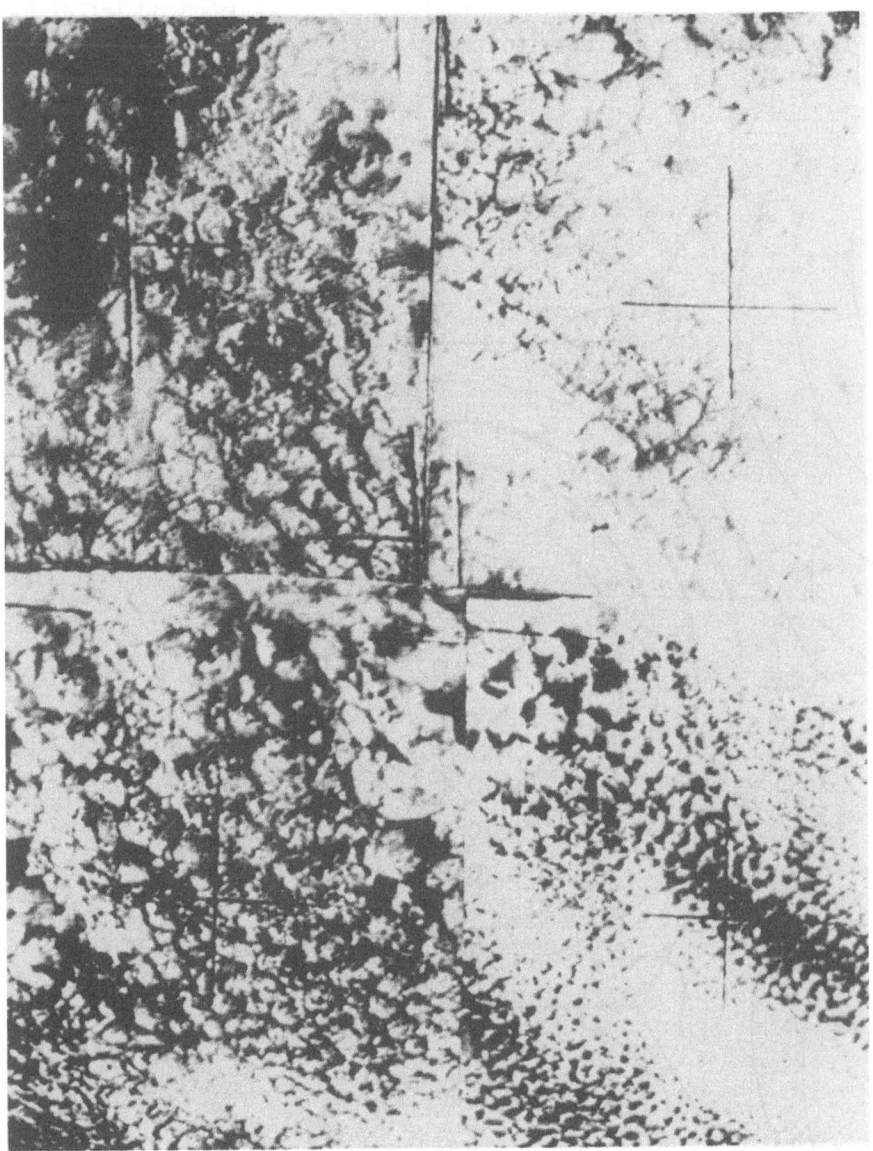

Figure 6. Satellite photograph of coexisting closed and open convection cells west of Baja California (near 23°N, 122°W) taken by COSMOS-144 at 0800 GMT 20 April 1967. (Photo provided through the coutesy of the Dr. V. A. Bugaev, former director, Hydrometeorological Research Center of the USSR). Representative cell diameter and depth are, respectively, 20 km and 1.5 km.

can be noted in Figure 8. These open hexagonal cells have cloud-free centers and enhanced cumulus congestus at their vertices, a truly remarkable identity to the classical solution.

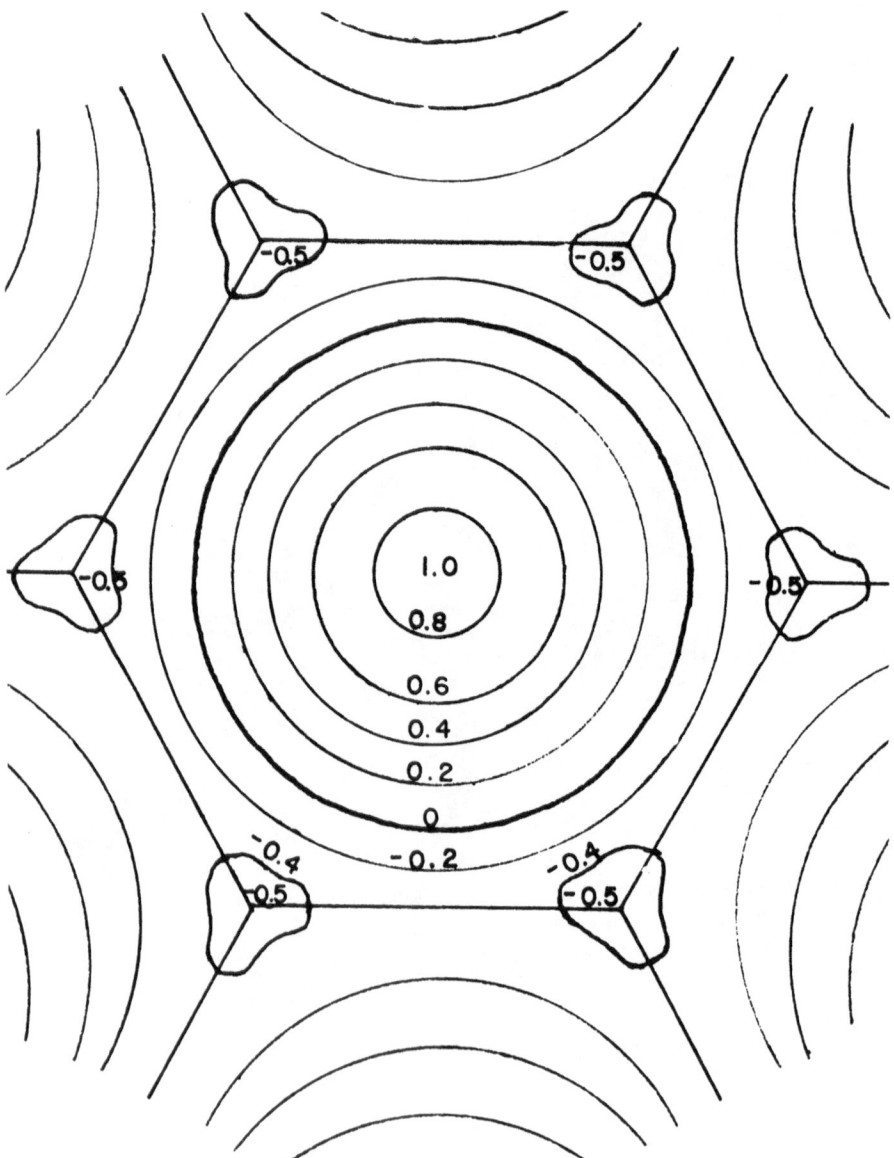

Figure 7. The classical linear solution of the relative vertical velocity field in a hexagonal convection cell, as obtained by Pellew and Southwell [15].

Figure 8. Hexagonal "open" cells north of Cuba, Photographed from Gemini 5 at 1827 GMT, 23 August 1965. Convective enhancement at the vertices of the hexagons can also be noted.

3. BOUSSINESQ EQUATIONS

In a manner similar to Speigel and Veronis [23], the total state density $\rho^*(x,y,z,t)$ can be expressed as

$$\rho^* = \rho_m + \rho_0 (z) + \rho'(x,y,z,t) \tag{7}$$

where $\rho_m \equiv$ constant space average, $\rho_0 \equiv$ density variation in the absence of motion and ρ' is the density fluctuation resulting from motion. Using (7) and the second Boussinesq assumption, the Navier-Stokes equations for an initially motionless fluid of constant molecular kinematic viscosity can be written in approximate perturbation form as

$$\frac{\partial u_i^{\prime}}{\partial t} + \vec{u}^{\prime} \cdot \nabla u_i^{\prime} = - \frac{1}{\rho_m} \frac{\partial p^{\prime}}{\partial x_i} - \frac{g\rho^{\prime}}{\rho_m} \delta_{i3} + \nu\nabla^2 u_i^{\prime} \quad . \tag{8}$$

Applying the premise of first-order perturbation theory to this nonlinear equation reduces (8) to Equation (1) in Rayleigh's model. Unsing the first Boussinesq assumption, one can write

$$\frac{\rho^{\prime}}{\rho_m} = \frac{T^{\prime}}{T_m} \tag{9}$$

Substituting (9) into (8) yields the conventional form of the Boussinesq momentum equation,

$$\frac{\partial u_i^{\prime}}{\partial t} + \vec{u}^{\prime} \cdot \nabla u_i^{\prime} = - \frac{1}{\rho_m} \frac{\partial p^{\prime}}{\partial x_i} + \frac{gT^{\prime}}{T_m} \delta_{i3} + \nu\nabla^2 u_i^{\prime} \quad . \tag{10}$$

The thermodynamic energy equation for the total state of a fluid can be written as

$$\frac{dh}{dt} = c_v \frac{dT^*}{dt} + p^* \frac{d\alpha^*}{dt} \tag{11}$$

or alternately as

$$\rho^* c_v [\frac{\partial T^*}{\partial t} + \vec{v}^* \cdot \nabla T^*] + p^* \nabla \cdot \vec{v}^* = \kappa\nabla^2 T^* + Q^* + \Phi^*, \tag{12}$$

where κ is thermal conductivity, Q^* represents radiational heating (or cooling) rates and Φ^* represents the heating rate due to viscous dissipation of motion. Again in a fashion similar to Spiegel and Veronis [23] for a motionless basic state, Equation (12) in perturbation form becomes

$$\tag{13}$$

$$\rho_m c_v [\frac{\partial T^{\prime}}{\partial t} + \vec{v} \cdot \nabla(T + T^{\prime})] + (p + p^{\prime}) \nabla \cdot \vec{v}^{\prime} = \kappa\nabla^2 T^{\prime} + Q^{\prime} + \Phi^{\prime} \quad .$$

Since, for shallow convection

$$(p + p^{\prime}) \nabla \cdot \vec{v}^{\prime} \simeq \frac{\rho_m}{T_m} (\frac{\partial}{\partial t} + \vec{v}^{\prime} \cdot \nabla) (T + T^{\prime}) + u_3 g\rho_m, \tag{14}$$

then (14) substituded into (13) yields

$$\rho_m c_p (\frac{\partial}{\partial t} + \vec{v}^{\prime} \cdot \nabla)T^{\prime} + u_3 (\Gamma_d - \gamma) = \kappa\nabla^2 T^{\prime} + Q^{\prime} + \Phi^{\prime}, \tag{15}$$

where γ is the ambient temperature lapse rate and $\Gamma_d = g/c_p$ is
the dry adiabatic lapse rate. Equation (15) is the Boussinesq
energy equation for convective motion in a perfect gas which
differs from the comparable equation derived by Rayleigh which
holds for most liquids. These differences are that the temperature gradient is replaced by its excess over adiabatic and c_v
is replaced by c_p. It is important to note that the assumption
of constant κ and ν has been made (up to this point), but as
pointed out in Palm's study [24], as well as Segel and Stuart
[25], variations in these quantities can affect the circulation
direction and convective geometry.

Subsequent to the paper by Spiegel and Veronis, Ogura and
Phillips [26] attempted to remove the restriction placed on the
vertical scale by placing a limitation on the permitted frequencies of the motion. Specifically, they assumed that the percentage range in potential temperature is small and that the time
scale is set by the Brunt-Vaisala frequency. Their approximate
set of equations eliminated acoustic waves, an important feature
for numerical integration of the equations since longer time
steps are possible. Ogura and Phillips also showed that the system of equations reduces to the Boussinesq form if the vertical
scale is small compared to the depth of an adiabatic atmosphere.
Dutton and Fichtl [27] have developed the approximate equations
of motion for gases and liquids for both *deep* and *shallow* convection. For deep convection (i.e. the vertical scale of motion is
comparable to the height scale) they show that the approximate
continuity equation requires the momentum field to be solenoidal,
while for shallow convection the continuity equation yields a
non-divergent perturbation velocity field. Another difference
obtained is that in shallow convection the effects of perturbation
pressure are retained only in the pressure gradient force, but in
deep convection they also appear in the first law of thermodynamics and the equation of state. Finally, they remove the restriction present in the studies by both Spiegel and Veronis [23] and
Ogura and Phillips [26], namely that the equations apply only to
an ideal gas. In summary, Dutton and Fichtl state that the most
significant aspect of their results is that the linearized approximate equations of motion for both deep and shallow convection
and gravity waves in the atmosphere lead to wave motion equations
which are equivalent to those of an incompressible fluid. Therefore, the known results about the stability of flows in an incompressible fluid can be applied directly to atmospheric problems.
This is important because a major part of the interest in the
role of convection and gravity waves in atmospheric processes
centers on how the stability properties of these motions are related to the generation and maintenance of turbulence in the atmosphere.

4. SATELLITE OBSERVATIONS

Man's existence in a sea of air has permitted essentially a
continuous opportunity to observe convective cloud systems. This
observational capability was enhanced over the years with the
invention of hot air balloons, sail planes and aircraft, which
provided a different perspective of viewing cloud formations from
above. The development of radar technology during this century
has provided yet another means of study, by remotely examining
convective circulations and cloud developments. Interestingly,
Benard-Rayleigh convection theory and laboratory experimentation
was viewed by most meteorologists as having little or no reference to atmospheric convection. This view was simply supported,
by the seemingly mere absence of hexagonal cloud systems or geometric arrays of convective cloud cells. In fact the period
following Rayleigh's original work did not seem to attract the
interest of most experimental and mathematical physicists, and
it was nearly 50 years before nonlinear problems were pursued.
Whitehead [28] expresses the view that this absence of interest
was due to the emergence of new fields of study such as quantum
mechanics and relativity that caught the interest of most physical scientists.

Full recognition of the relevance of Benard-Rayleigh convection to atmospheric convection had to await the advent of the
meteorological satellite program. This technological achievement
and advancement in atmospheric observational systems seemingly
warrants some treatment in this introductory section. During
the 1950's, Dr. Harry Wexler of the U.S. Weather Bureau worked
extensively in the planning and development of the concepts of a
meteorological satellite program. In the Soviet Union, similar
efforts were being made by Dr. V. A. Bugaev, which in later years,
led to considerable interaction between Wexler and Bugaev in
planning a global satellite program. On October 4, 1957 the
first earth-orbiting satellite was launched by the USSR, Sputnik
I, which followed a polar orbit that circled the Earth once
every 96 minutes. Efforts by American space scientists and engi-
neers culminated in the successful launching of Explorer I on
January 31, 1958. With these events the dreams and efforts by
Bugaev, Wexler and others could now become a reality. The concept
of a meteorological satellite program in the USA subsequently
led to the development of a program, named TIROS (Television and
Infrared Observational Satellite). On 1 April 1960, TIROS I was
launched successfully from Cape Canaveral, Florida, thus marking
the entry of the field of meteorology and weather observation
into the space age. The first nephanalysis was prepared at Fort
Monmouth, New Jersey, and was transmitted to the National Meteorological Center (NMC) at Suitland, Maryland. During the following year the efforts of an international committee appointed by
the World Meteorological Organization (WMO) succeeded in extending

the TIROS meteorological services to 21 countries. A thorough treatment and discussion of early weather satellites has been given by Hubert and Lehr [29]. An operational goal of the meteorological satellite program was to provide automatic picture transmission of cloud systems and weather patterns to countries with little or no weather service. All meteorological satellites launched prior to 1967 were polar orbiting satellites with nearly sun synchronous orbits. However, on 7 December 1966 and 5 November 1967, successful launches were made of two communications satellites, which marked a new system that was earth synchronous rather than sun synchronous, thus maintaining a fixed satellite subpoint. These efforts paved the way for the Geostationary Operational Environmental Satellite (GOES) system, that was placed into operation in preparation for the Global Atmospheric Research Program (GARP).

4.1 Early Observations of Convection

As first reported by Krueger and Fritz [30], one of the cloud features revealed by TIROS I was a cellular pattern having horizontal diameters as large as 50 to 80 km and consisting of clear centers bounded by ring- or U-shaped cloud elements about 15 to 25 km wide. Such an organization was too large to be discerned by a ground observer, and yet too small to be detected by the standard synoptic observational network. For the first time man was able to observe and study cloud patterns through the aid of the satellite and its camera system, which provided the perspective for detecting atmospheric manifestations of Bénard-Rayleigh convection cells. This single event alone revived the interests of meteorologists in the classical experimental and theoretical studies of convection, which had essentially been dormant for twenty years. The study by Krueger and Fritz was successful in drawing a relationship between the geometric array of cells seen in the atmosphere and those produced by Bénard in the laboratory; however, several differences were noted. Their study selected three satellite pictures of open cellular cloud formations (similar to that in Figure 1) and accompanying synoptic observations for analysis. Findings were: (1) a layer of moist air about 1.5 km deep was heated at the ocean surface resulting in a near adiabatic lapse rate, (2) superimposed over this layer was another of greater stability which served to inhibit the convection, and (3) throughout the convective layer there appeared to be little variation in wind speed and direction above that portion influenced by surface friction. Notable differences between atmospheric cells and Bénard cells were the degree of cell flatness, and the role played by entrainment and radiative processes, and the release of latent heat in the atmospheric case. The aspect ratio of laboratory cells is typically three to one, while that for the atmospheric cells was an order of magnitude larger (thus considerably flatter cells). Further differences noted were in the roles

played by heat conduction and viscosity, where molecular properties
dominate in laboratory fluids while the scale of convective motions in the atmosphere suggest that the use of eddy coefficients
for the diffusion of heat and momentum is required. In fact,
Priestly [31] argued that the extreme flatness of mesoscale convection cells in the atmosphere could be attributed to the anisot-
ropy of the eddy coefficients. The cloud photography from TIROS
I revealed an occurrence of cellular convection, primarily over
the oceans, of greater frequency and of larger areal extent over
the Earth's surface than anyone might have previously conceived.

A summary of the early satellite observations of convective
clouds was given by Hubert [29] in a report entitled "Mesoscale
Cellular Convection", which this author has used to adopt the
acronym MCC. Mesoscale cellular convection (MCC) is defined as
atmospheric manifestations of Bénard-Rayleigh convection in the
planetary boundary layer (thus shallow convection) that occur as
organized hexagonal arrays of convection cells, which can be
detected through satellite cloud photography. These cells typically have diameters of 10-100 km and can occur as either open or
closed circulation systems. Hubert's report further clarified
the frequent and extensive occurrence of MCC over the oceans
based on several years of satellite observations, and it was
emphasized for the first time that the patterns of MCC should
be regarded as a significant link in the study of the interactions
between the sea and the atmosphere. Hubert's work was particularly
helpful in drawing appropriate analogies between atmospheric and
laboratory convection, and the noting of various shortcomings and
inconsistencies between theory and atmospheric observations. One
of the early graphic examples of a hexagonal array of atmospheric
convection cells, examined by Hubert, was presented in Figure 8.
Hubert's work also documented that both open and closed cells
occur in the atmosphere, and that the aspect ratio of atmospheric
cells is approximately an order of magnitude larger than that for
laboratory cells. He also reported on radial patterns of convective clouds (named *actiniae*) as illustrated in Figure 9, which
were proposed as a possible transition mode between open and
closed cells. This figure clearly shows the actiniae located in
a region between coexisting open and closed cells.

Agee, et al. [32] have extended Hubert's satellite observations of MCC over the globe, which has resulted in a *global
climatology of* MCC as depicted in Figure 10. These findings show
that MCC occur primarily to the east of continents over warm ocean
currents and to the west of continents over cool ocean currents.
MCC can occur in any of these regions as open cells, closed cells,
or coexisting open and closed cells; however, open cells are statistically more favored over the warm water and closed cells over
the cool water. A critical aspect of this MCC climatology is recognition of at least two different types of convective mechanisms,

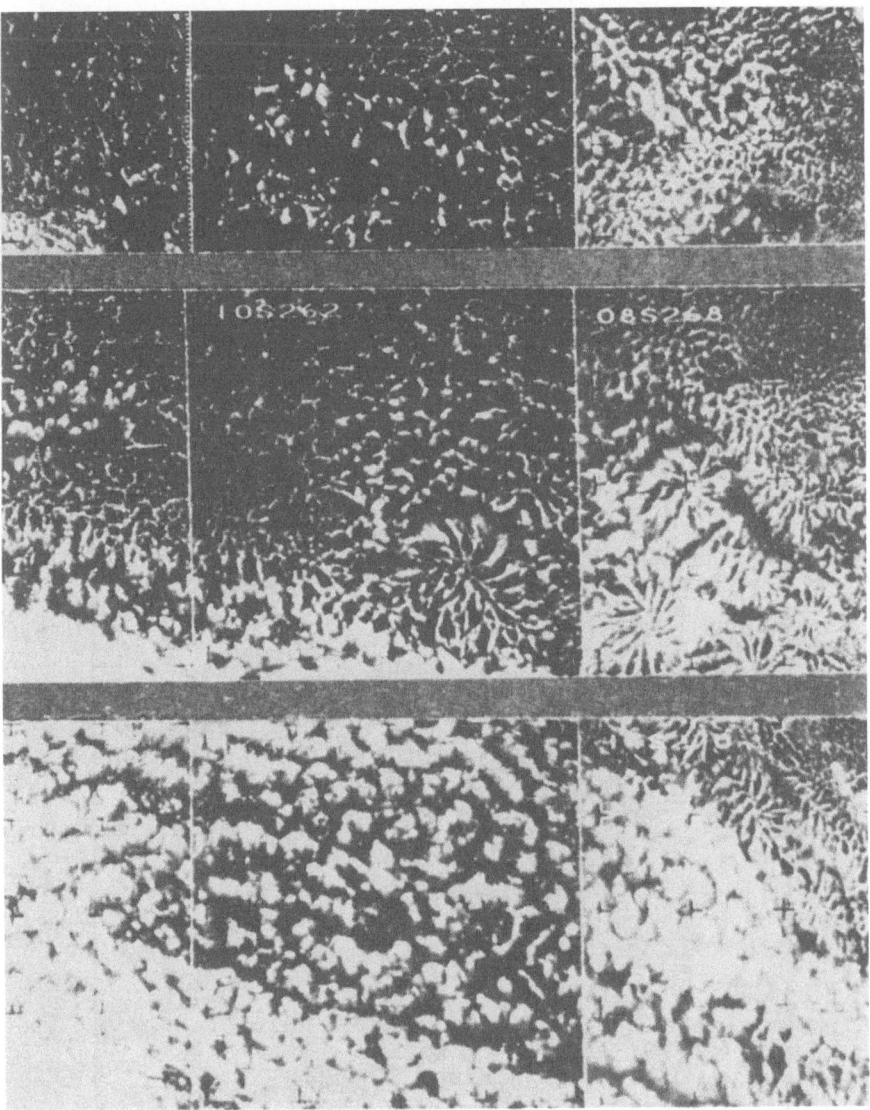

Figure 9. NIMBUS I imagery of coexisting open and closed cells over the Peru current (10°S, 95°W) at 1813 GMT on 15 September 1964. Radial arms of convective cloudiness, called actiniae, can be seen between the open and closed cell regions, suggestive of a transitional mode between open and closed MCC. Laboratory convection has also been seen to display a similar mode.

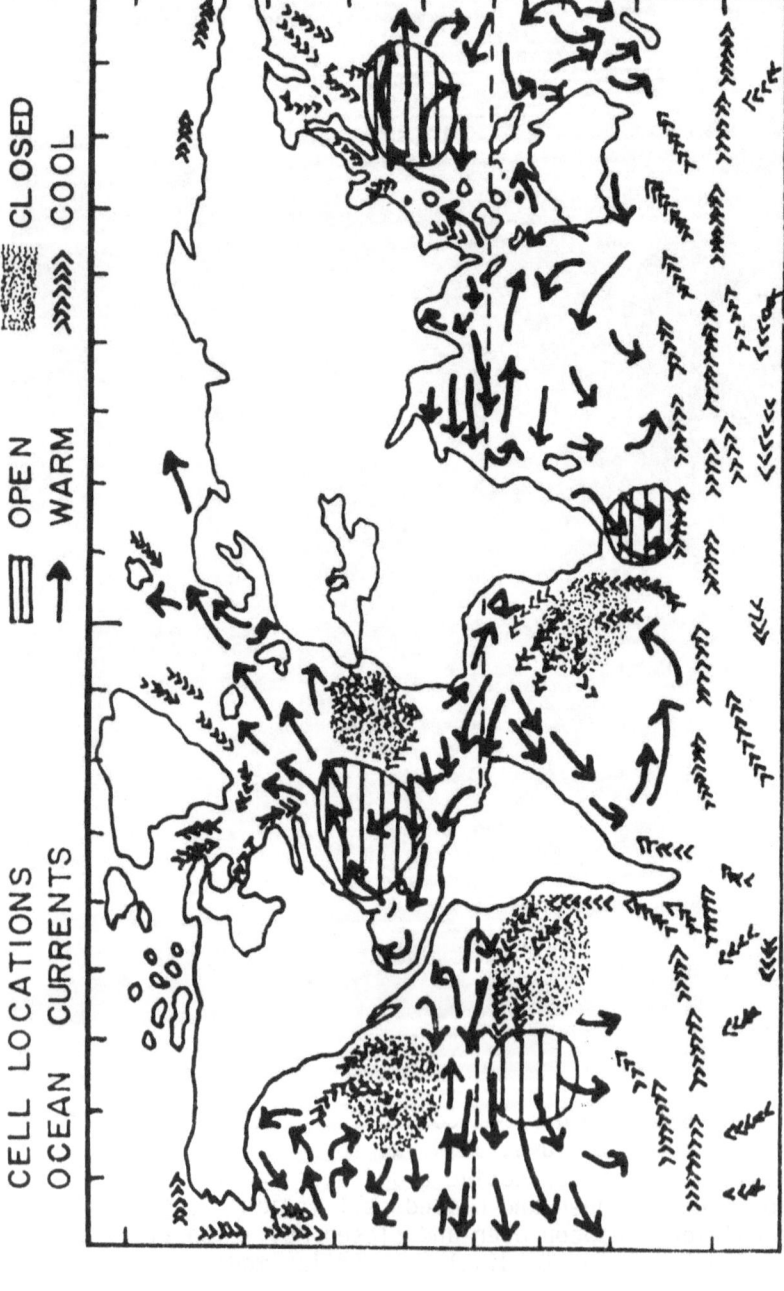

Figure 10. Global climatology of mesoscale cellular convection (MCC), after Agee, et al. [32], depicting the most favored regions of MCC over the oceans.

namely heating from below and cooling from above. To the east of
continents the wintertime cold dry air advects over warmer water,
and heat and moisture buoyancy initiates the convection from below. According to Agee and Howley [33], and Sheu and Agee [34],
this heat flux can exceed 800 W m^{-2} under optimum conditions.
Sheu and Agee have also determined two conditions *necessary* for
the occurrence of MCC due to heating from below: 1) sea surface
temperature at least 5°C greater than air temperature, and 2) surface wind speed in excess of 5 m s^{-1}. To the west of continents,
decks of stratocumulus clouds persist over the cooler ocean surface and frequently are observed to develop MCC formations in
response to cloud top cooling and infrared destabilization. Open
MCC are also observed frequently over the tropical oceans, and
manifest themselves through towering cumuli that develop in the
most convergent regions. This pattern may represent a third
mechanism or type of MCC largely due to convective overturning
gradually initiated by moisture buoyancy over a period of several
days. Such formations may be partly recognized from aircraft observations, as shown in Figure 11. However, the observation of a

Figure 11. Photo of suspected open MCC with accompanying cumulus
towers over the tropical Pacific, taken by E. Agee from commercial
aircraft on 25 February 1975.

similar convective pattern from a space platform, given in Figure 12, reveals more clearly the open array of MCC with a tendency for convective enhancement at the vertex of the hexagons. Such observations and photography from manned spacecraft missions have provided an enormous amount of extraordinary photographs of convective cloud systems. Another example is given in Figure 13, showing stratocumulus off the west coast of the USA. The appearance of penetrative convective elements in this shallow convection system are particularly striking.

Other examples of shallow convection over water have been reported by Lyons and Pease [35] in association with a steam fog

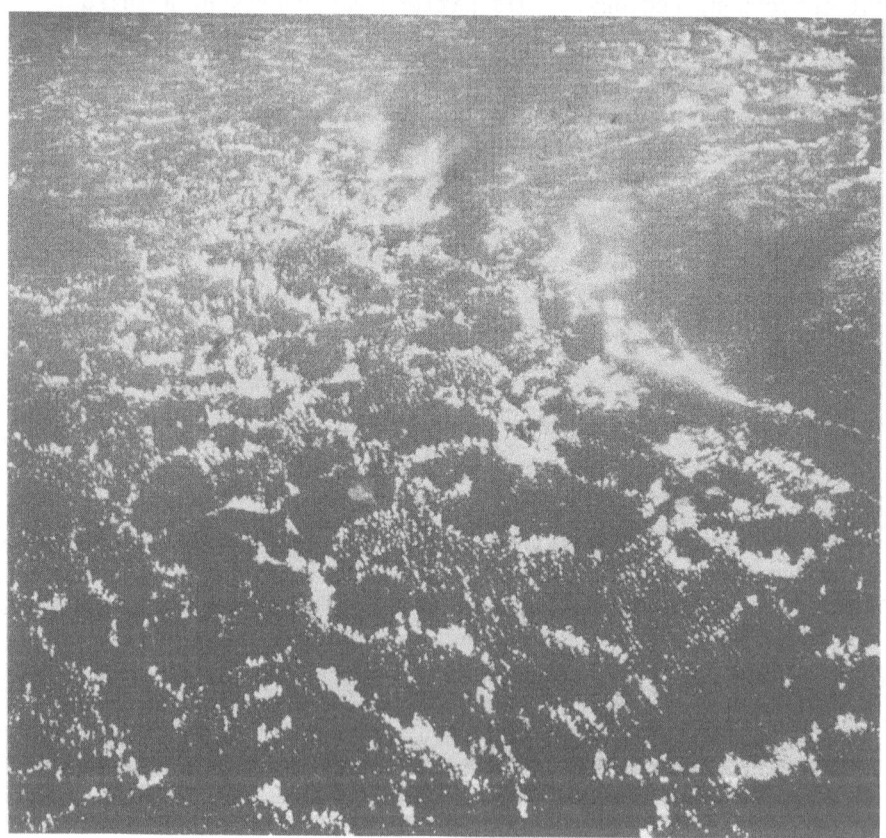

Figure 12. Gemini 10 photograph showing open MCC with accompanying cumulus towers over the tropical Pacific on 20 July 1966. As expected, convective towers tend to be enhanced at the vertices of the hexagons. (NASA photo S-66-45965).

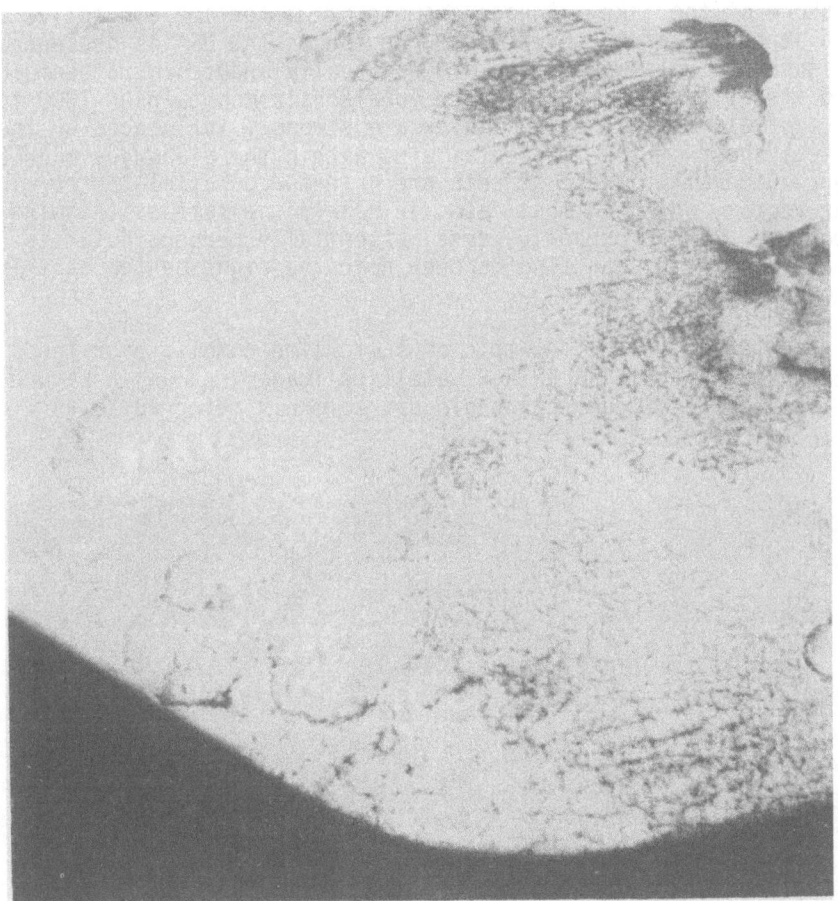

Figure 13. Photograph from the Apollo 9 mission on 8 March 1969
showing stratocumulus clouds off the west coast of the USA. Particularly striking in these shallow convective systems are the
penetrative convective elements and the surrounding u-shaped
boundaries (or "wakes") of subsiding cloud-free air.

over Lake Michigan, produced by the wintertime advection of cold
polar air (-6°F) over the much warmer lake surface (33°F). In
this report, Lyons shows photographs of hexagonal open cells within the steam fog that contain vortices called *steam devils*, at
the vertices of the hexagons. These steam devils in some instances
become as tall as 500 m and actually become associated with the
base of shallow convective clouds that are generated. Such cloud
systems can become well-organized into streets during strong cold
air outbreaks, resulting in rather intense snow squalls on the

lee shore of the lake. A good example of wintertime convective
cloud streets over the Great Lakes region of the USA is presented
in Figure 14. Also shown are orographically induced stratocumulus
cloud streets, appearing over the Appalachain Mountains (∿1000 to
2000 m of elevation), all of which are strongly influenced by the
vertical shear of the horizontal wind within the planetary boundary
layer. Generally, these streets are oriented parallel to the wind
shear vector, which in most cases in nearly the same as the wind
direction. Some orographic waves, essentially perpendicular to
the cloud streets, can also be seen near the right center of this
photograph.

Finally, a typical example of summertime cumulii over land
is captured in the NASA ERTS E satellite imagery shown in Figure
15. This field of convective clouds, commonly referred to as

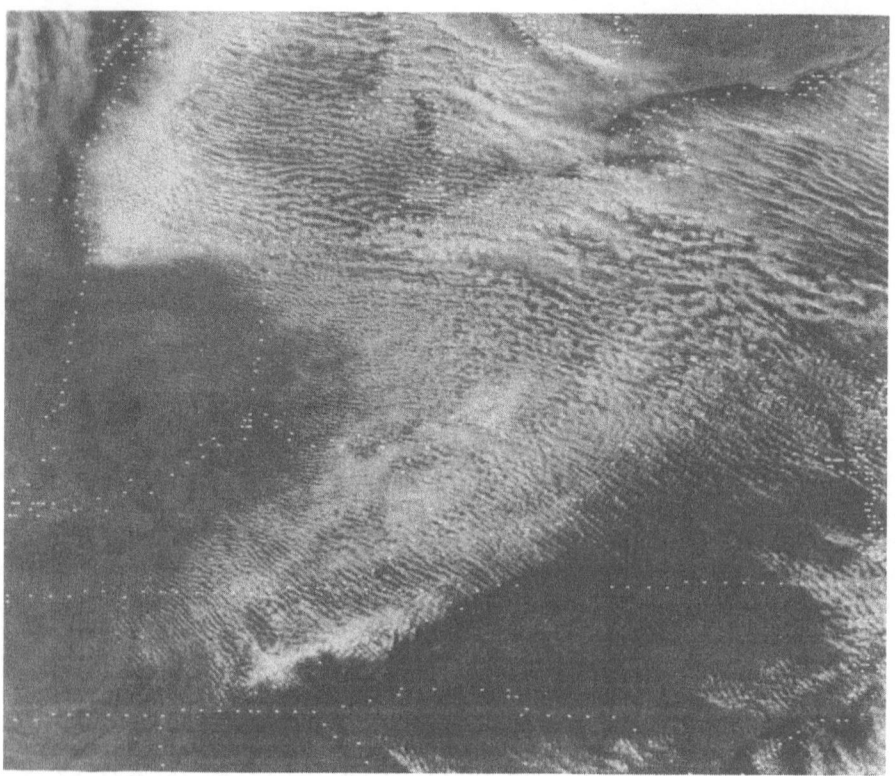

Figure 14.   Cloud streets over the Great Lakes region (USA) at
2000 GMT, 18 December 1975, induced by cold air over warmer water.
Orographic induced cloud streets and waves can also be noted.

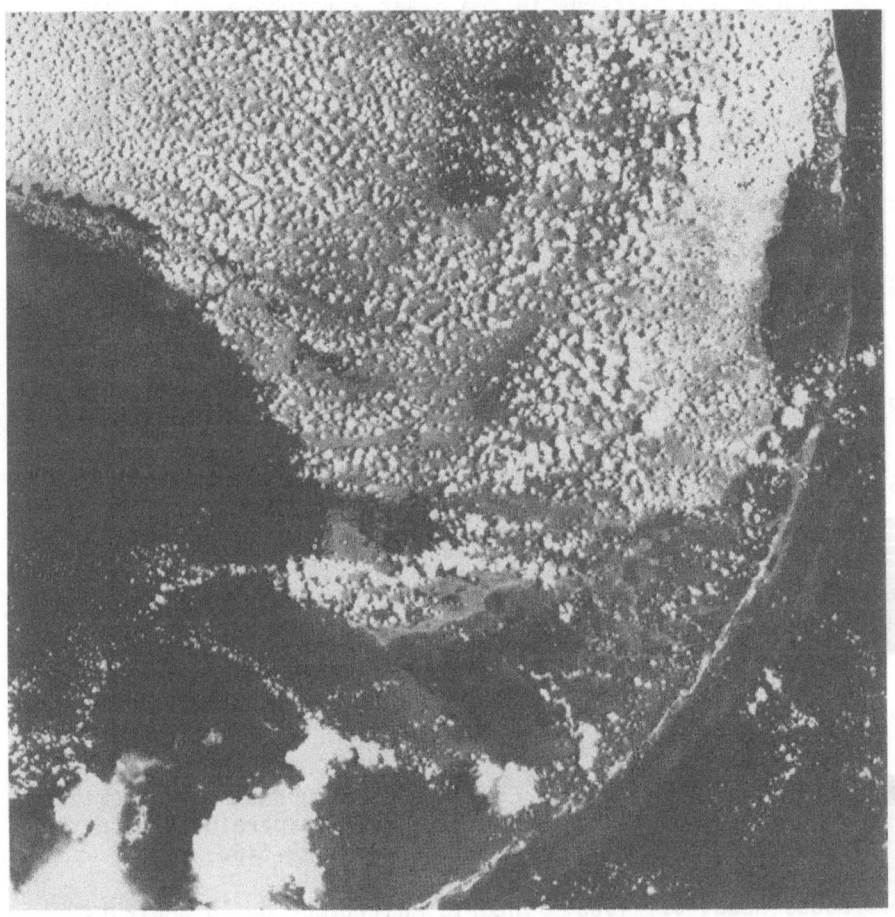

Figure 15. NASA ERTS E imagery of popcorn cumulii over the south tip of the Florida peninsula at 1526 GMT on 18 August 1972.

"popcorn cumulii", is situated in the planetary boundary layer over the south tip of Florida. The light winds accompanying this event on the morning of 18 August 1972, permitted the non-uniformly heated land surface to form a rather random pattern of thermals and cumulus clouds. With stronger winds and boundary layer shear, this random field would likely organize into cloud streets.

REFERENCES

1. Hardy K. and H. Ottersten, 1969: Radar investigations of

convective patterns in the clear atmosphere. J. Atmos. Sci., 26, 666-672.

2. Konrad, Thomas G., 1970: The dynamics of the convective process in clear air as seen by radar. J. Atmos. Sci., 27, 1138-1147.

3. Chandrasekhar, S., 1961: Hydrodynamic and Hydromagnetic Stability. Oxford Press, London, 652.

4. Rumford, Count, 1870: Of the propagation of heat in fluids. Complete Works, 1, American Academy of Sciences, Boston, 239.

5. Prout, W., 1834: Bridgewater Treatises, 8, edited by W. Pickering, London, 65.

6. Brunt, Sir David, 1951: Experimental cloud formation. Compendium of Meteorology, American Meteorological Society, 1255-1262.

7. Bénard, H., 1901: Les tourbillons cellulaires dans une nappe liquids transportant de la chaleur par convection en regime permanent. Ann. Chim. Phys., 23, 62-144.

8. Brunt, D., 1939: Physical and Dynamical Meteorology. Cambridge University Press, 219-220.

9. Rayleigh, O. M., 1916: On convection currents in a horizontal layer of fluid, when the higher temperature is on the underside. Phil. Mag. Ser., 6, 32, 529-546.

10. Boussinesq, J., 1903: Theorie Analytique de la Chaleur, 2, 172.

11. Chandrasekhar, S., 1957: Thermal convection. Daedalus, 86, No. 4, 325-339.

12. Platzman, G. W., 1967: Linear Theory of Marginally unstable convection motion. NCAR Technical Notes, 24, 89-148.

13. Jeffreys, H., 1926: The stability of a layer of fluid heated below. Phil. Mag., 2, 833-844.

14. Jeffreys, H., 1928: Some cases of instability in fluid motion. Proc. Roy. Soc. (London), A, 118, 195-208.

15. Pellew, A., and R. V. Southwell, 1940: On maintained convection motion in a fluid heated from below. Proc. Roy. Soc., A, 176, 312-343.

16. Low, R. A. and D. Brunt, 1925: Instability of viscous fluid motion. Nature, 115, 299-301.

17. Bénard, H., 1927: Hydrodynamique Experimental Sur Les tourbillons cellulares et la theorie de Rayleigh. Acedemie Des Sciences, 185, 1109-11, 1256-59.

18. Bénard, H., 1928: Sur les tourbillons cellulaires, les tourbillons en bandes, et la theorie de Rayleigh. Bull. Soc. franc. phys., 266, 112-225.

19. Hubert, L. F., 1966: Mesoscale cellular convection. Meteor. Satellite Laboratory, Rep. No. 37, U.S. Department of Commerce, Washington, D.C., 68 pp.

20. Chandra K., 1938: Instability of Fluids Heated from Below. Proc. Roy. Soc., (A)164, 231-242.

21. Christopherson, D. G., 1940: Note on the vibration of membranes. Quart J. Math., 11, 63-65.

22. Krishnamurti, R., 1975: On cellular cloud patterns - Part I: Mathematical Model. J. Atmos. Sci., 33, 1353-1363.

23. Spiegel, E. A. and G. Veronis, 1960: On the Boussinesq approximation for a compessible fluid. Astrophy. J., 131, 442-447.

24. Palm, E., 1960: On the tendency toward hexagonal cells in steady convection. J. Fluid Mech., 8, 183-192.

25. Segel, L. A. and J. T. Stuart, 1962: On the question of the preferred mode in cellular thermal convection. J. Fluid Mech., 13, 289-306.

26. Ogura, Y. and N. A. Phillips, 1962: Scale analysis of deep and shallow convection in the atmosphere. J. Atmos. Sci., 19, 173-179.

27. Dutton, J. A. and G. H. Fichtl, 1969: Approximate equations of motion for gases and liquids. J. Atmos. Sci., 26, 241-254.

28. Whitehead, J. A., Jr., 1971: Cellular convection. Amer. Sci., 59, 444-451.

29. Hubert, L. F. and P. E. Lehr, 1967: Weather Satellites. Blaisdell Publishing Company, 119 pp.

30. Krueger, A. F. and S. Fritz, 1961: Cellular cloud patterns

revealed by TIROS I. _Tellus,_ _13_, 1-7.

31. Priestly, C. H. B., 1962: Width-height ratio of large con-
    vective cells, _Tellus,_ _14_, 123-124.

32. Agee, E. M. and T. S. Chen and K. E. Dowell, 1973: A review
    of mesoscale cellular convection. _Bull. Amer. Meteor. Soc._
    _54_, 1004-1012.

33. Agee, E. M. and R. P. Howley, 1977: Latent and sensible
    heat flux calculations at the air-sea interface during
    AMTEX 74. _J. Appl. Meteor.,_ _16_, No. 4, p. 443-447.

34. Sheu, P. J. and E. M. Agee, 1977: Kinematic analysis and
    air-sea heat flux associated with mesoscale cellular
    convection during AMTEX 75. _J. Atmos. Sci.,_ _34_, 793-801.

35. Lyons, Walter A. and Steven R. Pease, 1972: Picture of the
    Month; "Steam Devils" over Lake Michigan during a January
    artic outbreak. _Mon. Wea. Rev.,_ _100_, No. 3, 235-237.

AN OBSERVATIONAL STUDY OF CONVECTIVE CLOUD STREETS

K.J. Weston

Department of Meteorology, University of Edinburgh, U.K.

VHRR satellite photographs and radiosonde information are used to study occasions of cloud streets over land. Over the British Isles streets appear to occur most frequently in spring and autumn in anticyclonic or straight flow and weak cold air advection. The streets were found to have aspect ratios between 2:1 and 4:1 and to be closely aligned along the mean wind direction in the layer which they occupied. In common with the work of Kuettner, a curved wind profile is found to favour streets, but his explanation of its origin, that of a decreasing geostrophic wind with height, is not confirmed.

1.   INTRODUCTION

It has long been recognised that the motions of the atmosphere occur on a whole spectrum of sizes from the global to the turbulence scale, but it was not until the advent of meteorological satellites that the full extent of cloud patterns on the medium scale – the mesoscale – became apparent. Satellite imagery is ideally suited for studies of mesoscale phenomena (say from 5 to 100 km) – a scale too large to be easily studied from a fixed point and too small to be adequately resolved by data from the conventional synoptic network.

Cloud streets are a familiar mesoscale phenomenon. Viewed from the ground they appear as lines of cumulus and stratocumulus, sometimes composed of discreet elements and at other times forming almost unbroken lines. When viewed from satellites these cloud streets are sometimes seen to cover very large areas, perhaps many thousands of square kilometres. Figure 1 shows a

31

E. M. Agee and T. Asai (eds.), Cloud Dynamics, 31–41.
Copyright © 1982 by D. Reidel Publishing Company.

K. J. WESTON

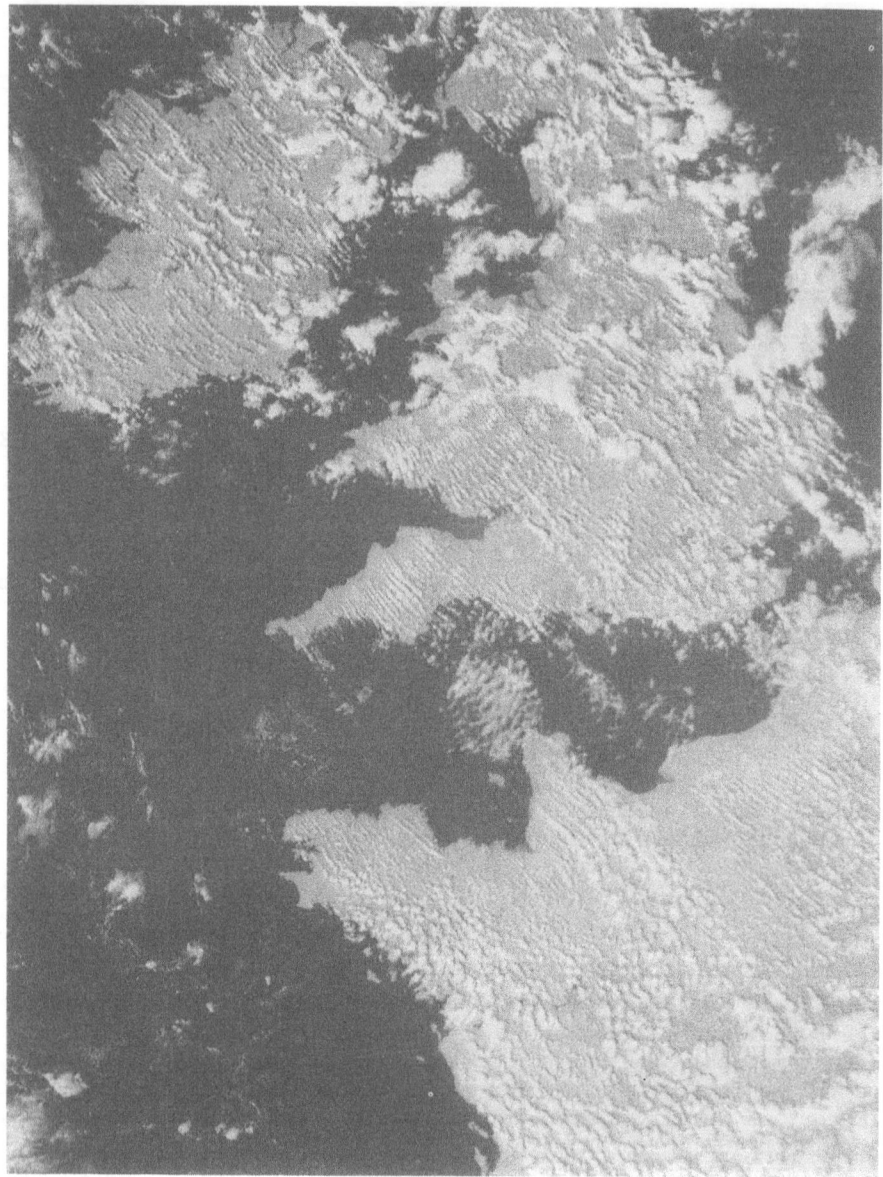

Figure 1.   TIROS-N satellite picture, 1518 GMT, 2 April 1980.
    Cloud streets cover much of England, Ireland and northern
    France in a north-westerly flow between an anticyclone
    centred in Biscay and low pressure centres over Denmark and
    the Norwegian sea.   Picture received at Dundee University,
    Electronics Laboratory.

striking example of cloud streets over the British Isles and
northern France.   The streets are particularly well-defined
over south-west England despite the ground being hilly, rising
to over 600 m in places: thus the mechanism producing the cloud
streets is sufficient to overcome the mesoscale circulations and
flow irregularities normally associated with high ground.

The aim of this study is to look at the various features of
cloud streets - their frequency of occurrence, aspect ratios,
alignment - and to look at the wind profiles on cloud street
occasions and in particular their relationships to the profiles
of geostrophic wind.

2.   PREVIOUS INVESTIGATIONS

Kuettner (1959, 1971) carried out observational and
theoretical studies of cloud bands.   Using data from a variety
of sources, including special observations taken during the
BOMEX project, he found that the length of these lines varied
from 20 to 500 km, with spacings of between 2 and 8 km.   He
measured the ratio of width to depth, called the aspect ratio,
and found it to be between 2 and 4.   The occurrence of lines
appeared to be associated with vigorous cold air outbreaks with
the cold centre to the right of the flow direction leading to a
decrease of the geostrophic wind with height and a curved wind
profile - one in which the direction changed little with height,
but in which the speed reached a maximum somewhere within the
convective layer.   He found an average profile curvature on
these occasions of 10^{-5} m^{-1} s^{-1} and attributed this to a
frictional wind increase with height in the lower part of the
boundary layer and a thermal wind decrease in the upper part.
According to the mechanism discussed by Kuettner the curved
profile leads to barotropic instability and a circulation pattern
sufficient to organize the convection into streets.   He found
that the cloud streets were aligned quite closely along the
direction of the surface wind.

These findings were in contrast to those of Plank (1966)
who studied convection using photo-reconnaissance data from
flights over Florida.   He concluded that there were no consistent
differences between the wind profiles on occasions of patterned
and non-patterned convection;   and, although the average
alignment of streets was only slightly to the left of the wind
at cloud base, that on more than 30% of occasions the deviation
was greater than 30°.

Le Mone (1973) determined the characteristic structure and
dynamics of cloud streets using tower and aircraft data.   She
found aspect ratios and orientations to be in broad agreement

with the results of Kuettner and that roll structure resembled
that predicted by theoretical models invoking instability due to
the presence of an inflection point in the wind profile in the
plane normal to the roll axis, as is present in an Ekman layer
(Brown, 1972).

Nicholls (1978) described aircraft measurements of boundary
layer parameters on an occasion of cloud streets over the sea.
He showed that such motions strongly influence the transport of
water vapour and momentum above the surface layer.

3.   PRESENT STUDIES

To identify the preferred time of year for the occurrence of
cloud streets over the British Isles, TIROS-N satellite pictures
for the whole of 1980 were examined.   Streets occurred on 53
days, but were more frequent and more extensive in spring and
autumn.   Figure 2 shows the annual variation of the area of
land covered by streets for 1980.   Because of the limited data
set used the results should be considered with caution, but they
are broadly similar to those presented by Kuettner (1959) for
cloud streets over the Boston area.   The low frequency of
streets in winter is probably due to the relatively small number
of occasions of cumulus convection over land at a time of low
solar input, while the summer minimum may be associated with the
lighter winds during the summer months - streets do not normally
occur on days of light winds.

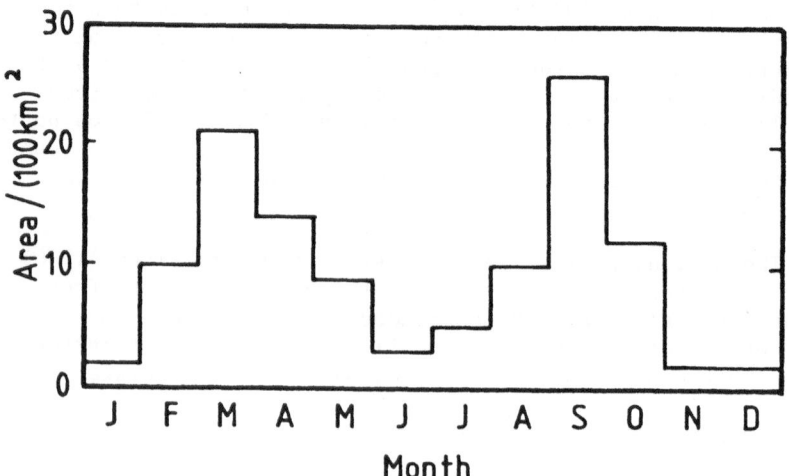

Figure 2.   Total area of British Isles covered by cloud streets
      during each month of 1980, as deduced from TIROS-N satellite
      pictures.

For the remainder of the study the data used were enlarged visible and infrared VHRR photographs of the British Isles for the months of March and April 1977 from the NOAA 5 satellite, and midday upper air ascents from British Isles land stations. March and April 1977 were months of particularly high occurrence of cloud streets, twenty-one cases being identified.

3.1 Aspect ratio

For the determination of the spacing-depth ratio of cloud streets we require the depth of the convective layer and the average spacing between streets.   Since no direct measurements of cloud tops were available, the depth of convection was inferred from radiosonde soundings.   Perhaps the most reliable method for estimating the level of the top of a convective layer is to identify this with the level at which the saturation wet-bulb potential temperature has a minimum value - ie the level at which the lapse-rate changes from greater to less than the wet-adiabatic. In almost all of the cases studied here this coincided with a level at which there was a distinct fall in humidity with increasing height, supporting the estimation.   The spacing of streets was measured directly from satellite photographs taken during the late morning and an average value obtained over the area covered.   The depth of convection was estimated from the nearest available radiosonde sounding or an average taken in the event of more than one station lying within the area of cloud streets.   Figure 3 shows values of spacing and depth for the twenty-one cases considered.

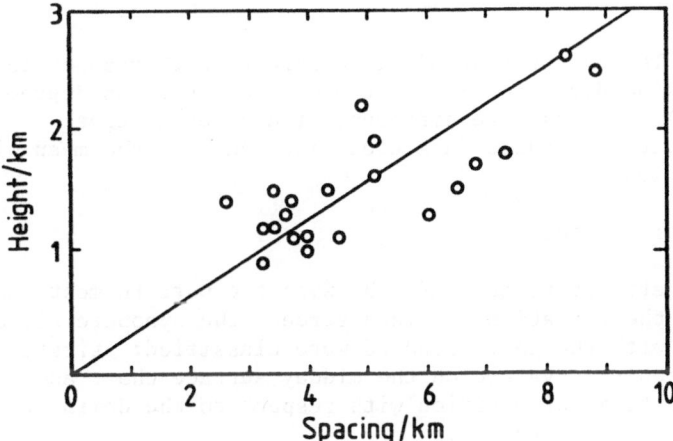

Figure 3. The relationship between the spacing of cloud streets
           and the height of the top of the convective layer.   The
           solid line represents the average aspect ratio of the 21 cases.

The results show considerable scatter but there is clearly an
increase in spacing as the depth increases.   The average aspect
ratio is 3.2 to 1,close to the value found by Kuettner.

3.2 Alignment

Figure 4 shows the relationship between the direction of
alignment of the cloud streets and the mean direction of the wind
in the convective layer.   In all cases the alignment of streets
was within 25° of the mean wind direction and three-quarters of
cases were within 10°.   In some cases there is a difference of 2
hours between the times of the satellite photograph and the
radiosonde measurements and it seems likely that concurrent
observations would show an even closer agreement.   Thus the
alignment of cloud streets is an excellent indication of the mean
boundary layer wind direction, though experience is needed in
distinguishing streets from wave clouds.

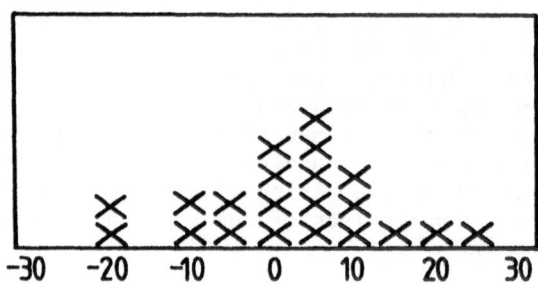

Figure 4.   The orientation of cloud streets with respect to the
      mean wind direction in the convective layer, in degrees (21
      cases).   A positive difference indicates a street
      orientation rotated in a clockwise sense to the mean wind
      direction.

3.3 Synoptic pattern

In an attempt to identify the synoptic pattern most likely
to lead to the formation of cloud streets the synoptic situations
associated with the cases studied were classified: firstly in
terms of isobar curvature on the midday surface chart and
secondly in terms of position with respect to the dominant
weather system over the area.

Of the twenty-one cases thirteen showed anticylonic isobar
curvature and the remaining eight had straight isobars.   This
may reflect the fact that streets tend to form only when convection is relatively shallow: the average depth of the convective

layer on the twenty-one occasions studied here was 1.5 km and in
no case did the depth exceed 2.6 km.   Further evidence in this
direction is found from satellite pictures of cold air outbreaks
over the North Atlantic.   As cold air flows off the ice sheet
over relatively warm ocean, well-defined cloud streets commonly
form; but these break up into convective cells further downstream.   The spacing of the streets in the neighbourhood of this
transition is typically about 5 to 10 km, and with an aspect
ratio of 3 to 1 this implies a convective layer depth of about
1.5 to 3 km.

The most common synoptic pattern amongst the cases studied
was the eastern flank of an anticyclone: ten of the twenty-one
cases were in this category.   Of the remainder, seven cases
occurred in a ridge, four of these immediately behind a cold front;
three cases occurred on the southern flank of a low; and one
case on the north-western flank of a high.

3.4 Mean wind profiles

For the study of the wind profile on cloud streets occasions
only those cases were selected in which streets were particularly
well-defined and widespread - nine cases.   For each day the wind
speeds at 12 GMT were resolved along the direction of alignment
of the streets and plotted as a function of height, normalised
with respect to the depth of convection as determined by the
method described in section 3.1.   All nine occasions were
combined to form an average profile and this is shown in Figure 5
together with the profile of geostrophic wind, constructed using
midday radiosonde temperature data to determine thermal winds,
and normalised with respect to height in the same way as for the
observed profiles.   The mean profile of observed wind shows a
well-defined maximum within the convective layer with a profile
curvature of about 10^{-5} m^{-1} s^{-1}, as found in Kuettner's work.
However, the mean geostrophic wind increases with height throughout the convective layer, so that on these occasions the profile
curvature cannot be explained by a decreasing geostrophic wind
with height combined with surface friction, as put forward by
Kuettner.

The observed wind approaches the geostrophic value above the
convective layer and in the region of the maximum wind, at about
0.7 or 0.8 times the depth of the convective layer, with
considerable sub-geostrophic flow near the ground, which is to be
expected, but also near the boundary layer top.

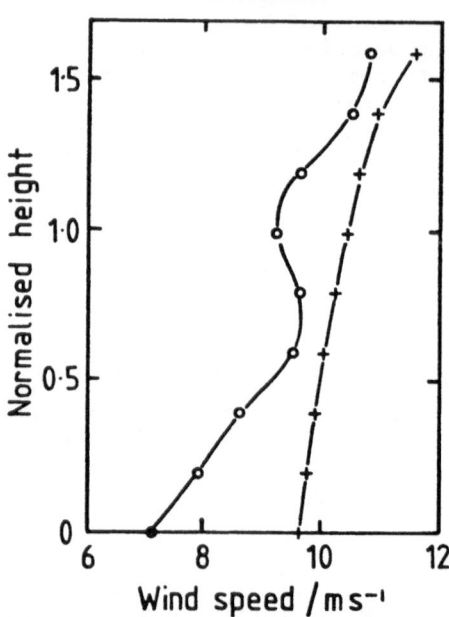

Figure 5. Mean values of observed wind speed (circles) and
   geostrophic wind speed (crosses) resolved along the direction
   of orientation of the cloud streets as a function of
   normalised height for nine occasions of well-defined cloud
   streets.

3.5 Thermal advection

     Kuettner listed cold advection as one of the necessary
conditions for cloud street formation.   This finding is borne
out by the present study.   Using midday radiosonde ascents from
British Isles stations it was found that in all nine cases of
well-defined streets there was cold advection within the
convective layer with an average magnitude of 0.15 deg. C hr^{-1},
leading to a backing of the geostrophic wind with height.
However, this value is about that expected solely from warming of
the boundary layer by the sensible heat flux from the ground,
leading to a downstream temperature increase.   Indeed, an
average sensible heat flux of only 60 Wm^{-2} would account for the
observed cold advection.   Thus, although all nine occasions of
cloud streets studied here exhibit cold advection, it appears to
be a result of the convection rather than a contributory cause.

4.   THE SIGNIFICANCE OF CONVECTIVE STREETS FOR FLUX MEASUREMENTS

     Many techniques have been employed for the measurement of
fluxes of sensible and latent heat at various levels in the

boundary layer and intercomparisons have been made between the
different methods.   Ground-based instruments may be used to
estimate fluxes by the flux-profile method (eg Wood, 1977) or by
eddy correlation techniques (Haugen et al., 1975).   The first of
these methods is used close to the surface where gradients of
temperature and humidity are large, while the second is limited
to the height of tower-mounted probes, say 100 m or so.   Balloon-
borne turbulence sensors have successfully measured fluxes to
greater heights (Readings and Butler, 1972), while Milford et al.
(1979) effectively used an instrumented powered glider as a
turbulence sensor, enabling heat and moisture fluxes to be
determined at several heights.   Cattle and Weston (1975) used a
budget technique to indirectly deduce fluxes as a function of
height.

Observational programmes aimed at comparing fluxes measured
by different techniques have been carried out by several workers.
Some investigators (Lenschow and Johnson, 1968; Milford, 1973)
found reasonable agreement between budget and eddy correlation
methods but Moores et al. (1979) in a comprehensive inter-
comparison showed discrepancies between methods which varied from
occasion to occasion.   It has been suggested that some of these
differences are due to fluxes carried on scales larger than about
a kilometre.   Eddy correlation methods from fixed sites are
particularly suspect when a significant part of the heat transfer
is on large scales;   and even fluxes measured by a relatively
slow-moving glider are liable to underestimate fluxes on such
occasions.

Le Mone (1973) and Nicholls (1978) showed the importance of fluxes
on relatively large scales.   Le Mone found that between a
third and a half of the vertical sensible heat flux was carried
by roll convection at heights of about 100 m, the proportion
increasing with height.   In view of the frequent occurrence of
convective streets, usually on space scales of about 6 km, the
presence of such motions must be carefully considered in any
interpretation of flux measurements within the boundary layer.

5.   SUMMARY AND CONCLUSIONS

Cloud streets over the British Isles appear to be most
frequent in spring and autumn.   In common with other workers
aspect ratios were found to be about 3 to 1 and streets were
aligned quite closely along the direction of the mean wind in the
boundary layer.   Most occasions of cloud streets occur when the
the flow is anticylonic or straight, with the eastern flank of a
high or a ridge behind a cold front being particularly favourable
synoptic patterns.   A curved wind profile was present in the
majority of cases, but in general this curvature could not be

explained in terms of a decreasing geostrophic wind with height. Cold advection was present in all the cases studied, with a magnitude of about 0.15 deg. C hr^{-1}.

ACKNOWLEDGEMENT

The author is grateful for the supply of satellite photographs by the Department of Electrical Engineering and Electronics, University of Dundee.

REFERENCES

Brown, R.A.: 1972, 'On the inflection point instability in a stratified Ekman boundary layer'. J. Atmos. Sci. 29, pp 850-859.

Cattle, H. and Weston, K.J.: 1974, 'Budget studies of heat flux profiles in the convective boundary layer over land'. Quart. J.R. Met. Soc. 101, pp 353-363.

Haugen, D.A., Kaimal, J.C., Readings, C.J. and Rayment, R.: 1975, 'A comparison of balloon-borne and tower-mounted instrumentation for probing the atmospheric boundary layer'. J. Appl. Met. 14, pp 540-545.

Kuettner, J.P.: 1959, 'The band structure of the atmosphere'. Tellus 11, pp 267-294.

Keuttner, J.P.: 1971, 'Cloud bands in the Earth's atmosphere: observations and theory'. Tellus 23, pp 404-425.

Le Mone, M.A.: 1973, 'The structure and dynamics of horizontal roll vortices in the planetary boundary layer'. J. Atmos. Sci. 30, pp 1077-1091.

Lenschow, D.H. and Johnson, W.B.: 1968, 'Concurrent airplane and balloon measurements of atmospheric boundary layer structure over a forest'. J. Appl. Met. 7, pp 79-89.

Milford, J.R.: 1973, 'Measurements of convective heat fluxes using a powered glider'. Quart. J.R. Met. Soc. 99, pp 768-771.

Milford, J.R., Abdulla, S. and Mansfield, D.A.: 1979, 'Eddy flux measurements using an instrumental powered glider'. Quart. J.R. Met. Soc. 105, pp 673-693.

Moores, W.H., Caughey, S.J., Readings, C.J., Milford, J.R., Mansfield, D.A., Abdulla, S., Guymer, T.H. and Johnson, W.B.: 1979, 'Measurements of boundary layer structure and development over S.E. England using aircraft and tethered balloon instrumentations'. Quart. J.R. Met. Soc. 105, pp 397-421.

Nicholls, S.: 1978, 'Measurements of turbulence by an instrumented aircraft in a convective atmopsheric boundary layer over the sea'. Quart. J.R. Met. Soc. 104, pp 653-676.

Plank, V.G.: 1966, 'Wind conditions in situations of pattern form and non-pattern form cumulus convection'. Tellus 18, pp 1-12.

Readings, C.J. and Butler, H.E.: 1972, 'The measurement of
     atmospheric turbulence from a captive balloon'.   Met. Mag.
     101, pp 286-298.
Wood, N.L.H.: 1977, 'A field study on the representativeness of
     turbulent fluxes of heat and water vapour at various sites
     in southern England'.   Quart. J.R. Met. Soc. 103, pp 617-
     624.

RADIATIVE INFLUENCE ON SMALL SCALE CONVECTION WITHIN STRATUS CLOUD LAYERS

Bakan, S.

Meteorologisches Institut der Universität Hamburg

Abstract:
Solutions of the radiative transfer equation show that strong cooling of the uppermost cloud layers by emission of thermal radiation leads within a few minutes to a negative potential temperature gradient that may give rise to cellular convective instability. The result of a linear stability analysis for a cloud layer is, that except for the definition of a moist Rayleigh number, the influence of latent heat of condensation and of the radiation field is very small. The critical Rayleigh and wave numbers deviate by less than 3% from those of the radiation free case.
A two layer model with negative wet bulb potential temperature gradient in the upper layer and neutral stratification below shows that convection penetrates into the neutral layer. Horizontal cell dimensions of 300 to 1200 m with convection depths of 100 to 400 m are calculated, which are comparable to observed values.

1. Introduction:

The upper boundary of stratus cloud layers frequently exhibits a fairly regular array of undulations or small cloud turrets (1, 2). The horizontal scale of these structures is of the order of several hundred to about 1000 meters.
Wegener (1) interpreted these structures as Helmoltz waves due to a wind and density discontinuity at the cloud top inversion. Several other authors (e.g. 3, 4) improved this theory by introducing a linear temperature profile and vertical wind shear and found acceptabel agreement between predicted and observed wavelengths.
Another possible reason for this phenomenon can be found in the destabilizing effect of cloud top cooling by longwave radiation. Möller (5) and Feigelson (6) showed, that longwave emission by stratiform clouds results in a strong cooling of the uppermost layers and an

43

E. M. Agee and T. Asai (eds.), Cloud Dynamics, 43–56.

almost vanishing net flux in the clouds interior. Therefore, a negative potential temperature gradient builds up below the cloud top inversion which could give rise to a Rayleigh Bénard type cellular convection. This would explain that the mentioned structures are found on top of almost any cloud layer irrespective of an overlying wind shear layer. Only when a second cloud layer above shields the lower one from radiating to space no such cellular structures are found.

In the present paper the role of radiation for the onset of cellular convection in a shallow cloud layer is investigated. In the second chapter heating rate calculations show the ability of radiative cooling to space to produce the mentioned layer with a negative lapse rate. In chapter 3 the results of a linear stability analysis of this cloud layer are presented, including radiation and phase flux between water vapor and droplets. Finally, in the fourth chapter a similar stability analysis is carried out for a two layer model.

2. Radiative transfer in clouds:

The monochromatic radiative transfer equation (RTE) is solved in the Eddington approximation for a horizontally homogeneous atmosphere with an embedded stratus cloud layer (7, 8). The wavelength integration of the resulting fluxes leads to vertical profiles of the heating rate. The full line in Fig. 1 shows a typical result for longwave radiation in an atmosphere with a cloud layer between 1000 and 1500 meters. The dashed line, for comparison, gives the more exact values of a matrix operator (i.e. adding and doubling) method for the solution of the RTE (9) and the numbers represent the difference between the two results in percent. The good agreement between both results allow to use the approximate solution scheme, which is by a factor of 100 faster to compute.

The temperature and moisture profiles are modelled according to the US standard atmosphere for 45° N while in the cloud layer a moist adiabatic lapse rate and between 1500 and 1600 m an inversion of 3K are assumed. A typical vertical distribution of cloud droplet water is chosen after Feigelson (6). In the clear air part of the model atmosphere cooling rates of about 1 K/d are found. The uppermost parts of the cloud layer exhibit strong cooling to space with a maximum of about 170 K/d at 25 to 50 m below cloud top. Below this maximum the cooling reduces rapidly to virtually zero at about 150 m below cloud top. Therefore, cooling to space results in the rapid development of an approximately 100 m deep layer with negative potential temperature gradient. The depth of this layer and the change in the vertical temperature lapse rate depend mainly on the amount and vertical distribution of droplet water mass. Keeping the latter constant but multiplying the whole droplet density profile by a constant factor leads to the results in Fig. 2. The abszissa gives the maximum liquid water density which is located 50 meters below cloud top. The layer depth d is defined by the distance between the maximum cooling and a reduction of the cooling rate to 10% of this maximum value. The second curve gives the time change of the average temperature lapse rate dH/dt of this layer. For the case of

Fig. 1 ρ_w = 0.30 gm^{-3} with d = 60 m and dH/dt = 3.3 Km^{-1} d^{-1}, so that radiation would double an initially moist adiabatic lapse rate within about 3 minutes. Therefore, any stratus is expected to be highly unstable in the upper 150 meters, except when underlying another cloud layer so that the net radiation flux is very small and cooling to space is inhibited.

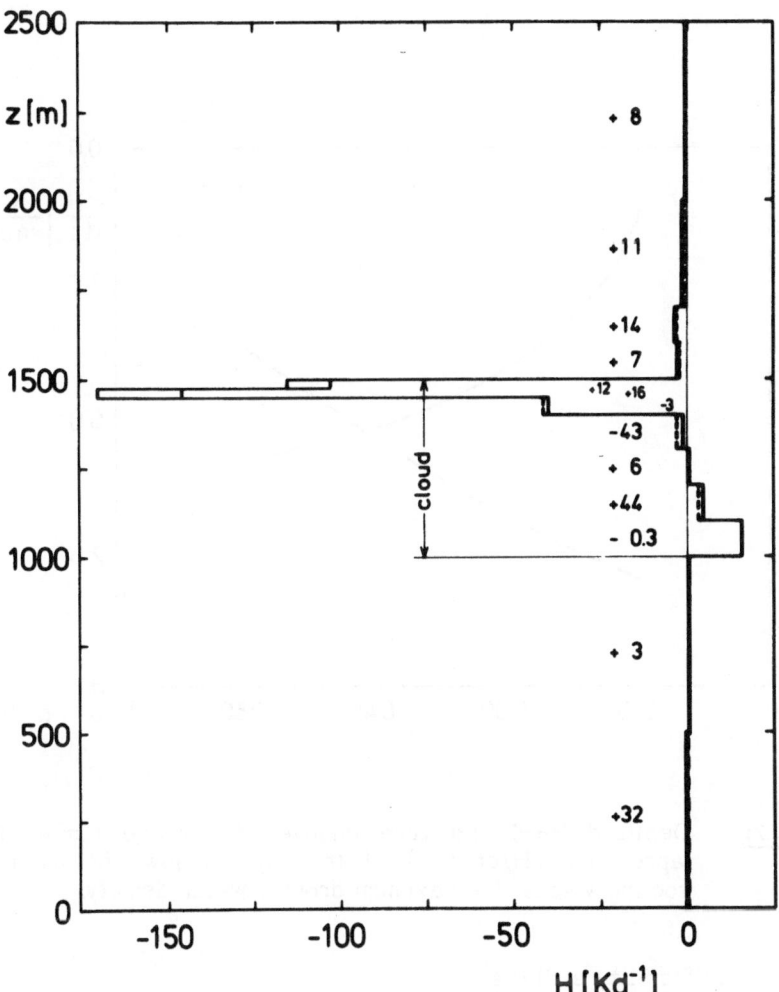

Fig. 1: Heating rate in the lower troposphere with a cloud layer. Numbers give the deviation between Eddington approximation (---) and matrix operator method (——) results.

Till now only longwave radiation has been considered which is typical
of nighttime conditions. During daytime solar heating could change the
picture. Solar heating rates turn out to be in the order of 20 K/d in
the upper part of the cloud layer with a vertical gradient much
smaller than that of the longwave radiation. Therefore, solar inso-
lation can be neglected as it is not of big influence on the stability
problem.

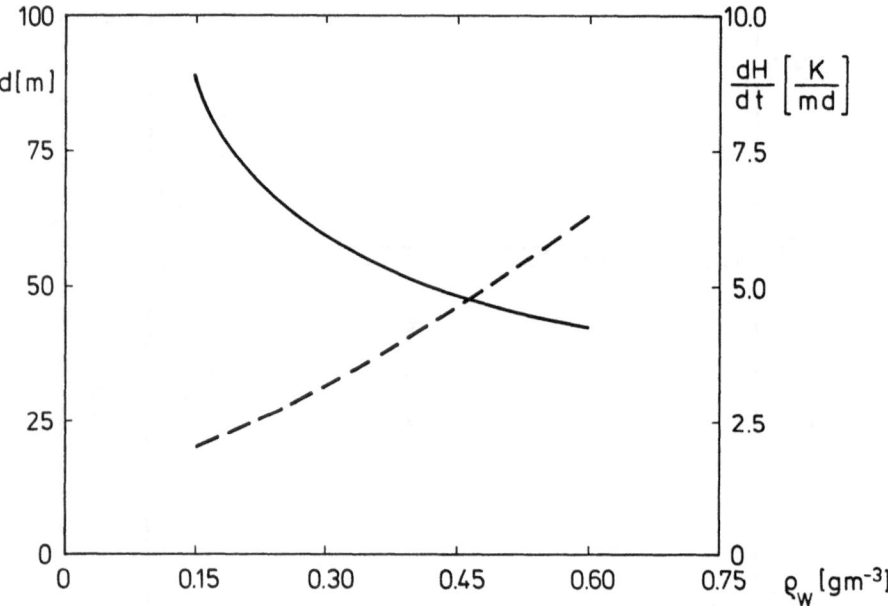

Fig. 2:   Depth d (——) and time change of average temperature
           lapse rate dH/dt (---) of the layer below the maximum
           cooling versus the maximum droplet water density.

3. Stability analysis
From a lot of studies it is well known that a horizontally homogeneous
fluid layer will become unstable and exhibit cellular structures if the
temperature gradient or rather its dimensionless form, the Rayleigh
number, exceeds a certain minimum value (10). Goody (11) found that
longwave radiation of such a fluid layer as an additional heat
transport mechanism is stabilizing, i.e. the critical Rayleigh number
increases. In this chapter the stability analysis is carried out for a

compressible layer of dry air, water vapor and water droplets, also considering radiative heat exchange and phase fluxes between water vapor and droplets. The Rayleigh number in the present paper is defined as

$$R = \frac{g \alpha \Delta \Gamma d^4}{\nu R}$$

with the acceleration of gravity g, the thermal expansion coefficient α, the excess of the actual over the moist adiabatic lapse rate $\Delta \Gamma$, the layer depth d, the turbulent heat excahnge coefficient R, and the turbulent viscosity ν.

For such a cloud layer the set of equations in the Boussinesq approximation is given by:

a - continuity equation: $\vec{\nabla} \cdot \vec{V} = 0$

with the wind vector $\vec{V} = (u, v, w)$

b - momentum equation:

$$\frac{d\vec{v}}{dt} = -\vec{\nabla} \frac{p}{\rho_o} + \nu \nabla^2 \vec{v} - \frac{\rho}{\rho_o} g\vec{k}$$

with an average density ρ_o of the base state, the actual density ρ, the pressure p, a unit vector $\vec{k}$ in the vertical direction and the total time derivative $d/dt = \partial/\partial t + \vec{v} \cdot \vec{\nabla}$

As boundary conditions for the velocity the usual vanishing of the vertical velocity w and free slip conditions for the horizontal flow (resulting in $\partial^2 w/\partial z^2 = 0$) are applied at both boundaries.

c - energy equation:

$$\overline{\rho c} \frac{dT}{dt} = \frac{dp}{dt} + k_T \nabla^2 T - L J_v - \vec{\nabla} \cdot \vec{F_N}$$

with temperature T, the heat capacity of a unit volume $\overline{\rho c}$, the turbulent heat diffusion coefficient k_T, the latent heat of condensation L, the phase flux J_v between water vapor and liquid water and the divergence of the net radiation flux $\vec{\nabla} \cdot \vec{F_N}$

d - water vapor balance:

$$\frac{d\rho_v}{dt} = k_v \nabla^2 \rho_v + J_v$$

with the water vapor density ρ_v and the turbulent exchange coefficient k_v for water vapor. This equation is used to specify the phase flux J_v in the energy equation. As we deal with a cloud layer it is always assumed, that the water vapor density ρ_v is given by the saturation vapor density ρ_s and, thus, by the temperature.

e - equation of state:

$$\rho = \rho_R (1 + \alpha (T_R - T))$$

with a reference temperature T_R and a reference density ρ_R

f - radiative transfer equation in the Eddington approximation

$$\vec{\nabla} \cdot \vec{F_N} = -4 \pi \beta (I_o - B(T))$$

$$\nabla^2 I_o = 3 \left(\frac{\beta}{\chi_s}\right)^2 (I_o - B(T))$$

with the radiance I_o as the isotropic part of the radiation field, the Stefan-Boltzmann expression for the energy source $B(T) = \frac{\sigma}{\pi} T^4$, the volume absorption coefficient ß and a measure for the importance of scattering χ_s. For practical calculations in clouds χ_s is almost 1 and does hardly influence the following arguments.

The RTE is written in a wavelength integrated form, as radiation calculations for clouds show, that it is possible to treat a cloud layer to a first approximation as a grey radiator (12).

At both boundaries constant diffuse radiation flux into the layer is assumed

$$F_{\downarrow T \atop \uparrow B} = \varepsilon_{T \atop B} \sigma T_{T \atop B}^4$$

where $F_{\downarrow T}$ is the downward flux at the top and $F_{\uparrow B}$ the upward flux at the bottom, T_T and T_B are the respective temperatures and ε_T and ε_B appropriate emissivities of the adjacent layers. This means that radiation from outside the considered layer does not react at an eventual perturbation of the radiation field inside the layer. In terms of I_o these boundary conditions are

$$I_o \pm \frac{2}{3} \frac{\chi_s^2}{\beta} \frac{\partial I_o}{\partial z} = \frac{1}{\pi} F_{\downarrow T \atop \uparrow B}$$

As usual, all variables are splitted up into a horizontally homogeneous base state and a perturbation, which leads to two sets of equations. The base state equations are not solved explicitly. It is assumed, that such a state exists in accordance with the equations, that it is motionless, that the wet bulb potential temperature gradient is constant with height, that the liquid water density is constant throughout the layer, and that the time dependence of this state is small compared to the time changes of the starting convection. The base state radiation distribution is calculated from the RTE with the assumed values for temperature and droplet water.

The continuity equation for the perturbations is $\vec{\nabla} \cdot \vec{v}' = 0$ and the momentum equation is

$$\frac{\partial}{\partial t} \nabla^2 w' = \nu \nabla^4 w' + g\alpha \left(\frac{\partial^2}{\partial x^2} + \frac{\partial^2}{\partial y^2} \right) T'$$

In the perturbed state the variation of the saturation vapor density with temperature is linearized

$$\rho_s(T) = \rho_{sR} + \phi(T - T_R)$$

with $\phi = \frac{\partial \rho_s}{\partial T}\big|_R = \frac{\rho_s}{T_R}\left(\frac{L}{R_v T_R} - 1 \right)$; ρ_{sR} being a reference saturation density and T_R a reference temperature. With the definitions

$$\Phi = \frac{L \phi}{\rho c} , \quad \hat{k} = \frac{k_T + \Phi k_v}{\rho c(1 + \Phi)} , \quad \Delta \Gamma = \frac{\partial T_o}{\partial z} - \frac{\partial T}{\partial z}\big|_{m.a.}$$

where $\frac{\partial T}{\partial z}\big|_{m.a.}$ represents the moist adiabatic lapse rate, the energy equation is

$$\frac{\partial T'}{\partial t} - \frac{1}{\bar{\rho}c(1+\Sigma)} \frac{\partial p'}{\partial t} = -w'\Delta\Gamma + \hat{\kappa}\nabla^2 T' - \frac{1}{\bar{\rho}c(1+\Sigma)}\vec{\nabla}\cdot\vec{F}$$

As the radiative cooling in the cloud is almost solely due to water droplets, the volume absorption coefficient is approximately given by ß = $\kappa\,\rho_w$, where κ is the mass absorption coefficient of droplet water. A perturbation of the droplet water mass yields ß = $\kappa(\rho_w^o + \rho_w')$. With the assumption that the water vapor density equals the saturation value, the water mass perturbation can be related to the temperature perturbation and, thus, ß = $ß_o - \kappa\phi T'$. Linearizing the source function with respect to temperature yields for the RTE

$$\vec{\nabla}\cdot\vec{F}' = -4\pi ß_o\left[I_o' - \left(4\frac{\sigma}{\pi}T_o^3 + (I_o^o - ß_o)\frac{\phi}{\rho_w^o}\right)T'\right]$$

$$\nabla^2 I_o' - 3\left(\frac{ß_o}{\chi_s}\right)^2 I_o' = -3\left(\frac{ß_o}{\chi_s}\right)^2\left[2(I_o^o - ß_o)\frac{\phi}{\rho_w^o} + 4\frac{\sigma}{\pi}T_o^3\right]T' - \frac{\partial I_o^o}{\partial z}\frac{\phi}{\rho_w^o}\frac{\partial T'}{\partial z}$$

The boundary conditions at the top and bottom of the layer are

$$w' = \frac{\partial^2 w'}{\partial z^2} = T' = I_o' \pm \frac{2}{3}\frac{\chi_s^2}{ß_o}\left(\frac{\partial I_o'}{\partial z} + \frac{\partial I_o^o}{\partial z}\frac{\phi}{\rho_w^o}T'\right) = 0$$

The equations are nondimensionalized with the layer depth d as a length scale and the diffusion time scale $\hat{\kappa}/d^2$ and the following separation of variables is applied

$$\begin{pmatrix}T'\\w'\\I_o'\end{pmatrix} = \begin{pmatrix}\theta(z')\\\frac{\hat{\kappa}}{d^2\Delta\Gamma}w(z)\\4\frac{\sigma}{\pi}T_o^3 J(z)\end{pmatrix}e^{\tilde{\sigma}t' + i(a_x x' + a_y y')}$$

$\tilde{\sigma}$ is the growth rate of the developing perturbations and a_x , a_y are dimensionless wave numbers that appear only in the combination $a^2 = a_x^2 + a_y^2$. Dimensionless time and space coordinates are characterized by a dot as an upper index (e.g. $z^{\bullet} = z/d$). Treating only the case of marginal instability $\tilde{\sigma} = 0$, which characterizes the exchange of stability between pure diffusion and growing convective instabilities, and eliminating the temperature leads to the following two equations

$$\left[D^2 - a^2 - \frac{3\chi\tau_o}{1+\Sigma}\left(1 + \frac{A}{4\tau_o}(\tilde{I}_o^o - \tilde{ß})\right)\right](D^2 - a^2)^2 w + Ra^2\left(w + \frac{3\chi\tau_o}{1+\Sigma}J\right) = 0$$

$$(D^2 - a^2 - \tilde{\tau}^2)J + \frac{\tilde{\tau}^2}{Ra^2}\left(1 + \frac{A}{2\tau_o}(\tilde{I}_o^o - \tilde{ß}_o)\right)(D^2 - a^2)^2 w + \frac{A}{4\tau_o}\frac{\tilde{I}_o^o}{\partial z^{\bullet}}\frac{\partial}{\partial z}(D^2 - a^2)^2 w = 0$$

with the boundary conditions

$$W = D^2 W = D^4 W = J \pm 2\frac{\tau_o}{\tilde{\tau}^2}\frac{\partial J}{\partial z} = 0$$

where $\tilde{I}_o^o = I_o^o/(\frac{\sigma}{\pi}T_o^4)$, $\tilde{ß}_o = ß_o/(\frac{\sigma}{\pi}T_o^4)$ and $D^2 = \frac{\partial^2}{\partial z^2}$. The parameter $\chi = 16 d\sigma T_o^3/(3 k_T)$ measures the relative importance of radiation and heat diffusion, $\tau_o = ß\cdot d$ is the optical depth of the base state, $\tilde{\tau} = \sqrt{3}\,\tau_o/\chi_s$ is a modified optical depth that accounts for scattering, and the parameter A = $\kappa\phi d/\alpha$ measures the change of the optical depth of the layer due to phase transitions connected with a temperature change.

An approximate solution of the form $W = \sum_i W_i \sin(i\pi z)$ is used in accordance with the boundary conditions for W. After some tedious algebra an analytic expression for the Rayleigh number can be found

$$R = \frac{1}{2a^2}\left[(N_1^6 R_1 + N_2^6 R_2) + (N_1^6 R_1 - N_2^6 R_2)\sqrt{1 + 4\frac{R_3\, N_1^4 N_2^4}{N_1^6 R_1 - N_2^6 R_2}}\,\right]$$

where R_1 , R_2 , R_3 are complicated functions of the wavenumber and the various radiation parameters.

$$N_1^2 = \pi^2 + a^2, \qquad N_2^2 = (2\pi)^2 + a^2$$

In the radiation free case ($R_3 = 0$, $R_1 = R_2 = 1$) this expression reduces to $R = N_1^6/a^2$ which is the solution to the classical Rayleigh problem yielding $R_{c_0} = 657.5$ and $a_{c_0} = 2.22$. In the limit, that radiative heat transport is of no importance in damping or amplifying perturbations, the solution for a moist layer is the same as for a dry layer but with a different definition of the Rayleigh number.

Fig. 3 shows the critical Rayleigh and wave number versus the relative importance χ of radiative and diffusive heat transport. In case of a dry radiating gas (full line, $\Phi = A = 0$) increasing importance of radiation leads to larger R_c . It has to be mentioned that in this dry case the $\Delta\Gamma$ in the definition of R is the normal potential temperature gradient. Radiation as an additional heat transfer mechanism acts in the same direction as heat diffusion and stabilizes the layer by damping perturbations. The critical wave number, too, grows with χ , so that the horizontal dimensions of the starting convection are smaller than in the radiation free case. Allowing for the latent heat of condensation (A = 0 but $\Phi \neq 0$) gives essentially the same results (broken line), but with somewhat reduced deviations from the classical result (thin line). If radiation, for example, cools a volume, latent heat of condensation is set free to reduce this cooling. The dotted line results when the droplet perturbation is allowed to influence the radiative source terms as a perturbation of the layer optical depth ($A \neq 0$). When a volume is cooled by a negative temperature perturbation the emission perturbation $4\frac{\sigma}{\pi}T_0^3 T'$ will be negative due to the temperature dependence of the Stefan-Blotzmann expression. But at the same time condensation produces a positive perturbation of the liquid water density and, thus, the volume extinction coefficient. This second effect turns out to be more important and, therefore, the influence of radiation leads to a destabilization of the layer, i.e. smaller values of R_c .

It is interesting to note that radiation influences the stability mainly via cooling to space and less by the radiative exchange between perturbations of opposite sign. This is documented by the decrease of radiatively induced deviations from the classical stability analysis with increasing ε_T, i.e. with a decreasing possibility of cooling to space.

In Fig. 4 R_c and a_c are plotted versus the temperature of the lower boundary as a measure for the average layer temperature. In the dry case, the result is independent of the absolute temperature and indicates again stabilization by radiation. Introduction of phase fluxes reduces R_c and a_c the more, the higher the temperature of the layer is. This reduction is related to the greater amount of droplet water that is set free per unit temperature change at higher temperatures according to the Clausis Clapeyron equation.

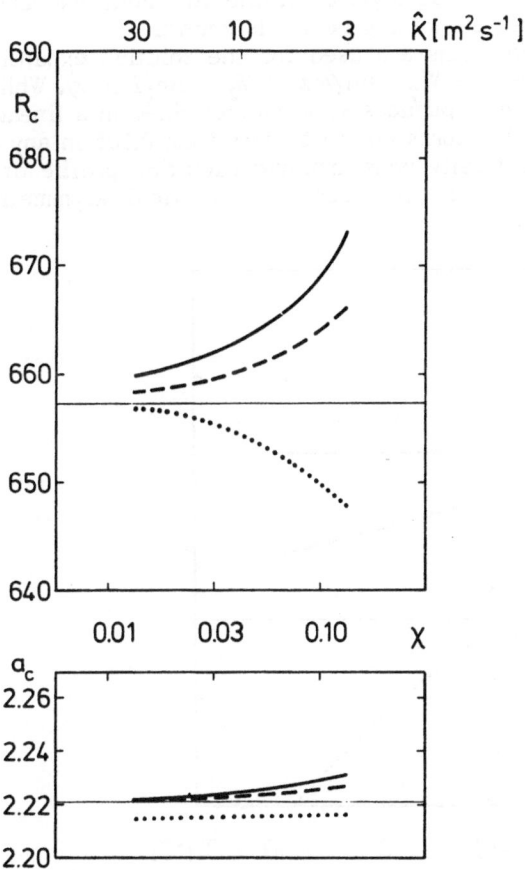

Fig. 3: Critical Rayleigh number (upper part) and wave number (lower part) versus the relativ importance of the radiative and diffuse heat exchange with (---) and without (——) considering latent heat of condensation and with the additional influence of droplet water perturbations on optical depth (. . .). The upper abszissa scale shows the corresponding value of the turbulent exchange coefficient.

These results show, that thermal radiation by cloud droplets does have
a certain influence on the stability problem. But while radiative
cooling is the main reason for the destabilization of the base state
temperature profile, it has a very weak influence on the critical
values of the Rayleigh and the wave number. In all the tested cases
the deviations of R_c and a_c from the classical values $R_{co} = 657.5$
and $a_{co} = 2.22$ never exceed 3%.
The stability analysis is performed only for parameter combinations
that are possible in cloud layers. All the turbulent exchange coefficients are set equal, as nothing better is known.
In praxis, only two terms are used for the solution expansion of the
vertical velocity $W(z) = W_1 \sin(\pi z) + W_2 \sin(2 \pi z)$. While the absolute values of the amplitudes stay undetermined in a linear problem,
the relation W_2 / W_1 turns out to be less than 0.005 in any case. This
means that the vertically nonsymmetric radiation profile of the base
state does not produce too much of a vertical asymmetry of the
perturbation solution.

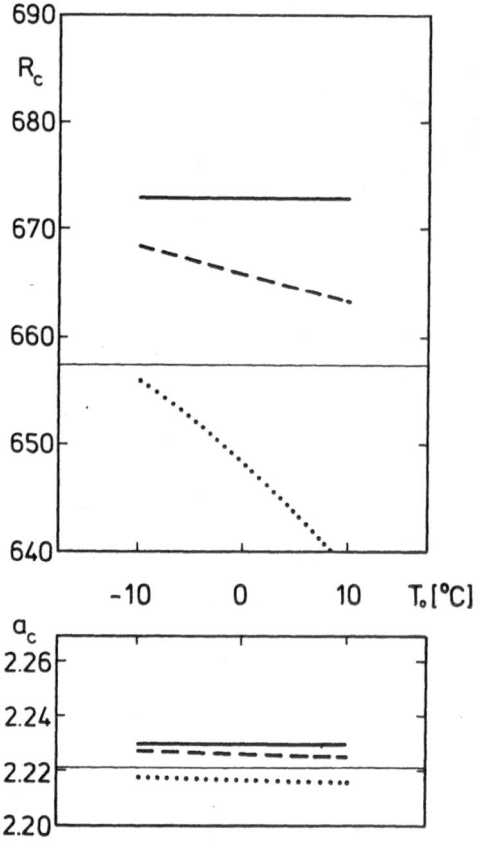

Fig. 4:   As Fig. 3, but versus the temperature of the lower boundary.

4. Two layer model

Several simplifications and approximations have been used to get to the results of the previous chapter. One of the most restricting conditions seems to be the choice of constant temperature ($T' = 0$) and vanishing vertical velocity ($w' = 0$) at the lower boundary of the considered layer. Such rigid plate conditions might be acceptable at the upper boundary which, in nature, is usually a strong inversion that damps vertical motion. But inside the cloud layer no such stable zone can be found. Therefore, in the following the lower boundary is replaced by an additional layer with neutral stratification in the base state. Hence, developing convection is not limited to the negative lapse rate layer but may extend deeper into the cloud. For simplicity this lower layer is taken to be infinitely extended. Both layers are assumed to be filled with cloud droplets, but radiation is not taken into account as it turned out to be of minor influence for the stability analysis. The perturbation equations for this model are:

layer 1: momentum equation

$$\frac{\partial}{\partial t} \nabla^2 w_1' = \nu \nabla^4 w_1' + g\alpha \left(\frac{\partial^2}{\partial x^2} + \frac{\partial^2}{\partial y^2} \right) T_1'$$

energy equation

$$\frac{\partial T_1'}{\partial t} = -w_1' \Delta \Gamma + \hat{K} \nabla^2 T_1'$$

layer 2: momentum equation

$$\frac{\partial}{\partial t} \nabla^2 w_2' = \nu \nabla^4 w_2' + g\alpha \left(\frac{\partial^2}{\partial x^2} + \frac{\partial^2}{\partial y^2} \right) T_2'$$

energy equation

$$\frac{\partial T_2'}{\partial t} = \hat{K} \nabla^2 T_2'$$

Separation of variables in analogy to chapter 3 leads in the marginal case $\partial/\partial t = 0$ to two equations for the amplitude of the temperature perturbation

$$(D^2 - \alpha^2) \, \Theta_1(z') = -R\alpha^2 \, \Theta_1(z')$$

$$(D^2 - \alpha^2) \, \Theta_2(z') = 0$$

The solution of the second equation is $\Theta_2(z') = \Theta e^{\alpha z'}$. A trial solution for $\Theta_1(z')$ is given by $\Theta_1(z') = \sin c(1-z')$ which solves the differential equation and fulfills the upper boundary conditions. These two solutions have to be matched at the interface where it is required that

$$T_1' = T_2' \qquad \text{and} \qquad \frac{\partial T_1'}{\partial z} = \frac{\partial T_2'}{\partial z}$$

This leads to the solution condition $c \cdot \text{ctg} \, c = -a$, which holds for a spectrum of c's due to the periodicity of the ctg. For the Rayleigh number the introduction of these solution yields

$$R = \frac{(c^2 + a^2)^3}{a^2}$$

In analogy to the classical result for one layer any value of c corresponds to a critical Rayleigh number R_c , the smallest of which appears with the smallest value of c:

$$R_c = 133.8 \quad , a_c = 1.025, \qquad c_c = 2.04$$

This reduction of R_c and a_c is due to the effective increase in the convection depth by allowing penetration of the perturbations into the neutral layer. Fig. 5 shows the vertical variation of the amplitude of the temperature variation as compared to the rigid boundary case. In a cloud it is expected that radiatively driven cellular convection is not only confined to the radiative cooling layer but may penetrate deeper into the cloud. The following table shows the critical potential temperature difference in °C between the layer boundaries with separation d depending on the turbulent exchange coefficient K.

d in m	50	100	150
K in m^2/s			
3	0.30	0.04	0.01
10	2.98	0.37	0.11

There is very little data on the value of the exchange coefficient in cloud layers. After Vinnichenko et al. (13) 10 m^2/s seems to be an upper limit.

The calculated cooling rates lead to the onset of instability within a few seconds to several minutes. Therefore, convective instability has to be expected on the top of any layer cloud that is able to radiate freely to space. In the case of larger exchange coefficients turbulent mixing in clouds could be faster than radiative cooling. This would deepen the unstable layer as compared to the case with negligible turbulent mixing. But as shown in the table, deepening of the layer leads to a smaller critical temperature difference, and again convective instability sets in, but with somewhat larger scales.

The maximum depth of the convectively unstable layer that one would expect from these arguments is 150 to 200 m. With the critical wave number of 1.025 this corresponds to a range of horizontal wavelengths between 300 and 1200 m. According to Fig. 5 these cells may penetrate as deep as about 400 m into the cloud layer.

Unfortunately, there seems to exist almost no quantitative material on the discussed phenomena, especially nothing is known about the depth of the observed structures. Therefore, a rigorous test of the theoretical results is impossible. At least, the observed horizontal scales of several 100 to 1000 m fit well into the theoretically predicted range of wavelengths.

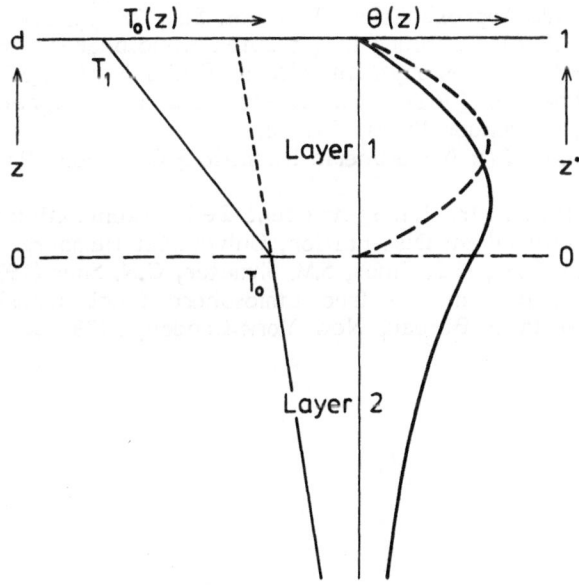

Fig. 5:   Vertical variation of the base state temperature profile
T_o (z) and the temperature perturbation amplitude Θ(z). Two
layer case (———), classical one layer case (---).

Literature:

1. Wegener, A. (1906): Studien über Luftwogen; Beitr. z. Phys. d. fr.
         Atmos., 2, pp. 55-72.
2. Borovikov, A.M. et al. (Ed.: Khrgian, A.K., 1961): Cloud Physics;
         Gidromet. Izdatelstvo, Leningrad (engl. translation by Israel
         Program for Scientific Translations, Jerusalem, 1963).
3. Solberg, H. (1936): Schwingungen und Wellenbewegungen in einer
         Atmosphäre mit nach oben abnehmender Temperatur; Astro-
         phys. Norv., 2, pp. 123-172.
4. Sekera, Z. (1948): Helmholtz Waves in a Linear Temperature Field
         with Vertical Wind Shear; J. Met., 5, pp. 93-102.
5. Möller, F. (1943): Labilisierung von Schichtwolken durch Strahlung;
         Meteorol. Zeitschr., 60, pp. 212-213.
6. Feigelson, E.M. (1964): Light and Heat Radiation in Stratus Clouds;
         Izdatelstvo "Nauka", Moskow (engl. translation by Israel
         Program for Scientific Translations, Jerusalem, 1966).
7. Joseph, J.H., W.J. Wiscombe, J.A. Weinman (1976): The Delta-
         Eddington Approximation for Radiative Transfer; J. Atmos.
         Sci., 33, pp. 2452-2459.

8. Shettle, E.P., J.A. Weinman (1970): The Transfer of Solar Irradiance Through Inhomogeneous Turbid Atmospheres Evaluated by Eddington's Approximation; J. Atmos. Sci., 27, pp. 1048- 1055.
9. Grassl, H. (1978): Strahlung in getrübten Atmosphären und in Wolken; Hamb. Geophys. Einzelschr., Reihe A, Heft 37.
10. Chandrasekhar, S. (1961): Hydrodynamic and Hydromagnetic Stability; Clarendon Press, Oxford.
11. Goody, R.M. (1964): Atmospheric Radiation; Clarendon Press, Oxford.
12. Bakan, S. (1982): Strahlungsgetriebene Zellularkonvektion in Schichtwolken; Dissertation, Universität Hamburg.
13. Vinnichenko, N.K., N.Z. Pinus, S.M. Shmeter, G.N. Shur (1973): Turbulence in the free atmosphere (engl. translation by Consultants Bureau, New York-London, 1973, Ed. J.A. Dutton).

INFRARED REMOTE SENSING OF THE VERTICAL AND HORIZONTAL DISTRIBUTION OF CLOUDS

Moustafa T. Chahine

Jet Propulsion Laboratory
California Institute of Technology
Pasadena, California 91109

Robert D. Haskins
Institute for Atmospheric Optics and Remote Sensing
Hampton, Virginia 23666

An algorithm has been developed to derive the horizontal and vertical distribution of clouds from the same set of infrared radiance data used to retrieve atmospheric temperature profiles. The method leads to the determination of the vertical atmospheric temperature structure and the cloud distribution simultaneously which provides information on heat sources and sinks, storage rates and transport phenomena in the atmosphere. Experimental verification of this algorithm was obtained using the 15 μm data measured by the NOAA-VTPR temperature sounder. After correcting for water vapor emission, the results show that the cloud cover derived from 15 μm data is less than that obtained from visible data.

1. INTRODUCTION

Clouds play a major role in the radiative processes in the Earth's atmosphere both in the absorption and reflection of solar radiation and in the emission of thermal energy. A knowledge of the vertical location, the horizontal distribution and the optical properties of clouds is critical in determining the driving mechanisms for motions in the atmosphere and oceans.

Infrared atmospheric temperature sounding data permit the determination of the horizontal and vertical distribution of clouds provided that all the sounding channels observe the same field of view at the same time. The requirement of simultaneous observations is necessary because clouds are usually inhomoge-

57

E. M. Agee and T. Asai (eds.), Cloud Dynamics, 57–71.
Copyright © 1982 by D. Reidel Publishing Company.

neous in space and time. Chahine (1970) investigated the problem
of extracting information about the horizontal distribution, N,
of clouds and their pressure level, p_c, from radiance data used
to recover atmospheric temperature profiles, T(p). A simple
minimization approach was used by Chahine to determine the
combination $[N,p_c]$ which gives the best agreement between the
measured and calculated radiances. Limited verifications of this
approach were subsequently carried out by Shaw *et al* (1970),
McCleese and Wilson (1976), Chahine *et al* (1977b), Smith and
Platt (1978) and Wielicki and Coakley (1981). In this paper we
will apply the method to retrieve global cloud data from the 15
μm data measured by the Vertical Temperature Profile Radiometer
(VTPR) sounder flown on the NOAA weather satellite.

2.   APPROACH

The upwelling radiance from a planetary atmosphere is a
function of the thermal state of the atmosphere, the concentration of radiatively active gases, and the extents, heights, and
radiatively transfer properties of clouds and aerosols. Thus, in
principle, it should be possible to recover useful information
about the physical and chemical structure of an atmosphere from
analysis of the upwelling radiance. However, the problem in
analyzing such data lies in finding ways to uncouple the effects
of these variables and retrieve the true values of each unknown
separately. *By treating the cloud effects as short term
oscillations over the clear-column radiance*, an analytical method
was developed by Chahine (1974) to retrieve clear-column vertical
temperature profiles from radiance measurements made in the
presence of clouds. The method requires radiance data measured
in two spectral regions and over two adjacent fields of view
having different amounts of clouds. The uncoupling of the
effects of clouds is carried out without any a prior information
about the amounts, heights and optical properties of the clouds
in the fields of view. Once the clear-column temperature
profiles are determined the same radiance data could then be used
to determine the heights, amounts, and radiative transfer
properties of clouds.

Formulation of this approach is straightforward. It will be
carried out here in two steps:

First, we consider the radiance $\tilde{I}(\nu)$ measured at frequency ν
in the presence of clouds, and define the corresponding clear-
column radiance $I(\nu)$ as

$$I(\nu) = B(\nu,T_s) \ \tau(\nu,p_s) + \int_{\ln p_s}^{-\infty} B[\nu,T(p)] \ \frac{\partial \tau(\nu,p)}{\partial \ln p} \ d \ln p \qquad (1)$$

where $\tau(\nu,p)$ is the transmittance of a clear-column of gaseous absorbers between pressure level p and the sounder. B is the black body Planck function and T(p) is the clear-column vertical temperature profiles. Next we express the difference between $I(\nu)$ and the measured $\tilde{I}(\nu)$ in terms of an expansion function $G(\ ,p)$ and an expansion coefficient N as

$$I(\nu) - \tilde{I}(\nu) = NG(\nu,p_c, \ldots) + N'G'(\nu,P_c,\ldots) \tag{2}$$

In this first step we don't need to define the form of $G(\nu,p)$ because we aim to eliminate it. Detailed discussion of this approach is given by Chahine (1977a). For simplicity we take here one expansion term only and consider observations made over two adjacent fields of view (subscripts 1 and 2) having different amounts of clouds, such that $\tilde{I}_1 \neq \tilde{I}_2$ and write Eq. (2) as

$$I_1(\nu) - \tilde{I}_1(\nu) = N_1 G(\nu,p)$$

$$I_2(\nu) - \tilde{I}_2(\nu) = N_2 G(\nu,p). \tag{3}$$

Now, if the true fields of view are contiguous we can assume that $I_1(\nu) \simeq I_2(\nu)$ and drop their subscripts. Equation (3) becomes then

$$I(\nu) = \tilde{I}_1(\nu) + \eta[\tilde{I}_1(\nu) - \tilde{I}_2(\nu)] \tag{4}$$

$$\text{where} \quad \eta = \frac{N_1}{N_2 - N_1} = \text{constant}$$

Determination of T(p) and $I(\nu)$ will be discussed in Section 3 and it follows the methods described by Chahine (1974, 1975, 1977).

In the second step we substitute the value of $I(\nu)$ into Eq. (3) and write

$$N_k G(\nu,p_c, \ldots) = I(\nu) - \tilde{I}_k(\nu) \tag{5}$$

for k=1,2. The left-hand side of Eq. (5) represents the radiative transfer effects of the clouds and the terms on the right-hand side are all known. The determination of the cloud amount N_k and cloud top height P_c from Eq. (5) will require an assumption of a cloud model as described in Section 4.

3.   DETERMINATION OF THE CLEAR-COLUMN RADIANCE

The determination of clear-column vertical temperature profile T(p) automatically gives the value of the clear-column radiance I(ν) according to Eq. (1). However, in order to solve Eq. (4), to get T(p) one needs to know the value of η. Since for every value of η one can obtain a corresponding temperature profile $T_\eta(P)$, we need an additional piece of information, such as some appropriate microwave data, in order to determine the value of η which satisfies this property. Unfortunately no microwave data were available in 1975 in connection with the VTPR. Therefore, in order to select the correct value of η for $T_\eta(p)$ we assumed a priori knowledge of the surface temperature T_s and developed an iterative scheme which solved simultaneously for η and T(p) and gave a value of T(p) at the surface P_s which agreed with T_s or

$$T(p_s) = T_s.$$

We obtained T_s from the NOAA surface analyses and proceeded to recover T(P) according to the method of Chahine (1974). Once T(p) is determined, the determination of the corresponding clear-column radiance I(ν) can be calculated directly according to Eq. (1).

4.   DETERMINATION OF THE CLOUD PARAMETERS

Once I(ν) is determined we substituted its value on the right-hand side of Eq. (15) and consider the left-hand expression.

In order to determine the cloud parameters we will assume that the difference between the clear-column radiance and the radiance measured in a given field of view is solely due to clouds. The values of the parameter N, and p_c will depend on the specific mathematical form of the expansion functions $G(\nu,p_c)$ i.e., on the cloud model adopted.

To deduce the properties of clouds one is faced, in principle, with the problem of solving the complete equation of transfer in a Mie scattering medium. Obviously, this approach is not practical for use with temperature sounding data, mainly because of the nature of the spectral measurements used. However, we find temperature sounding data to be very useful in cloud studies when used in connection with simple radiative transfer cloud models.

For the case of the VTPR sounding data, which cover the spectral range from 12-15 μm, we will assume that the cloud effects expressed in Eq. (5) are due to the presence of a single layer of non-reflecting clouds. In this case we will conveniently rewrite Eq. (5) as

$$N_k[\varepsilon_c(\nu) \; \tilde{G}(\nu,P_c,\ldots)] = I(\nu) - \tilde{I}_k(\nu) \qquad (6)$$

where $G(\nu,P_c) = \varepsilon_c(\nu)\tilde{G}(\nu,P_c)$ and $\varepsilon_c(\nu)$ is the cloud emissivity. If we assume that

$$\varepsilon_c(\nu) + \tau_c(\nu) = 1$$

where $\tau_c(\nu)$ is the cloud transmissivity, Chahine (1982) showed that $\tilde{G}(\nu,P)$ can be expressed as

$$\tilde{G}(\nu,p_c) = B(\nu,T_s) \; \tau(\nu,p_s) + \int_{\ln p_s}^{\ln p_c} B[\nu,T(p)] \frac{\partial \tau(\nu,p)}{\partial \ln p} \, d \ln p \qquad (7)$$

$$- B[\nu,T(p_c)] \; \tau(\nu,p_c) \; .$$

The only unknown on the right-hand side is the cloud top pressure level p_c. Determination of p_c can be accomplished by considering two sounding frequencies, say ν_1 and ν_2 for which $\tau(\nu,p_s)$ is preferably small in order to minimize the effect of changes in earth surface emissivity. From Eq. (6) and Eq. (7) we assume $\varepsilon_c(\nu_1) = \varepsilon_c(\nu_2)$ and write for say the first field of view

$$\frac{G(\nu_1,p_c)}{G(\nu_2,p_c)} = \frac{I(\nu_1) - \tilde{I}_1(\nu_1)}{I(\nu_2) - \tilde{I}_1(\nu_2)} \qquad (8)$$

and solve for p_c by minimization in a manner similar to the approach described by Chahine (1982) in Section 4.a.

The values of the expansion coefficient N in this case cannot be separated from the cloud emissivity. We will therefore define the cloud opacity $N_k\varepsilon_c(\nu)$ as the apparent infrared cloud cover and denoted it as

$$\tilde{N}_k(\nu) = N_k \; \varepsilon_c(\nu) \; .$$

We obtain the value of $\tilde{N}_k$ directly as

$$\tilde{N}_k = N_k \quad \varepsilon_c(\nu) = \frac{I(\nu_1) - I_k(\nu_1)}{\tilde{G}(\nu_1, p_c)} . \tag{9}$$

In practice, because of the instrument noise and uncertainties in the atmospheric temperature profiles $T(p)$, we chose as a solution the set $[\tilde{N}, p_c]$ which minimizes not only Eq. (8) but also the radiance residuals of all of the temperature sounding frequencies which are affected by the clouds.

The minimization approach adopted here requires taking different values, $p_c^{(i)}$ for p_c

$$p_c = p_c^{(1)} < p_c^{(2)} < p_c^{(3)} , \dots < p_c^{(n)} \tag{10}$$

then computing the corresponding values $\tilde{N}^{(i)}$ from Eq. (9). We take as solution $[\tilde{N}^{(i)}, p_c^{(i)}]$ which absolutely minimizes

$$\tilde{N}_k \tilde{G}(\nu_j, p_c) = I(\nu_j) - \tilde{I}_k(\nu_j) \tag{11}$$

for all the temperature sounding channels which are effected by clouds, $j = 1, 2, \dots J$. This method leads to unique solutions provided that the temperature profile $T(p)$ is monotonic (or not isothermal) between $p_c^{(1)}$ and $p_c^{(n)}$.

5.   APPLICATION TO VTPR DATA

We applied the single-layer non-reflecting cloud model to analyze 15 μm temperature sounding data from the NOAA-VTPR sounder. The sounding frequencies of the VTPR shown in Fig. 1 cover the range between 668.5 and 747.0 cm^{-1}.

The determination of clear-column temperature profiles from the VTPR data requires *a priori* knowledge of the surface temperature T_s. We obtained our T_s from the NOAA surface analysis and proceeded to recover $T(p)$ according to the method of Chahine (1974). We investigated the effects of errors in the assumed surface temperature on the accuracy of the values of $\tilde{N}$ and p_c and concluded that the effects of $\pm$ 2K error in T_s on $\tilde{N}$ and p_c are small, especially for $p_c < 700$ mb. We investigated also the sensitivity of the VTPR sounding frequencies to low-level clouds, with $p_c > 700$ mb, and concluded that such clouds cannot be determined from the 15 μm frequencies of the VTPR. These conclusions are in agreement with results shown in Fig. 8 of Chahine (1974), McCleese and Wilson (1976) and Wielicki and Coakley (1981). Corrections for the effects of water vapor on the transmission functions $\tau(\nu,p)$ were made before generating the clear-column radiances.

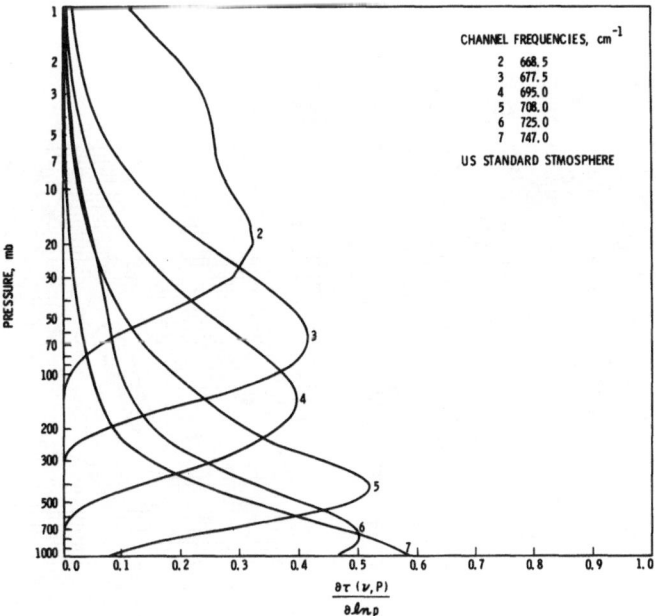

FIG. 1: Weighting functions of the VTPR temperature sounding channels.

It should be reemphasized here that while the determination of the clear-column radiance is obtained without assumptions about the types of the clouds, the determination of the amount and heights of the clouds requires the use of cloud models. In the case of this verification, we assumed that the difference between reconstructed clear-column radiance and the measured radiance in a given field of view is due to water vapor effects and to the presence of a single layer of non-reflecting clouds. We accounted for the effects of water vapor through the atmospheric transmission functions and proceeded to calculate $\tilde{N}$ and p_c.

The VTPR cloud maps illustrated in this section correspond to one week of data averaged from January 1 through 7, 1975. The cloud distributions were originally calculated in 1977 for the VTPR footprint size of 55 x 55 km in the nadir. These results were subsequently averaged to a grid size of 4° latitude and 5°

longitude. Only the averaged results were then stored for later comparison with other cloud maps to be obtained from different sources. A typical comparison of these results is shown in Figures 2, 3 and 4.

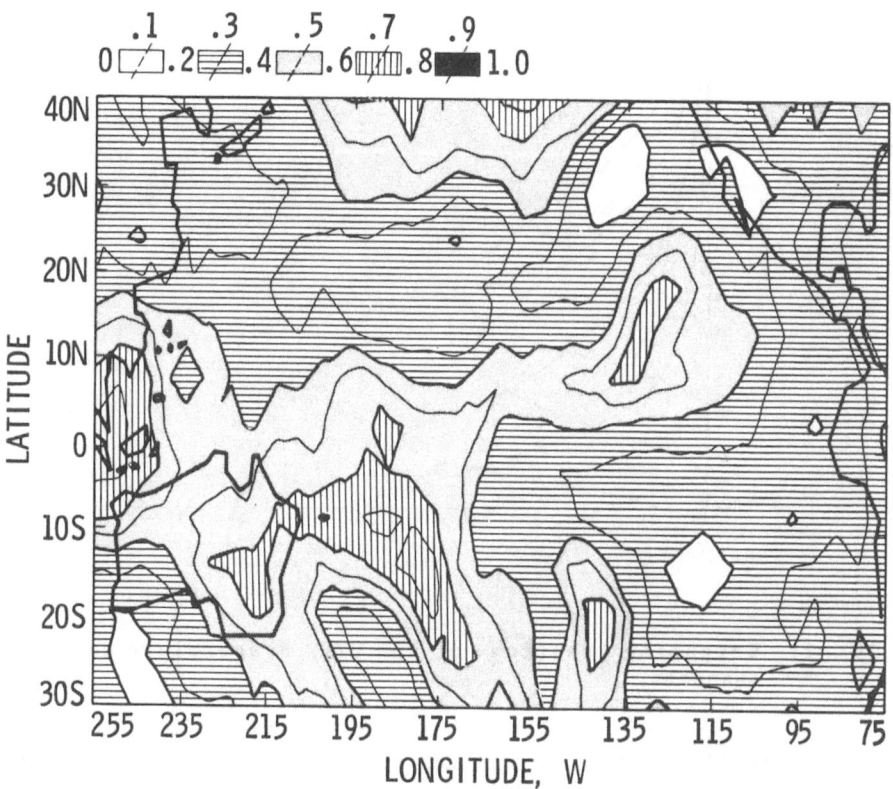

FIG. 2:   Contours of the effective cloud amounts derived from VTPR μm temperature sounding data for the period of January 1 through 7, 1975.

Comparison with Visible Data

Figure 2 shows contours of cloud amounts for a region across the Pacific Ocean between 40°N–30°S and 75°W–255°W, corresponding to one week of data averaged from January 1 through 7, 1975. Fig. 3 shows contours of cloud amounts derived by J. Sadler (1979), for the same region and period of time, from photographs obtained from the vidicon cameras of NOAA's satellite.

Sadler, however, uses a smaller grid size of 2.5° x 2.5° and a scale of 0 to 8 for the cloud amount as in Sadler (1968). A simple linear rectification algorithm was employed to change Sadler's data from the 2.5° x 2.5° grid size to 4° x 5° grid. To obtain the corresponding cloud cover values in decimals we simply divided Sadler's values by 8.

The results shown in Figs. 2 and 3, therefore, compare infrared cloud maps obtained sequentially from an orbiting sounder with visual cloud maps derived from pictures obtained instantaneously. Consequently, only persisting cloudiness appears to be common between the two types of cloud data. Therefore, the zonally averaged values shown in Fig. 4 offer a more realistic comparison between the amount of clouds observed in the infrared and visible parts of the spectrum.

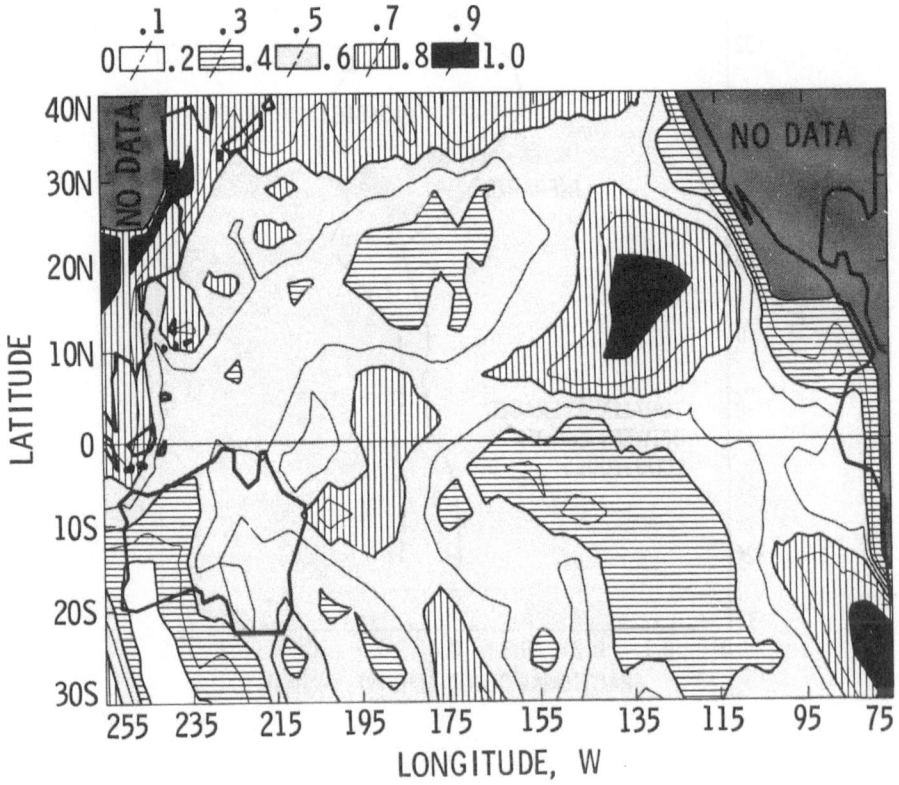

FIG. 3:   Contours of cloud amounts derived from visible data by J. Sadler for the period of January 1 through 7, 1975.

The conclusion that the apparent infrared cloud amount $\tilde{N} = N\varepsilon$ is smaller than the cloud amount observed in the visible is due to the fact that (1) the VTPR 15 µm sounding channels are not sensitive to low-level clouds discussed earlier, and (2) the cloud emissivity in the 15 µm region is less than unity as shown by Yamamoto *et al* (1970). The second property may be substantiated experimentally also by noting that in the region around 15°N and 130°W where low clouds are obscured by persistant high and dense layers of clouds, the 15 µm effective cloud cover remains smaller than the corresponding cover in the visible.

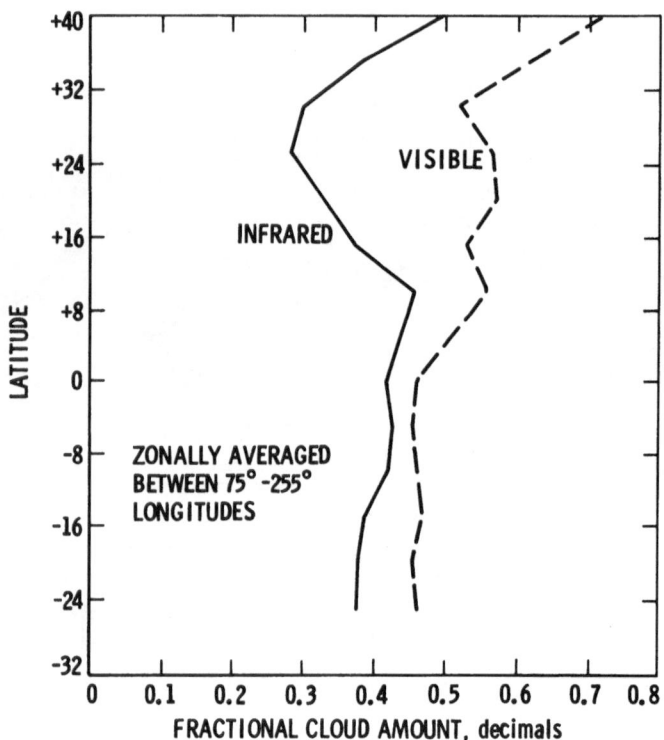

FIG. 4: Meridional profiles of zonally averaged distribution of fractional cloud cover corresponding to the data shown in Figs. 2 and 3.

The average cloudiness for the region shown in Figs. 1 and 2 is 0.39 for the 15 μm and 0.52 for the visible data. The ratio of the 15 μm cloud cover to the visible cloud cover is ~ 0.75.

Global 15 μm Cloud Maps

The complete global distribution of clouds in the 15 μm is given in Figs 5 and 6. Figure 5 shows the distribution of the effective cloud amounts in steps of 0.1. The retrieved cloud amounts varied between 0 and 0.89 on the grid size of 4° x 5°. A simple examination of some of the cloud details in Fig 5 shows that the major earth cloud formations are retrieved. We notice also that the VTPR 15 μm data did not detect the persistent low-level stratus off the west coasts of South America and Africa where a relatively large area with 0.5 to 0.8 cloud cover is usually observed in the visible. This is a direct consequence of the inherent insensitivity of the VTPR 15 μm data to clouds near the surface.

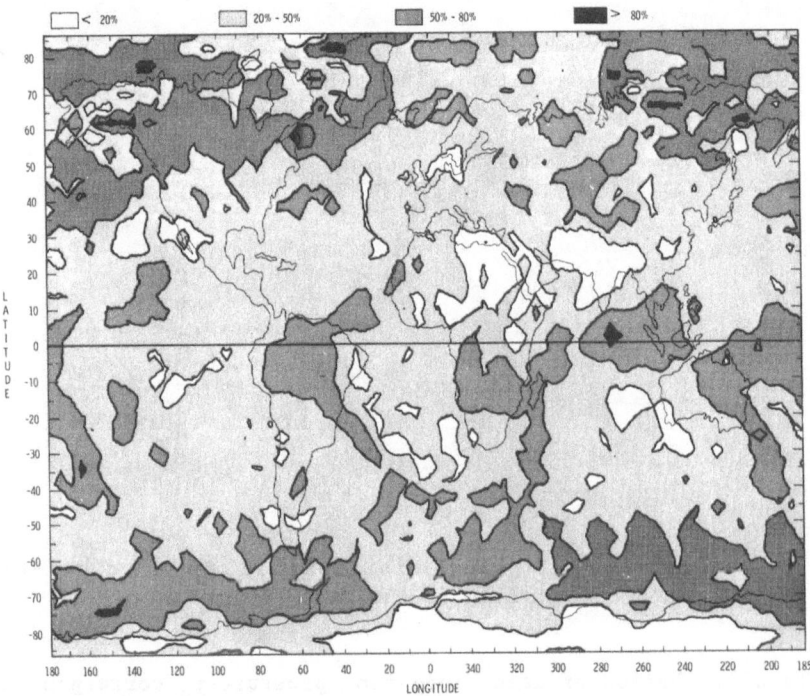

FIG. 5: Contours of the global distribution of the effective cloud cover derived from VTPR 15 μm temperature sounding data for the period of January 1 through 7, 1975.

Examination of the zonally averaged global distribution in Fig. 5 shows the expected maximum of cloudiness in the equatorial regions where horizontal convergence is strong. The predominance in the low latitude of convective clouds, which are local in character and cover only a part of the sky, is the reason for the minimum cloudiness shown in the latitudes of the subtropical anticyclones. From the subtropics cloudiness tends to increase polarward up to the ± 70th parallels.

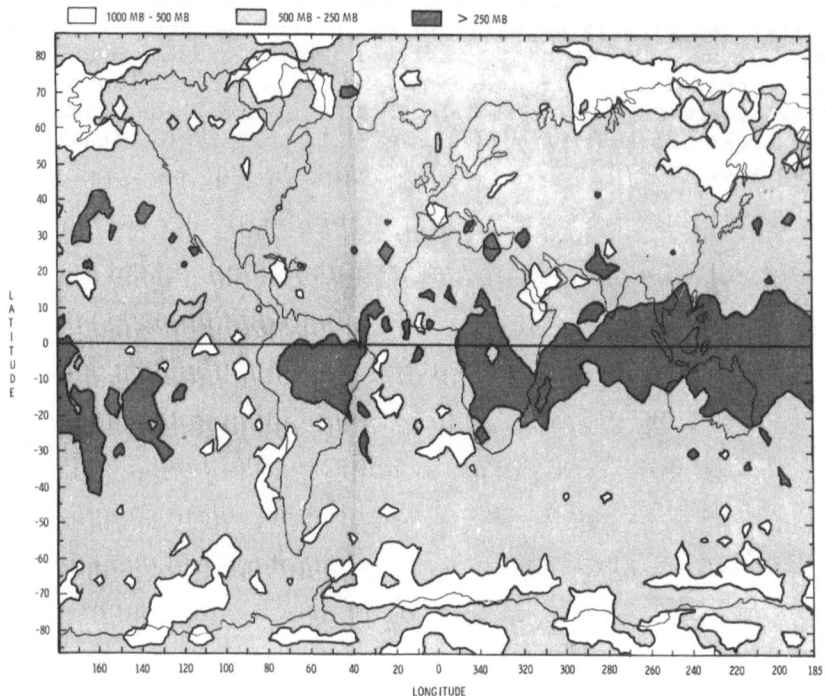

FIG. 6:   Contours of the global distribution of mean cloud-top heights corresponding to the case of Fig. 4.

The distribution of mean cloud-top pressure p_c corresponding to Fig. 5 is given in Fig. 6. On the grid size of 4° x 5° the derived values of p_c ranged between the surface (no clouds) and 145 mb. A close examination of the contours in Fig. 7 reveals that the results possess the basic features of the global circulation pattern and compare well with the small amount of available statistical information, as in Paltridge and Platt

(1976). No other cloud height data exist at present for the time period of this test. However, we should note there that the polar cloud-top pressure levels shown in Fig. 6 appear to be uniformly higher than expected. This result is due to the difficulty of locating the cloud-top height within a nearly-isothermal layer. The minimization-search sequence employed assigns the cloud-top height to the first minimum encountered, and, in the case of the polar regime, this is the top of the tropopause layer.

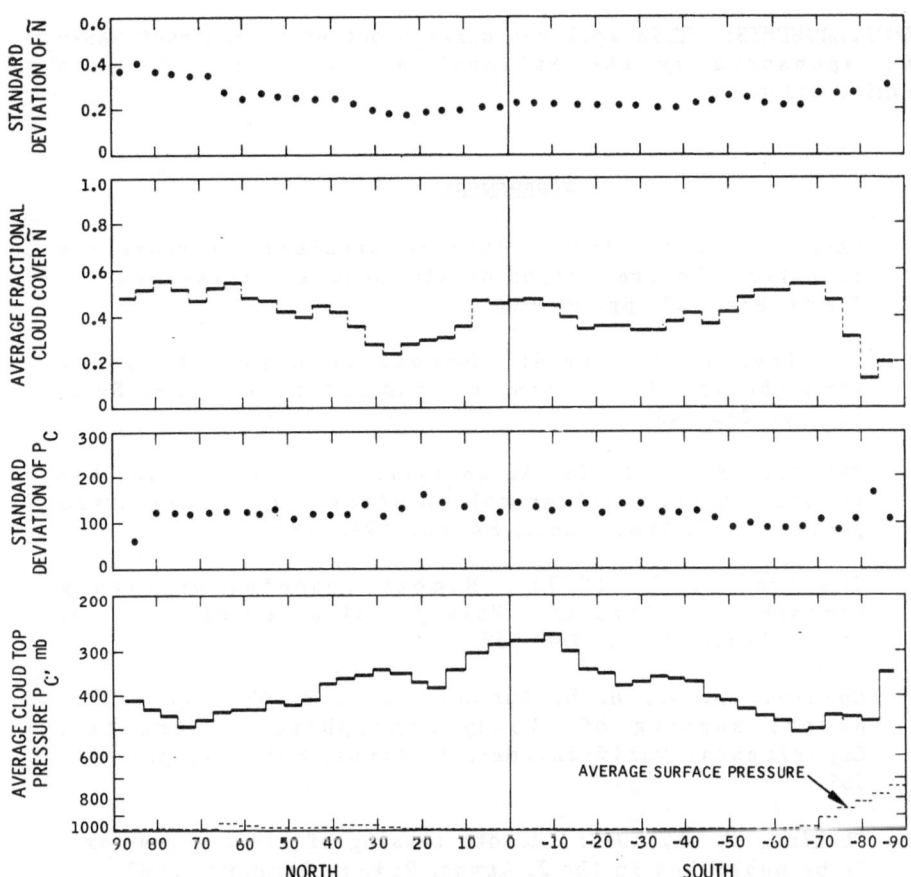

FIG. 7: Meridional profiles of zonally averaged distributions of the effective cloud amounts and mean cloud-top heights and their time-dependent standard deviations, corresponding to the period of January 1 through 7, 1975.

In Fig. 7, the global weekly average of the effective cloud cover as seen in 15 μm band is 0.41, corresponding to 0.40 for the northern hemisphere and 0.42 for the southern hemisphere. It is interesting to note here that the average annual climatological global cloud cover in the visible obtained by Sellers (1965) is 0.54, corresponding to 0.52 for the northern hemisphere and 0.57 for the southern hemisphere. The ratio of the global cloud covers in the 15 μm to the visible is still ~ 0.75 in spite of the difference in the time averaged used in Fig. 7.

ACKNOWLEDGMENTS: This work was carried out under contract NAS7-100, sponsored by the National Aeronautics and Space Administration.

REFERENCES

(1)   Chahine, M. T., 1970:   Inverse problems in radiative transfer: Determination of atmospheric parameters. J. Atmos. Sci., 27, pp. 960-967.

(2)   Chahine, M. T., 1974:   Remote sounding of cloudy atmospheres. I. The single cloud layer. J. Atmos. Sci., 31, pp. 233-243.

(3)   Chahine, M. T., 1975:   An analytical transformation for remote sensing of clear-column atmospheric temperature profiles. J. Atmos. Sci., 32, pp. 1946-1952.

(4)   Chahine, M. T., 1977a: Remote sounding of cloudy atmospheres. Part II. Multiple cloud formations. J. Atmos. Sci., 34, pp. 744-757.

(5)   Chahine, M. T., H. H. Aumann and F. W. Taylor, 1977b: Remote sensing of cloudy atmospheres. Part III. Experimental verifications. J. Atmos. Sci., 34, pp. 758-765.

(6)   Chahine, M. T., 1982:   Remote sensing of cloud parameters. To be published in the J. Atmos. Sci. in January, 1982.

(7)   McCleese, D. J. and L. S. Wilson, 1976:   Cloud top heights from temperature sounding instruments. Quant. J. R. Met. Soc., 102, pp. 781-790.

(8)   Paltridge, G. W., and C. M. R. Platt, 1976:   Radiative processes in meteorology and climatology. Elsevier Scientific Publishing Co. Amsterdam-Oxford-New York, pp. 1-34.

(9)   Sadler, J. C., 1968: Average cloudiness in the tropics
       from satellite observations. East-West Center Press.
       University of Hawaii, Honolulu, pp. 215.

(10)   Sellers, W. D., 1965: Physical Climatology. University of
       Chicago Press, Chicago, Ill. and London, 272 pp.

(11)   Shaw, J. H., M. T. Chahine, C. B. Farmer, L. D. Kaplan, R.
       A. McClatchey and P. W. Schaper, 1970: Atmospheric and
       surface properties from spectral radiance observations in
       the 4.3 micron region. J. Atmos. Sci., 27, pp. 773-780.

(12)   Smith, W. L., and C. M. R. Platt, 1978: Comparison of
       satellite-deduced cloud heights with indications from
       radiosonde and ground-based laser measurements. J. Appl.
       Meteor., 17, pp. 1796-1802.

(13)   Wielicki, B. A. and J. A. Coakley, 1981: Cloud retrieval
       using infrared sounder data: error analysis. J. of Appl.
       Meteor., 20, pp. 157-169.

(14)   Yamamoto, G., M. Tanaka and S. Asano, 1970: Radiative
       transfer in water clouds in the infrared regions. J.
       Atmos. Sci., 27, pp. 282-293.

LONG-LASTING PRECIPITATION CELLS IN STRATIFORM CLOUDS

H. SAUVAGEOT, R. AURIA and B. CAMPISTRON

Observatoire du Puy de Dôme
Centre de Recherches Atmosphériques, Campistrous
65300 Lannemezan, FRANCE.

ABSTRACT

    Distributions of reflectivity and air motion in a precipita-
ting stratiform cloud were observed by a recently developped
millimeter wave Doppler radar ($\lambda = 0.86$ mm). Precipitation trails
were found to originate near the cloud top in generating layer
exhibiting a cellular convective structure, with vertical velocity
up to 2.5 ms^{-1}. Each trail was generated in small and fairly compact
convective region exhibiting frequently a dynamical and microphysical structure of finer scale. The organization of this substructure appears strongly influenced by the vertical shear of the
environmental wind at the level of the convective layer. A conceptual
model for the relations between the vertical profile of the wind
and the substructure of the long-lasting precipitation generating
cells is tentatively proposed.

PRECIPITATIONS IN STRATIFORM CLOUDS

    Short range and high resolution radar observations show clearly
that pseudo continuous, steady and widespread precipitations from
stratiform clouds result in fact from spatially non uniform processes aloft. Such precipitations are generally produced by compact
generating elements located near the cloud top in a shallow generating layer. The resulting precipitation patterns are characteristic snow trails or streamers sloping and spreading under the
wind shear effect. Partial overlapping of the trails makes time and
space distributions of precipitation intensity near the ground more
uniform. Generating cells and associated precipitation trails or
fallstreaks are a proeminent characteristic of stratiform clouds.

E. M. Agee and T. Asai (eds.), Cloud Dynamics, 73–85.
Copyright © 1982 by D. Reidel Publishing Company.

In the past, generating cells and trails have been a topic of considerable experimental investigations (1 to 10). This kind of structure is that observed in thick stratiform clouds of extra-tropical cyclonic disturbances (11) particularly in warm frontal bands, in wide cold frontal bands and also sometimes in the other types of band during their decaying stage. Similar structure are observed too in cirrus clouds (12 to 16) and in thin stratiform clouds due to local effect and without direct relation with the cyclonic system of the general atmospheric circulation. For example, an association between convective cells and fallstreaks has been pointed out in altocumulus (17) and in roll shaped stratocumulus clouds (18). But generating cells and trails are also observed in some part of thick convective systems which can be locally and temporarily stratiform, for example near the top of decaying cumulonimbus clouds (19) or in tropical convective storm (20).

The main causes of development and persistence of instable layer aloft are a) a mesoscale upward velocity of about 10 cms^{-1} sufficient to release the potential instability necessary for cell development (6, 7) ; b) latent heat released when snow crystals grow at the expense of water vapor through deposition and the Bergeron-Findeisen effect in the cells (5) or in the precipitation trails (12, 18, 21) ; c) the cooling of cloud top and the heating of cloud bottom by radiative heat transfers (23, 16) and d) a differential advection of cold and dry air above the top of a warm saturated cloud layer (7). Several causes can be working simultaneously. As a whole, the concept of generating cells is applied to various structures showing between them some significant physical and dynamical differences. In addition the identification and delimitation of a generating layer and inside it of individual cells are sometimes imprecise.

Some precipitation generating cells in thick warm frontal clouds are apparently active for more than one hour and creates apparently continuous trails longer than 100 km. On the contrary, in thin stratiform clouds like some altocumulus or cirrocumulus clouds, smaller and shorter generating cells are generally observed. However a number of obscurities and uncertainties are remaining in the understanding of the microphysical and dynamical processes which, from combined actions, contribute to the production of long lasting precipitations : particle growth, radiative effects, regenerating mechanisms making the steadiness of the generating process, etc... In fact, the main problem is an experimental one linked to the difficulty to obtain significant data on the complete history of these processes, particularly on their dynamical aspects. Radars operating at millimeter wavelengths are able to observe the very small particles involved in the earliest stages of precipitation initiation and can therefore provide useful information on the subject. This paper discuss a set of millimeter wave Doppler radar

data where the generating cells appear to have a finer dynamical
structure probably not resolvable by most weather radar.

OBSERVATIONS AND RESULTS

The observations presented were made on 28 March 1981 above
the Plateau of Castilla, near Valladolid in Western Spain in the
frame of the Precipitation Enhancement Project (PEP) of the World
Meteorological Organization. The main tool used for the data
collection is the millimeter wave Doppler radar RABELAIS. The
technical characteristics of this radar as well as diverse informations on the measuring and data processing system are given in
annex. Other observation facilities included two 3 cm wavelength
radars located in the vicinity of RABELAIS and giving regularly PPI
informations on the evolution of the horizontal distribution of
precipitation echoes. Two instrumented aircrafts equiped with numerous atmospheric research instruments, including particle measuring
systems, were used for systematic local observations of cloud and
precipitation microstructure. A radiosounding station located near
the radars was operated 3 times a day during the working periods.

On 28 March 1981 a depression on Western Europe drove a
perturbed flow over Spain. Echo cells moved in the E-NE direction
above PEP site with velocity of about 50 kmh^{-1}.

Azimuthal scans for several values of radar beam elevation and
vertical beam Doppler radar measurements of clouds and precipitation
were alternately performed from 1030 to 1730 GMT[1]. Throughout all
the period of observations, an intermittently precipitating stratiform cloud deck with top around 2.5 km of altitude at 1030,
raising up to 3.2 km at 1200 was observed. Most of the time, the
echoes were continuous from the surface to the cloud top. The data
presented below correspond to a selected period of about 5 minutes.
Selection of this period for presentation in this paper resulted
from many considerations such as proximity in time to available
environmental upper air data (radiosounding, aircraft and other
panoramic radars), technical quality of tape recorded Doppler radar
data and trajectory of the precipitation just above the radar, in
the observation surface of the radar beam.

As pointed out in the preceding section, generating cells are
observed in various conditions and configurations. For that, each
case study is particular. In this point of view, the selected case
is not an archetype of precipitation cells and trails ; its main
quality is to show in relatively simple conditions   several major
features of this kind of structure.

Fig. 1a is a contoured time height cross section of the radar
reflectivity factor Z[2]. The main features of the pattern are a

precipitation trail falling from a region of strong reflectivity
located near the cloud top between 2700 m and 3000 m from 1145-1144.
The precipitation trail reach the 0 °C isotherm level around 1800 m
at about 1141. An equivalent space scale deduced from the mean wind
between 2000 and 3000 m is given at the top of Fig. 1a. Just below
the cloud top, the radar detect a non precipitating cloud medium
at a reflectivity level reaching - 35 dBZ. Temperature at the cloud
top is - 6 °C (fig. 2). Between the head of the precipitation trail
and the cloud top, the vertical gradient of the reflectivity factor
is 50 dBZ for 400 m. The Z values inside the trail are around
10-15 dBZ between 2700 and 1900 m (a careful examination of the
numerical data shows a small increase downward). The strong variation of the reflectivity factor below 2000 m with a maximum around
1800 m marks the position of the precipitation melting layer.

The slope of the precipitation trail between 3200 m and 2400 m
corresponds to a negative wind shear in the vertical profile of
horizontal wind (fig. 2). At low level, below 1500 m a reverse
slope is observed. At the ground, precipitation rate is less than
0.5 mmh^{-1}.

A contoured time-height cross section of air velocity for the
same echoes as in Fig. 1a is shown in Fig. 1b. As pointed out in
annex, mean Doppler velocity was measured by pulse pair processing
technic and Doppler velocity spectra were computed each 5 s from
time series processing. Usually, vertical air velocity $\overline{w}$ is calculated from measurements of the mean vertical Doppler velocity of
precipitation $\overline{V}_z$ by adding to $\overline{V}_z$ the mean terminal velocity in still
air of the scattering particles $\overline{V}_\ell$. For Doppler velocity positive
upward, $\overline{w}$ is given by :

$$\overline{w} = \overline{V}_z + \overline{V}_\ell$$

The mean velocity $\overline{V}_\ell$ is deduced from the radar reflectivity factor
using a relationship of the form $V_\ell = aZ^b$ where a and b are experimental constants characteristic of the observed precipitation (22).
The method is not applicable for V_z less than about 1 ms^{-1}. This
last case is frequently observed in stratiform clouds. In such
circumstances, if a radar able to detect non precipitating particles
is used, w can be directly obtained from the upper bound of spectra
of vertical Doppler velocity, which correspond to the non precipitating component of the target. For Fig. 1b this last method was
used above the 0 °C level (cloud basis was around 1400 m). Below,
$\overline{w}$ was calculated from $\overline{V}_z$ using $\overline{V}_\ell$ values estimated from the equation :

$$\overline{V}_\ell = 2.7 \, Z^{0.12}$$

This relationship was obtained from precipitation size distri-

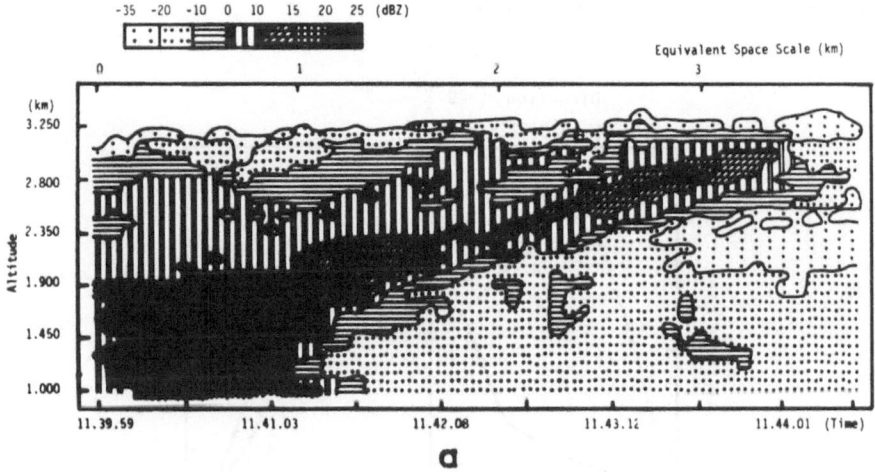

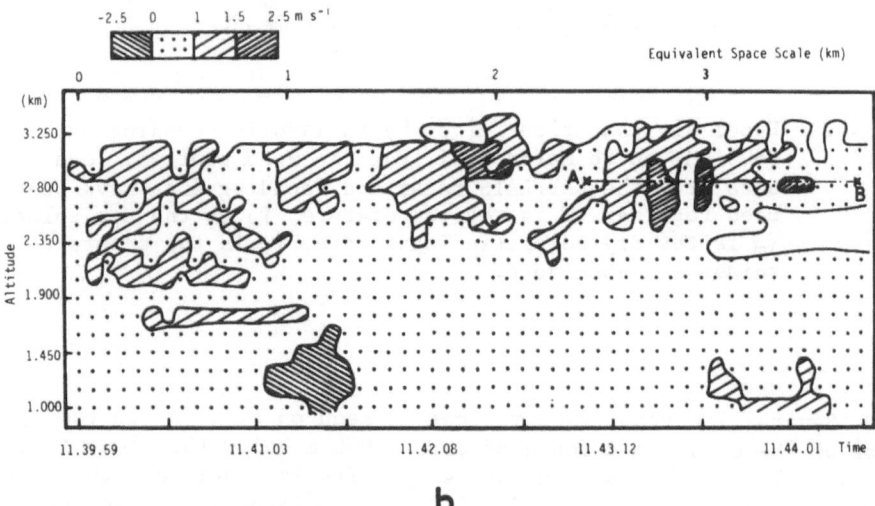

Fig. 1 – Contoured time-height sections of (a) radar reflectivity factor and (b) vertical air velocity on 28 March 1981 deduced from RABELAIS radar. Line AB shows the location of spectra presented in Fig. 3.

bution measurements performed at the ground on the radar site with
a spectrometer for raindrop during the radar observation (with a
correlation coefficient of 0.9).

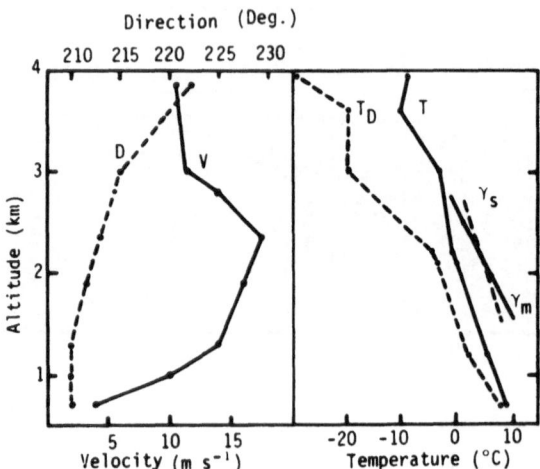

Fig. 2 - Upper air conditions given by Valladolid sounding at
         0800 GMT on 28 March 1981. For the wind profile, local
         data from the RABELAIS radar and sounding data have been
         used together. T is the temperature, T_d is the dew point,
         γ_m is the dry adiabatic lapse rate, γ_s is the saturated
         adiabatic lapse rate.

Several important features can be seen on Fig. 1b. In the upper
part of the cloud on a depth of about 1000 m below the cloud top,
a layer of small and discontinuous updrafts is observed ; the maximum value, reaching 2.0 - 2.5 ms^{-1}, is associated to the reflectivity core of the generating zone. Below this layer, the precipitation trail falls in a region of zero vertical air velocity up to
the melting level which is marked by a flattened and shallow ascending region. The generating zone exhibits several narrow updrafts
with a tendancy for a decrease in thickness and strength toward the
downshear side or the zone. Fig. 3 shows spectra from the 2800 m
level, in the generating zone, along line AB (Fig. 1b). On the left
of each curve spectral moments are given. The horizontal line is
the noise level. It can be seen in Fig. 3 that spectra corresponding
to convectively active parts of the generating zone are shifted

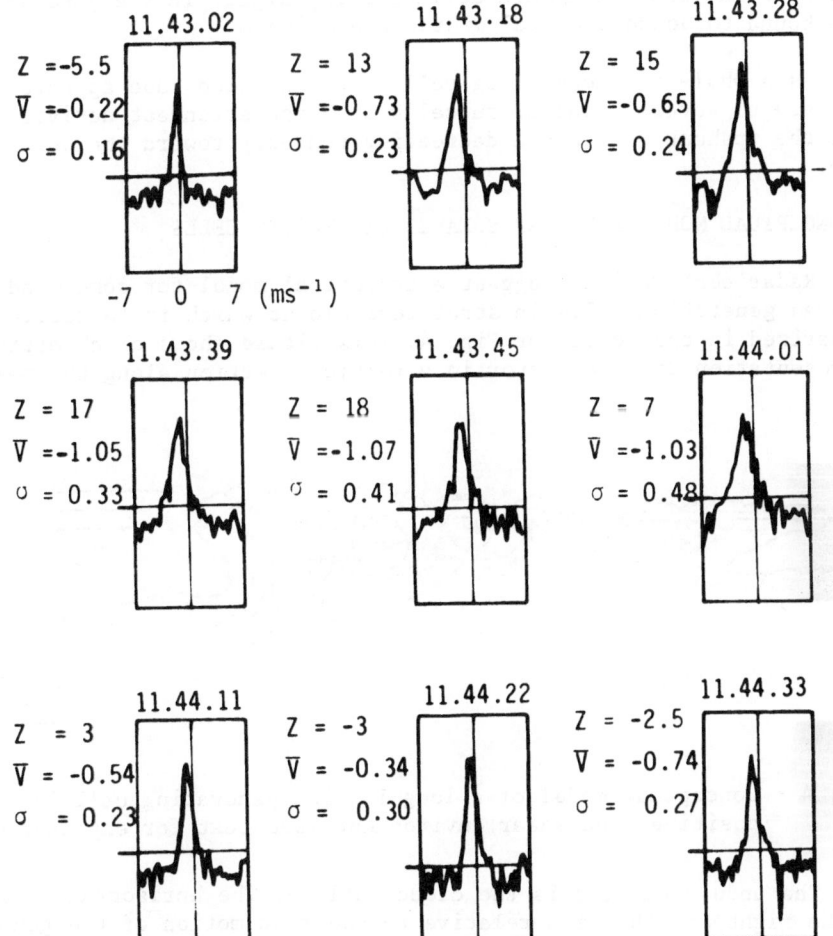

Fig. 3 - Doppler velocity spectra observed in the generating zone
along the line AB (see Fig. 1b). The spectral moments are
given on the left of each curve. The radar reflectivity
factor indicated is deduced from the non coherent channel
of radar receiver owing to saturation problem of the
coherent receiver. Doppler velocity are indicated along
the x-axis in ms^{-1}. Positive values (upward) are on the
right and negative values (downward) are on the left.

toward the right (upward velocity) relatively to those outside of
this zone and their standard deviation is larger. In the generating
zone bound to bound spectral widths reach 4-6 ms^{-1}.

As a whole the generating cell in the studied case appears as
composed of several smaller subcells with the strongest activity
near the upshear side and a decreasing tendency toward the downshear
side.

A CONCEPTUAL MODEL FOR WIND SHEARED GENERATING CELLS

Radar observations suggest a conceptual model for some wind
sheared generating cells in stratiform clouds which is tentatively
summarized in the scheme of Fig. 4. This figure shows a schematic
representation of a generator in a vertical section along the mean
wind.

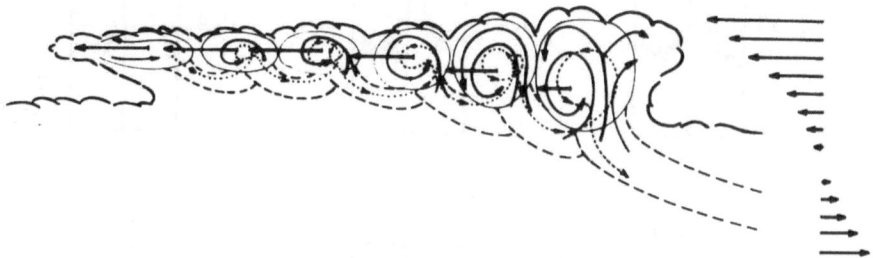

Fig. 4 - Conceptual model of a long-lasting generating cell in a
         positive wind shear environment (see text for explanation).

The undulated line is the cloud outline. The horizontal arrows
on the right are the wind relative to the mean motion of the generator. Curved arrows represent the air motion in the cloud, dotted
lines are the assumed trajectories of some growing precipitation
particles and dashed lines are the precipitation trails. Growing
processes in the cells will not be discussed here. Some aspects of
trajectory and growth of precipitation particles in such small wind
sheared convective generators have been discussed from a kinematic
model (8). Results, which will not be recalled in this paper suggest
that the wind shear influence on the convective circulation is to
concentrate the precipitation near the upshear side of the cell.

It is known that a convective cell shows some dynamical cohe-
sion and moves horizontally with the wind at mid cell level. In
presence of a positive wind shear the precipitation trail associated
to a cell slopes towards the rear and, if the latent heat released

in the precipitation trail is sufficient, some updrafts developt
above the trail and form one or several small new subcells upshear
of the former cell.

A particular cell is sustained by the heat released in the
precipitation trail coming from the cells situated downshear and
creates itself a stronger precipitation trail with larger precipitation particles and therefore a more vertical slope. Such a trail
is producing an updraft less spread out horizontally, more concentrate and coming from a level lower than for the cell downshear.
In these conditions the relative horizontal velocity of the cell,
which depends on the mean horizontal air velocity in the layer
where the updraft is acting, is smaller than for the downshear cell.
As a whole all these effects and perhaps others can explain the
observed structure.

This conceptual model can be extended to other patterns of
precipitation generator observed by radar. The main features are
summarized in Fig. 5 where are represented the corresponding wind
profile and reflectivity contour of the cell in the diverse
conditions.

1. In absence of wind shear the generator is symetrical. The
vertical mid section shows two lobes corresponding to the two parts
of the convective circulation in the plane.

2. With wind shear of the same sign and nearly same value at
the level of the generator and below it, the generator is S shaped.
If there is a regeneration mechanism the generator stretch out. The
more the environmental shear is strong, the more the generator is
stretched out.

3. With a change in the sign of the environmental wind shear
or a low wind shear just below the generator, a case frequently
observed, the generator shows a single cell with an outline S
shaped.

4. With an opposite sign of wind shear at the generator level
and below but with small shear intensity we have a C shaped pattern.
If there is a regeneration mechanism, and a strong shear, the generator stretch out.

Far below the generator, the trail can have any profile
following the particularity of the wind with respect to the generator velocity.

CONCLUSION

Observations of a mid level stratiform cloud with millimeter

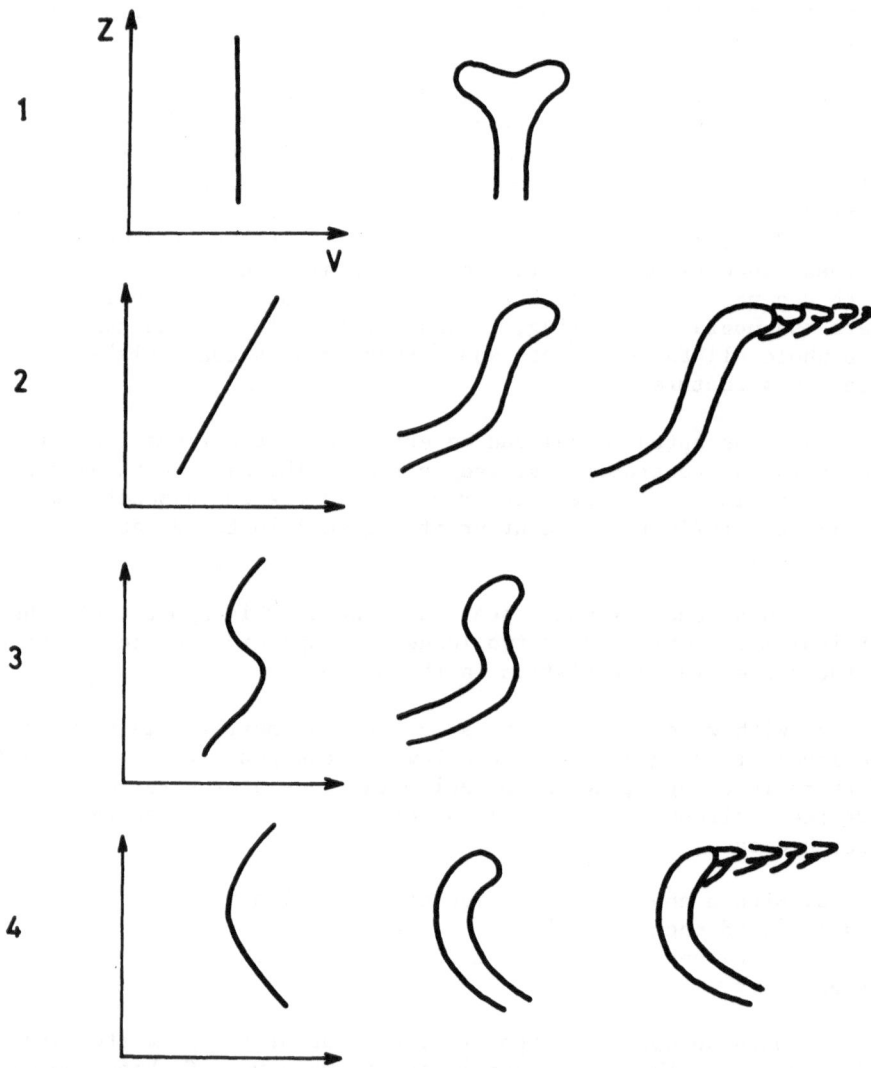

Fig. 5 – Conceptual model of generating cells. Environmental wind
          profiles are on the left and the reflectivity contours of
          the cells are on the right. (1) without wind shear (2)
          positive wind shear (3) S shaped profile (4) C shaped
          profile.

wave Doppler radar have shown that small convective generators responsible for precipitation generation can have a finer structure composed of several convective subcells. In the case studied the strongest convective activity was observed near the upshear side of the generator with in addition a decreasing tendency toward the downshear side. Vertical velocities up to 2.5 ms^{-1} were measured. The velocity maxima were associated with the highest reflectivity zones. The generating convective region was characterized by large bound to bound spectral widths reaching 4-6 ms^{-1}.

Based on the above radar data and on many other observations on the stratiform clouds, a conceptual model for the relation between the vertical profile of the environmental wind and the substructure of the long-lasting precipitation generating cells has been tentatively proposed. In this, wind shear is the main factor responsible for the substructure and for the stretching along the wind of the generator. The subcells tend to disappear near the downshear side where the generating zone flattens and to be regenerated near the upshear side where the main precipitation trail is located.

Certainly many other aspects of the precipitation generating cell mechanisms are still not understood and it appears that long lasting generator in stratiform clouds raise some challenging questions for modelisation and experimentation.

ACKNOWLEDGMENTS

This work was supported by the Institut National d'Astronomie et de Géophysique (A.T.P. Recherches Atmosphériques) and by the Direction des Recherches et Etudes Techniques (grant 80-190).

[1] all times are GMT.

[2] The radar reflectivity factor Z is the mean value on the radar pulse volume of ΣD^6 where D is the diameter of the equivalent spherical scattering particle and Σ the sum for a unit volume. Z is expressed in $mm^6 m^{-3}$ or in dBZ that is to say 10 log Z (see for example 24).

ANNEX - THE RABELAIS RADAR

RABELAIS is a mobile Doppler radar built for cloud and precipitation physic research by the Centre de Recherches Atmosphériques in collaboration with the Laboratoire Central de Télécommunications. The transmitter uses a magnetron Litton 4064A modified. The coherence is obtained from a STALO-COHO method. A dual pulse repetition frequency system is employed to extend the unambiguous velocity

limit. The coherent video signal is processed by pulse pair technic
for the mean Doppler velocities and through Fourier transform of
time series for calculations of the spectra and their moments. A
non coherent receiver is also available for the estimation of the
mean reflectivity. Preprocessed data can be visualized in real time
in the radar shelter and recorded on digital magnetic tape for
further processing on a multipurpose computer.

Characteristics of the RABELAIS radar :

- Antenna
  - diameter : 1.40 m
  - gain : 51.5 dB
  - angular width (3 dB, one way) : 0.415°
  - first secondary lobe : − 26 dB
  - elevation span : − 2° à 90°
  - azimuth span : ± 270°

- Transmitter
  - wavelength : 0.86 cm
  - frequency : 34 960 MHz
  - peak power : 70 kW
  - pulse width : τ = 0.3 μs (45 m)
  - pulse repetition frequencies (Hz) :
          mode 1 : 3125/2688
          mode 2 : 1344 not coherent

- Receiver
  - noise factor : at receiver input (with
    circulation and TRL losses) : 9 dB
  - adjustable receiver dynamic : Min 30 dB
                                  Max 54 dB
  - phase stability, ground clutter elimina-
    tion rate in the 120 Hz Doppler band at
    $\frac{Fr}{2}$ (1562 Hz) :
          at 2.8 km ..... 43 dB
          at 22 km ...... 35 dB

- Ambiguity
  (transmission mode 1)
  - non ambiguous range .... 48 km
  - non ambiguous velocity .... ± 40 ms^{-1}
  - resolution for 64 pairs ... 0.15 ms^{-1}

- Exploration cycle
  (transmission mode 1)
  - gate width : Min. 60 m
  - number of gate : Max. 256
  - measurement zone : adjustable

- Sensibility
  - After integration the min. detectable at
    10 km is : $Z(mm^6 m^{-3}) \simeq 2.7 \cdot 10^{-1}$ corres-
    ponding for water clouds to a liquid water
    content of 0.08 gm^{-3} and a mean volume
    diameter of 11 μm.

REFERENCES

1. Lhermitte, R.M., 1952 – C.R. Acad. Sci., 235, 1414–1416.
2. Marshall, J.S., 1953 – J. Meteor., 10, 25–29.
3. Gunn, K.L.S., M.P. Langleben, A.S. Dennis and B.A. Power, 1954 – J. Meteor., 11, 20–26.
4. Gunn, K.L.S. and J.S. Marshall, 1955 – J. Meteor., 12, 339–349.
5. Douglas, R.H., K.L.S. Gunn and J.S. Marshall, 1957 – J. Meteor., 14, 95–114.
6. Wexler, R. and D. Atlas, 1959 – J. Meteor, 16, 327–332.
7. Browning, K.A. and T.W. Harrold, 1969 – Quart. J. Roy. Meteor. Soc., 95, 288–309.
8. Sauvageot, H., 1974 – J. Rech. Atmos., 8, 213–219.
9. Carbone, R.E. and A.R. Bohne, 1975 – J. Atmos. Sci., 32, 1384–1394.
10. Harris, F.I., 1977 – J. Atmos. Sci., 34, 651–672.
11. Hobbs, P.V., 1978 – Rev. Geophys. and Space Physics, 16, 741–755.
12. Ludlam, F.H., 1956 – Quart. J. Roy. Meteor. Soc., 82, 257–265.
13. Heymsfield, A.J. and R.C. Knollenberg, 1972 – J. Atmos. Sci., 29, 1358–1366.
14. Konrad, T.G. and J.C. Howard, 1974 – J. Appl. Meteor., 13, 563–572.
15. Heymsfield, A.J., 1975 – J. Atmos. Sci., 32, 799–830.
16. Sauvageot, H., 1976 – J. Rech. Atmos., 10, 17–24.
17. Henrion, X. and H. Sauvageot, 1977 – Geophys. Res. Lett., 4, 360–362.
18. Sauvageot, H., 1976 – J. Rech. Atmos., 10, 119–122.
19. Sauvageot, H., 1976 – C.R. Acad. Sci., t. 283 B, 159–162.
20. Houze, R.A. and A.K. Betts, 1981 – Rev. of Geophys. and Space Phys., 19, 541–576.
21. Henrion, X., H. Sauvageot and D. Ramond, 1978 – J. Atmos. Sci., 35, 2315–2324.
22. Rogers, R.R., 1963 – J. Atmos. Sci., 20, 170–174.
23. Scorer, R.S., 1951 – Quart. J. Roy. Meteor. Soc., 77, 235–240.
24. Sauvageot, H., 1982 – Radarmeteorologie. Editions Eyrolles, Paris, 296 p.

LAKE-EFFECT SNOW STORMS ON LAKE MICHIGAN, USA.

R. R. Braham, Jr. and R. D. Kelly

University of Chicago, Cloud Physics Laboratory

ABSTRACT. Examples are presented of three different convective patterns in lake-effect snow storms over Lake Michigan, USA: wind-parallel bands of clouds and precipitation due to horizontal-roll convection; deep, shore-parallel cloud bands associated with lake-induced, mesolow pressure centers; and midlake cloud bands resulting from the convergence of land breezes from the lake shores.

1. INTRODUCTION

During fall and winter months in North America, cold arctic air masses from Canada often sweep across the Great Lakes causing strong convection in the planetary boundary layer (PBL). Under suitable combinations of lake-air temperature differences (ΔT), upstream air stability, wind shear, length of fetch, and regional pressure gradients, this convection becomes organized into lines and bands of cumulus and stratocumulus clouds capable of depositing considerable amounts of snow.

Locations along the downwind shores of the lakes receive 2-3 times the amounts of snow received at corresponding upwind locations. Several geographical and meteorological features combine to cause a high frequency of rather intense lake-effect snow storms on the North American Great Lakes. These lakes are large and deep, covering an aggregate area of 244,900 km² and extending over 1300 km in a general west-east direction. Located in the interior of a large continent, these lakes are subjected to strong insolational heating in summer and are open to the unhindered overpassage of air masses from the Canadian Arctic in winter. Their latitude, 42-48°N, insures that most of the winter

E. M. Agee and T. Asai (eds.), Cloud Dynamics, 87–101.
Copyright © 1982 by D. Reidel Publishing Company.

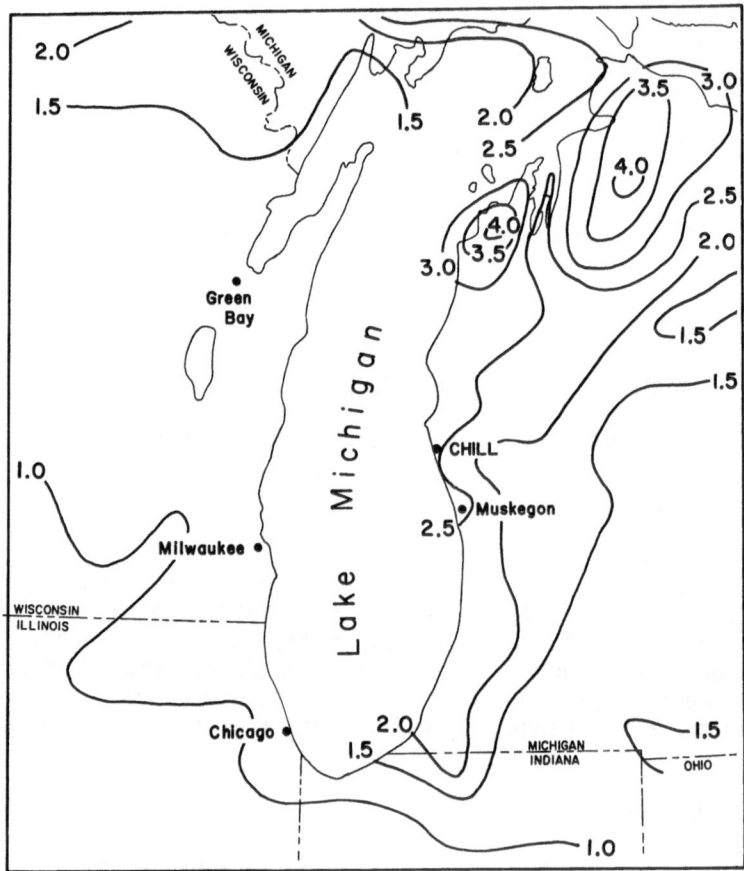

Figure 1. Map of the average annual snowfall, meters, around
Lake Michigan for the winters of 1970-71 through 1979-80.

precipitation falls as snow. They are near the preferred winter
cyclonic storm tracks, and as a consequence passing cyclones are
often followed by cold arctic anticyclones giving 2-4 days of
conditions favorable for lake-effect snows.

      With support from the National Science Foundation, in 1977
the University of Chicago, Cloud Physics Group, began a study of
boundary-layer convection and lake-snow storm microphysics over
Lake Michigan. Figure 1 shows the average annual snowfall around
Lake Michigan during the decade of the 1970's. Our project is
based at Muskegon, Michigan (see Figure 1), and makes use of
instrumented airplanes and Doppler radars.

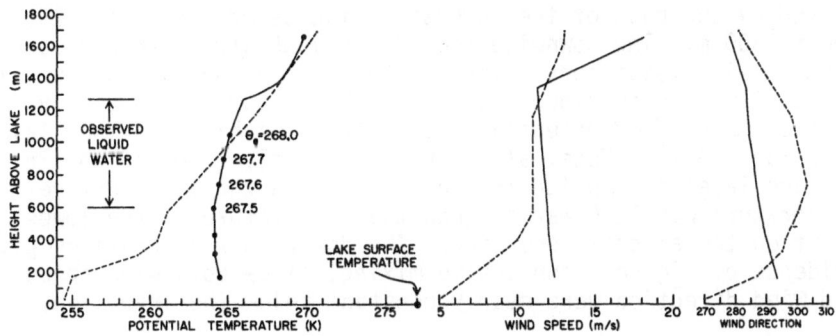

Figure 2. Soundings of potential temperature and wind speed and direction from Green Bay, Wisconsin (dashed lines) and near the CHILL radar (solid lines). Equivalent potential temperature (θ_e) values are included for the CHILL case.

With the radars we have observed several different modes of organization of precipitation bands attributed to lake-induced convection. During upwind conditions of strong winds and strong static stability, we find long, narrow lines of precipitation, suggestive of cloud streets. Under conditions of moderate winds and weak upstream static stability, we find a single major precipitation band roughly parallel to the downwind shore. With very weak regional pressure gradients (as in the center of an anticyclone) we find cloud lines over the middle of the lake as a result of land-breeze circulations from both shores. An example of each of these is given in the following sections.

2. WIND-PARALLEL BANDS OF 9 DECEMBER 1978

The Lake Michigan snow storm of 9 December 1978 was organized in a pattern often observed in lake-effect situations. The storm clouds as seen by satellite and the precipitation as detected by the CHILL radar occurred in long, parallel bands. These bands originated some 15 to 30 km off the western shore, extended all the way across the lake, and dissipated some 10 to 20 km inland of the eastern shore.

Figure 2 presents details of the boundary layer (BL) structure upwind of and over the lake. The dashed lines in these plots are data from the National Weather Service radiosonde, launched from Green Bay, Wisconsin (see Figure 1) at 1200 GMT. The solid lines are data from soundings taken by the research aircraft over the lake, near the CHILL radar site. Several effects of the lake can be seen in Figure 2. The BL depth, as

marked by the base of the inversion, increased from about 200 to
about 1300 m. The downwind (solid line) BL structure was consistent with moist convection, having nearly constant potential
temperatures below cloud base (i.e., below 600 m) and nearly
constant equivalent potential temperatures within the cloud layer
(600 to 1300 m). Note, also, that a strongly super-adiabatic
surface layer is implied in Figure 2, since the lake surface
temperature was 12°C warmer than the temperature at the lowest
point on the aircraft sounding. The downwind wind profiles give
evidence of vigorous convective mixing, since both wind speed
and wind direction were nearly constant below the inversion.

Figure 3 is a constant-elevation plot of the radar reflectivities measured in the storm, and shows the banded distribution
of precipitation. These bands were spaced at intervals of 3 to
7 km (average 5 km) and were generally oriented parallel to the
wind at the base of the inversion. Evidence is given below that
these bands resulted from horizontal-roll convection in the BL.

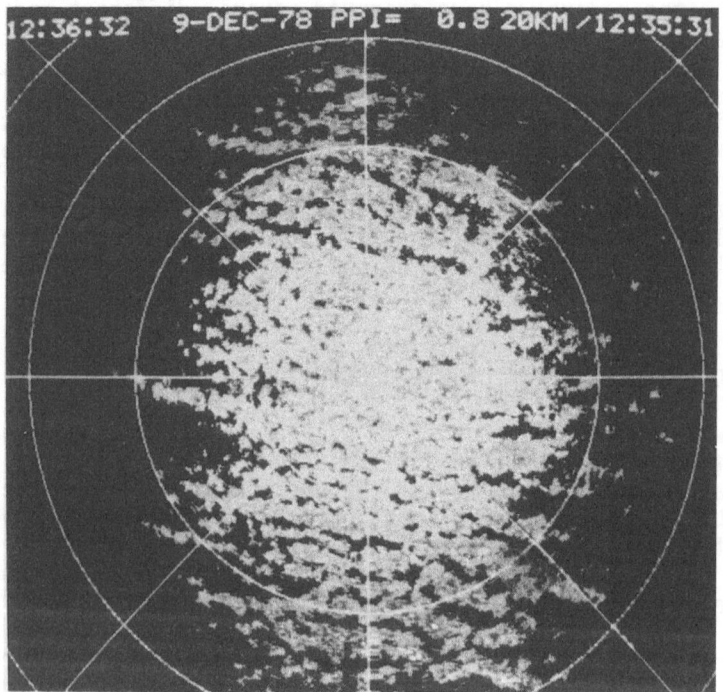

Figure 3. Radar reflectivity map from CHILL radar at 0.8°
elevation, 1736 GMT, 9 December 1978. Range circles are at
20 km intervals.

Since the convective depth was 1300 m, these rolls had an aspect ratio (ratio of roll spacing to depth) of 2.3 to 5.4 (average 3.8). Theoretical models of BL convection predict roll aspect ratios of 2.5 to 2.8.

Each wind-parallel echo band contained "cores," or "cells," of reflectivity maxima (up to 30 dBZ). These cores were spaced along the bands at intervals of 3 to 5 km, and moved along the bands at about 12 m s^{-1}. As seen in successive horizontal radar scans, and in vertical cross-sections, the cores were continuous in height through the BL depth.

Horizontal-roll convection, as predicted by theory and measured in the laboratory and in the atmosphere, is characterized by shallow, counter-rotating horizontal vortices [1]. Evidence of such circulations was found in the radar Doppler velocities measured in this storm. The Doppler velocities (V_D) recorded here are means of individual velocity spectra. They were obtained by the pulse-pair method, after integration over 128 consecutive radar pulses. Doppler measurements were enhanced in this storm by the widespread occurrence of precipitation, such that V_D could be determined over a large portion of the BL motion field. In the discussion to follow, positive V_D values correspond to target movement radially away from the radar.

The broad pattern of Doppler velocities measured in this storm was one of azimuth-dependent V_D values, as one would encounter in a horizontally uniform wind field. However, band-scale changes in V_D were detected. Figure 4 is a magnified view of the area southwest of the radar, from a horizontal scan at 1.7° elevation; the Doppler velocities were contoured and shaded according to the legend in the figure. Portions of four wind-parallel echo bands are visible in Figure 4, passing about 15, 18, 22, and 25 km south of the radar.

Moving along radial lines from the radar, we see that in most instances V_D increased with range across individual bands. This pattern is quite striking in the southern-most band, where we find that V_D increased as much as 4 to 6 m s^{-1} across the band. Since the radar measures only that component of target motion along radials from the radar, such cross-band increases in V_D are consistent with diffluent shifts in wind direction across the bands.

For purposes of illustration, such a diffluent wind field, along with a hypothetical radar echo band, is sketched in Figure 5. In this schematic the area of zero V_D (no movement along a radial) is stippled. The areas labeled "a," "b," and "c" represent increasing values of V_D. As one moves outward from the radar, across the band, the diffluent wind shift causes an

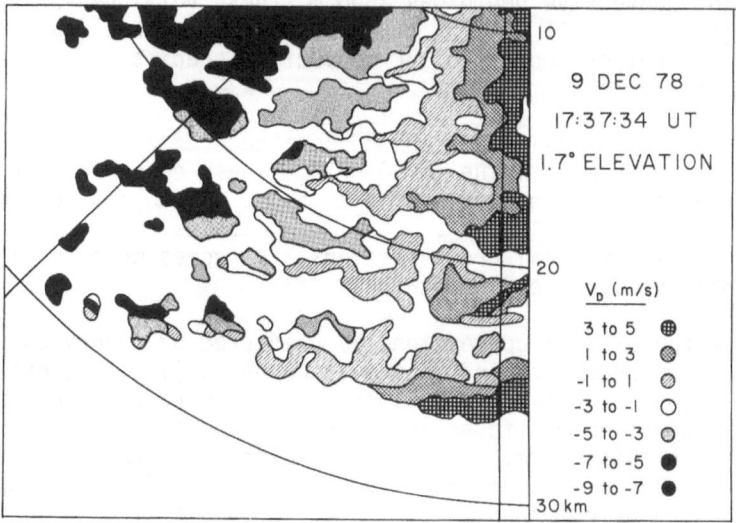

Figure 4.   Southwestern segment of constant-elevation map of Doppler velocities.

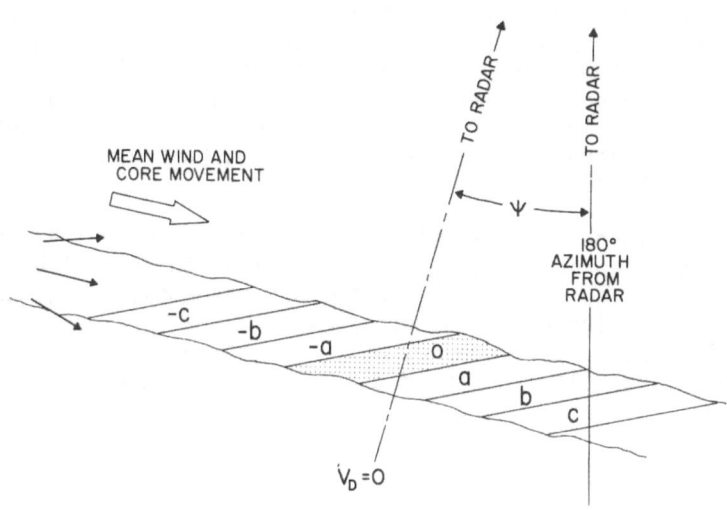

Figure 5.   Schematic view of Doppler velocity pattern in an echo band south of the radar.   The angle, ψ, is typically 5 to 15°.

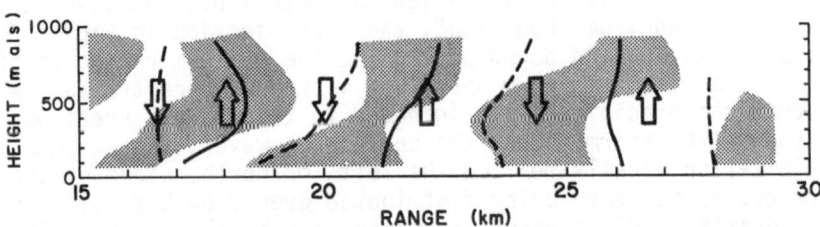

Figure 6. Vertical cross-section along 180° azimuth from CHILL radar. See text for details.

increasing outward component of target motion, and leads to an increase in V_D. Note that changes in wind speed could also cause changes in V_D. However, the fact that the $V_D=0$ area changes azimuth across the band, coupled with the two-dimensional nature of the pattern, precludes wind speed changes from being the major cause in this case. Similar reasoning shows that a confluent shift of wind direction within an echo band would lead to a decrease in V_D along a radial across the band.

Confluent and/or diffluent wind patterns were observed in echo bands in all quadrants and at different elevations in the area around the radar. Diffluent shifts within bands usually occurred at heights of 300 m or more. Confluent shifts usually occurred below 400 m. In some cases, confluent and diffluent shifts were observed side-by-side within individual bands, with confluence to the north and diffluence to the south side of each such band. This latter pattern usually occurred below 400 m. Overall, then, the Doppler measurements give evidence of the horizontally and vertically alternating areas of confluence and diffluence one would expect in a field of horizontal rolls.

In order to gain insight into the organization of precipitation within the storm, several vertical cross-sections were studied. One such section, along the 180° azimuth from CHILL, is presented in Figure 6. It was constructed by averaging data from three adjacent radials (a total of 3° azimuth) at each of the sequential elevation angles of 0.3, 0.7, 1.2, 1.7, and 2.7°. As described above, increases and decreases in V_D with range were interpreted as areas of diffluence and confluence, respectively. Diffluent areas are stippled in Figure 6, while the intervening confluent areas are left unshaded. The heavy, solid lines are the positions of maximum radar reflectivity values; the heavy, dashed lines are positions of minimum reflectivity values. The arrows mark the approximate locations of updrafts and downdrafts, as implied by vertical alternations in confluence and diffluence.

Given the position of the 180° azimuth line, and given the prevailing wind from about 280°, the cross-section in Figure 6 is equivalent to looking downwind. Since the maximum reflectivities represent areas of maximum snowflake size and concentration, we see that the precipitation originated in the updraft areas near the top of the storm. Near the surface, however, the maximum precipitation usually fell on the left, or north, sides of the updrafts, rather than being distributed evenly on both sides of the updrafts. This apparent asymmetry may be traced to at least two possible sources. First, tilted updrafts are predicted by theory and have been observed in other horizontal-roll cases [1]. Second, the vertical profile of BL winds measured by the aircraft (see Figure 2) has a northward pointing shear vector, since the winds back with height up to the inversion. Either case could lead to precipitation particle trajectories falling preferentially to one side of the roll updrafts.

3. SHORE-PARALLEL BAND OF 10 DECEMBER 1977

A major winter storm occurred in the midwestern part of the United States on 8-11 December 1977. It began with the formation of a low-pressure center over Oklahoma. This low moved very rapidly to the northeast, up the Ohio River Valley. The arctic air mass which moved in behind the low-pressure center carried freezing temperatures to the Gulf of Mexico and northern Florida. At Muskegon, Michigan, this storm deposited over 69 cm of new snow.

On 10 December we studied this storm with the CHILL Doppler radar and an NCAR cloud physics airplane. On this date, weather conditions on the west side of the lake were generally clear skies, temperatures about -20°C, dew points about -25°C, and the winds were from the northwest at about 10 kts. On the Michigan shore, skies were broken to overcast with low clouds, temperatures -8 to -12°C, dew point depressions 1-2°C, and winds were light easterlies.

Data from the GOES-East satellite (see Figure 7) showed several distinct lines of clouds oriented roughly NW-SE over the lake. These lines merged into a much larger cloud band located approximately along the eastern shoreline. At its southern end this cloud band curved inland across the shoreline about 100 km south of Muskegon.

Radiosonde data are available from Green Bay, Wisconsin (upwind of the lake) and from the CHILL site (on the downwind shore) for the morning of 10 December. Except for a shallow radiation inversion, the air upstream of the lake was only slightly more stable than dry adiabatic up to the subsidence inversion at 764 mb. The downwind sounding showed nearly saturated

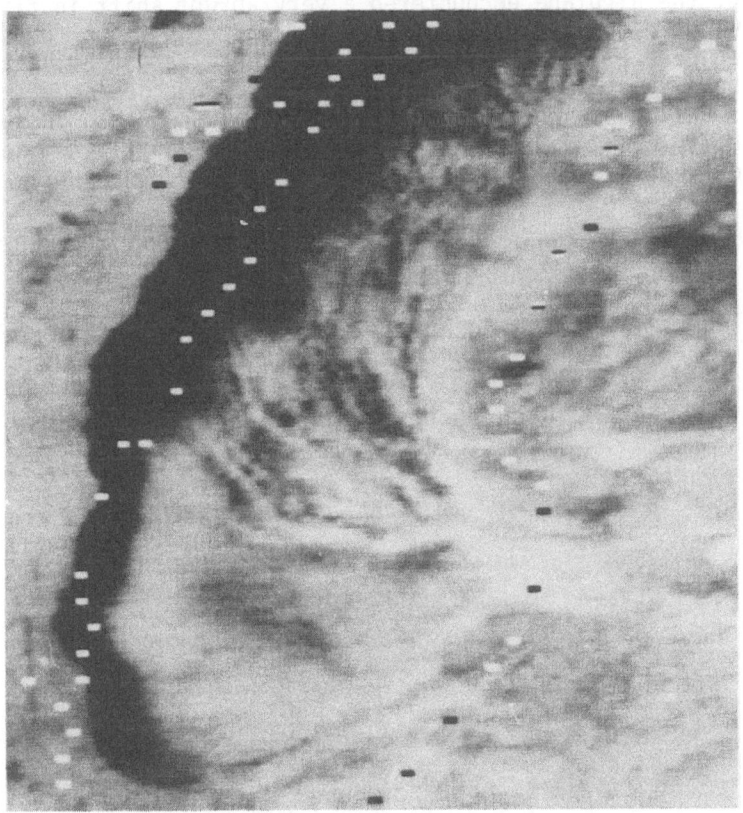

Figure 7. GOES-East visible image of Lake Michigan area on
10 December 1977, 1631 GMT.

adiabatic conditions up to 734 mb. Winds upwind of the lake were
northwesterly at all levels. At the CHILL site they were northerly between about 950 and 850 mb, with northwesterlies increasing
in strength above that height.

     Figure 8 shows the aircraft track for the flight of 10 Decem-
ber, the PPI radar image during the flight, and weather conditions
encountered along the flight track. At the time of take-off
(1745 GMT) surface winds at Muskegon and the CHILL site were
northeasterly. Between about 1800 and 1900 GMT, surface winds at
both locations switched to northwesterly. After take-off we flew
south along the eastern shore of Lake Michigan. Flight-level
winds were easterly. Steam arising from the lake was observed
about 16 km south of Muskegon, just prior to the time the plane
entered snow. Just before turning to the westerly heading, the
snow became very heavy. Shortly after turning to the westerly

heading, the airplane encountered a very abrupt shift in flight-
level winds. This shift in winds, from easterly to northwesterly,
occurred over a flight distance of 400 m and was directly in line
with the axis of the shoreline precipitation band as seen on the
radar. Heavy snow continued until after the plane had passed the
wind shift line, whereupon it decreased in intensity and, near
midlake, became showery in character.

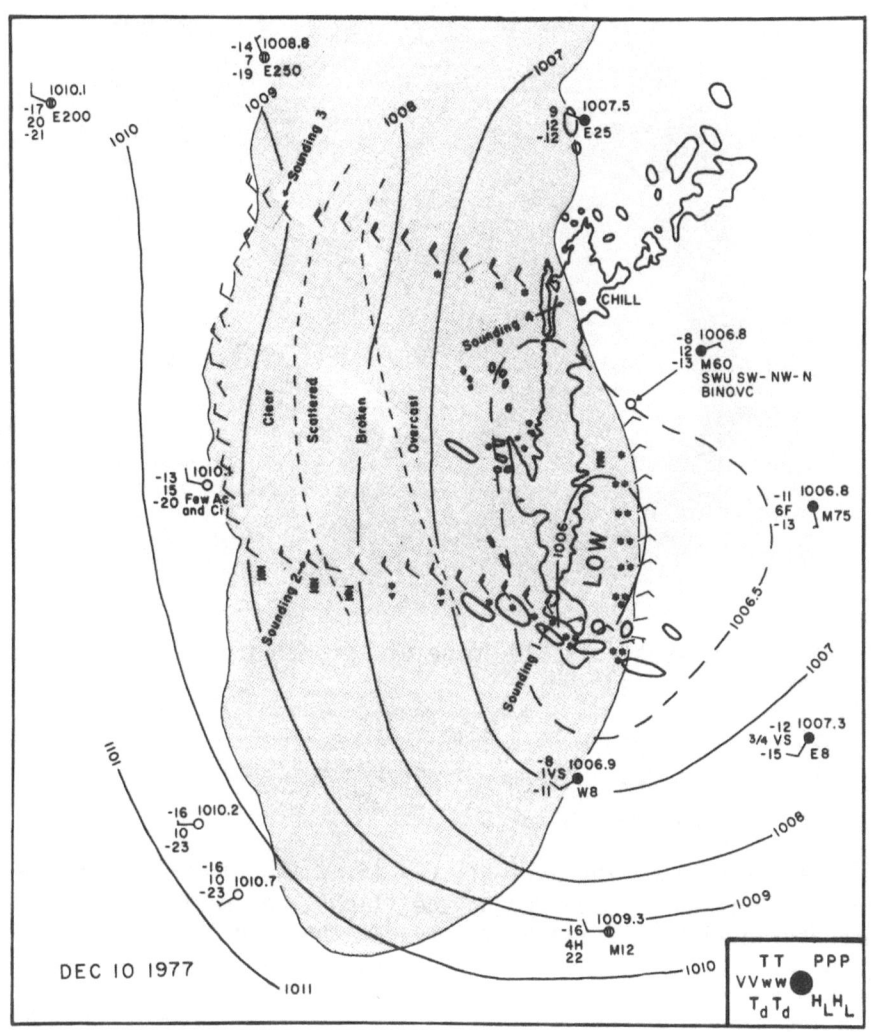

Figure 8. Mesoscale analysis of the lake-effect snow storm of
10 December 1977 on Lake Michigan. Contours of pressure (mb)
refer to the 210 m surface, see text for details. The solid
outline depicts the 20 dBZ radar reflectivity contour.

About halfway across the lake the cloud bases became very ragged and steam devils were observed rising from the lake surface to the clouds. In Figure 8 these observations have been denoted by combining standard weather symbols for fog and dust devils. This area of intense steam devils was several kilometers wide and extended tens of kilometers north-south.

Within a few kilometers of the Wisconsin shore, rafts of pancake ice were observed floating on the lake surface. Floating ice was observed all along the western shore of the lake.

On the eastbound (northern) leg of the flight, we again observed steam devils, but they were not as intense as those observed farther south. About midlake we entered a cloud of small ice crystals which reduced visibility to about 2 km. Crystal aggregates were first encountered about 50 km west of CHILL. Flight-level winds remained northwesterly throughout the eastbound leg of the flight.

Profiles of temperature, dew point, and equivalent potential temperature for the westbound leg of the flight are given in Figure 9. The left and right edges of the figure correspond to the Wisconsin and Michigan shorelines, respectively. Note that air was warmed and moistened rapidly after leaving the Wisconsin shore. There was a change in the rate of warming at flight level, in the general vicinity of the most intense steam devils. The maximum flight-level temperatures and θ_e were found in a broad zone on the west side of the convergence line. In the convergence line the airplane measured sustained upward air flow. It exceeded 1 m s^{-1} along 4 km of the flight track and reached a peak value of 4.3 m s^{-1} near the location of the wind shift. If we assume that conditions along the band are two-dimensional, these values give a computed flight-level convergence of 4×10^{-3} s^{-1} averaged over a horizontal distance of 5 km. Low-level soundings of temperature, humidity, and wind were made near each of the four corners of the "box-like" flight pattern. These soundings show, as one would expect, that the sub-cloud layer was very nearly dry-adiabatic, whereas within the cloud, conditions were very nearly saturated adiabatic.

The airplane measurements in the subcloud layer allow one to compute pressure values at the lake surface or at any intermediate level. To do this one uses the measured flight-level pressure, temperature, radio-altimeter height above the lake surface, and the known sub-cloud lapse rate. Similar calculations over land are less clearcut because the appropriate lapse rates are more uncertain. In order to study the mesoscale pressure distribution associated with this storm, we have reduced flight-level pressures to a level of 210 m, which is the approximate mean height of land stations near the lake and about 32 m above the lake surface.

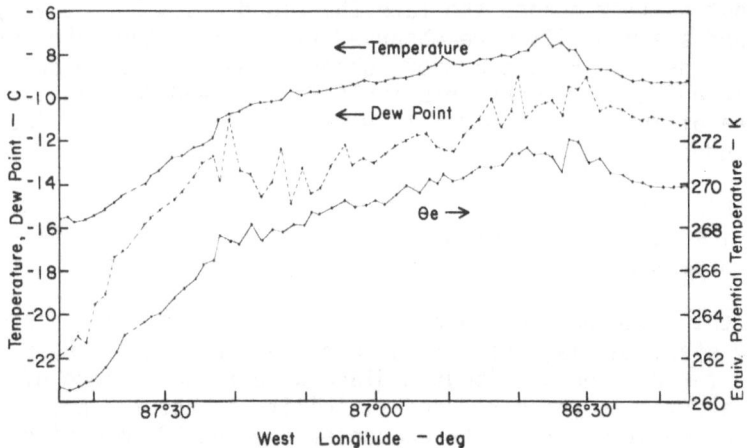

Figure 9. Horizontal profiles of temperature, dew point, and
equivalent potential temperature for the south leg of Flight 19,
10 December 1977 across Lake Michigan. Refers to a sub-cloud
flight level of about 250 m above the lake surface.

These calculations have assumed a dry adiabatic lapse rate
over the lake, and an isothermal lapse rate at measured station
temperatures over land. Results of these calculations are shown
in Figure 8. Isolines of pressure in millibars and half-millibars
are shown. The 1006.0 mb minimum pressure observed near the wind
shift line and along the eastern shore is 3 mb below pressures
observed along the western shore and 0.5 to 1.0 mb below those
along the eastern shore.

This analysis clearly shows that a warm-core mesolow was
responsible for the easterly winds observed along the Michigan
shore and for the southerly winds observed further inland. Once
easterlies (counter gradient winds) have been set up along the
downwind shore, there would be a sharp line of convergence over
the lake. Obviously, the mesolow is limited in areal extent and
location by the size and shape of the lake. It appears that in
this case the mesolow had its maximum intensity just north of the
point where the airplane made its east-to-west pass. South of
the mesolow one would expect, and we observed, westerlies. This
leads to an eastward turning of the cloud line, as was observed
by the satellite.

Once such a system is set up, it is self-sustaining until it
is disturbed by a change in large-scale weather conditions. Such
a disturbance occurred in the storm of 10 December while the airplane flight was in progress. We previously pointed out that the

surface wind at CHILL and Muskegon shifted to the northwest while
the flight was in progress. At the same time, the radar band
drifted onshore and heavy snow was observed at Muskegon. About
five hours later the snow decreased in intensity and winds again
became easterly -- apparently the mesolow had been reestablished.
Muskegon winds then remained easterly for another 12 hours, until
the approach of the next large-scale weather system made conditions unfavorable for lake-effect snows.

4. MIDLAKE SNOW BANDS

Our third example of a lake-induced snow band is one which
forms as a direct consequence of land-breeze circulations. Favorable conditions for such circulations are weak regional pressure-
gradients and substantial lake-land temperature differences.
These frequently occur when the region is under the influence of
a cold arctic anticyclone. Under these conditions a counter-
gradient land breeze may form along the downwind shore, while
along the upwind shore a land breeze may augment the otherwise
weak large-scale flow. The result is a line of convergence with
cumulus clouds and snow where the land-breeze fronts meet over
the lake. Passarelli and Braham [2] describe these findings and
give several examples, one of which is given here.

Figure 10 shows a portion of the GOES-East visible satellite
image on 7 November 1978. On this day Lake Michigan was on the
eastern side of a surface high-pressure ridge. Land stations in
Michigan and Wisconsin were reporting clear skies and temperatures
around 0°C to -3°C. The lake surface temperature was about 10°C.
Away from the lake, surface winds were light and variable with a
tendency for calm conditions in Wisconsin and weak northerly flow
in Michigan. Along the western shoreline of Lake Michigan, winds
were from the northwest, while along the eastern shore they were
northeasterly.

The research airplane conducted measurements at several
levels above the lake along an E-W flight track extending from
shore to shore between about the CHILL site and Milwaukee. These
data clearly show two land-breeze circulations meeting over the
western part of the lake. The most intense land-breeze flow was
confined to the lowest 3-400 meters, while the return flow
occurred between 1 and 2 km above lake level. The National
Weather Service station at Milwaukee reported "towering cumulus
NE-SE over the lake." Towering cumulus clouds were also observed
by one of the project scientists flying over the southern end of
the lake. But at the time the airplane made its measurements,
clouds were not encountered or reported. There are two reasons
for this -- the flight was made before daylight and the flight
track was made across the lake in the region where the GOES image

shows a break between the NE-SW oriented midlake band over the northern part of the lake and the much larger N-S band over the southern part of the lake.

In midlake situations, such as that of 7 November 1978, we often see the midlake band increase in size from north to south as a result of continued feeding of lake-warmed air to the convergence line. With a slow drift from north to south, bands of this type account for the high frequency of lake-snow storms around the south end of Lake Michigan.

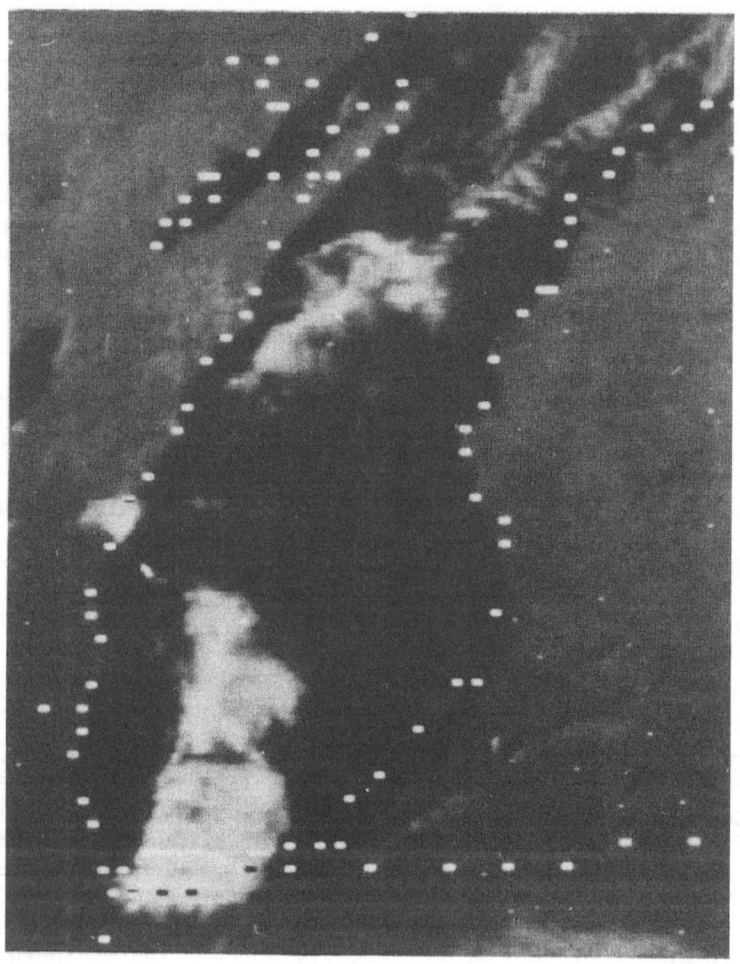

Figure 10. GOES-East visible image of Lake Michigan area on 7 November 1978, 1601 GMT.

5. DISCUSSION AND CONCLUSIONS

Lake-effect snow storms over Lake Michigan are comparatively shallow, with cloud tops typically 1.5 to 2.5 km above the surface. The cold temperatures and shallow depths in combination lead to modest snowfall production rates. The maximum observed radar echoes suggest melted equivalent precipitation rates of 0.5 mm hr^{-1}. As a result, the contribution of these storms to local snow accumulations stems primarily from their longevity and geographic persistence.

The storms are generally organized in one of several convective patterns. In some cases boundary layer processes dominate, as in the horizontal rolls and wind-parallel precipitation bands of 9 December 1978. In other cases mesoscale features dominate, as in the mesolow-centered, shoreline snow band of 10 December 1977, and in the land-breeze induced, midlake cloud band of 7 November 1978.

6. ACKNOWLEDGEMENTS

The authors would like to thank the Illinois State Water Survey, which operated the CHILL radar, and the Research Aviation Facility, National Center for Atmospheric Research (NCAR), which operated and reduced data from the research aircraft. Much of the computer-based analysis used in this study was performed at the NCAR Scientific Computing Division and Field Observations Facility. Thanks are also due the National Weather Service, in particular the personnel at Muskegon, Michigan, for their help and cooperation.

We deeply appreciate the help, advice, and ideas offered during this study by Drs. R. C. Srivastava, R. E. Passarelli, Jr., and M. R. Hjelmfelt. Thanks are also due M. J. Dungey for her efforts in reducing and analyzing the data, and to J. Jacobs for careful editing and typing.

The research described in this paper was supported by the National Science Foundation, Experimental Meteorology and Weather Modification Program.

REFERENCES

[1] Brown, R. A., Rev. of Geophys. and Space Physics, 18(3), 683, 1980.

[2] Passarelli, R. E., and R. R. Braham, Bull. Amer. Meteor. Soc., 62(4), 482, 1981.

DISCUSSION AND CONCLUSIONS

A SIMILARITY THEORY FOR UNSATURATED DOWNDRAFTS WITHIN CLOUDS [1]

Kerry A. Emanuel

Department of Meteorology and Physical Oceanography,
Massachusetts Institute of Technology

ABSTRACT

Observations of small and moderate cumulus clouds strongly
support the contention of Squires (1) that the bulk of the mixing in such clouds is attributable to unsaturated downdrafts
initiated at the cloud tops. The scale and dynamics of downdrafts of this type suggest that they may be treated using
similarity theory. Results of this application imply that
downward-propagating unsaturated thermals may be pervasive
throughout most convective clouds.

1. INTRODUCTION

Early aircraft observations of small cumulus clouds (e.g.,
Malkus [2]) firmly put to rest any notion that the properties
of such clouds could be explained merely by the pseudo-adiabatic
ascent of air from cloud base. The observations indicated that
the liquid water content and buoyancy of the clouds were far
smaller than their respective adiabatic values; indeed, the
temperature lapse rate within the clouds closely approximated
that of their environment. It was natural to assume, as did
Stommel (3), that as in the case of dry thermals and plumes,
most of the discrepancies were due to mixing through the sides of
the cloud. Observations (e.g., Warner [4]) show, however, that
little systematic variation in cloud properties occurs across
the cloud, in contrast to laboratory plumes and thermals in which
the time-mean quantities roughly conform to Gaussian distributions.
These and other considerations led Squires (1) to propose that

E. M. Agee and T. Asai (eds.), Cloud Dynamics, 103–116.
Copyright © 1982 by D. Reidel Publishing Company.

the bulk of the mixing in such clouds is due to unsaturated downdrafts initiated at the cloud tops and driven by evaporative cooling. Such downdrafts differ fundamentally from classical dry or
moist convection in that they rely on turbulent mixing to provide
simultaneously the liquid water and dry air necessary for evaporative cooling. The phenomenon is therefore peculiar to clouds.
Squires (1) was able to show, using a simple model employing
constant eddy mixing, that such downdrafts are capable of penetrating to great depths within typical clouds.

Squires' idea is particularly attractive as it accounts for
many of those observed properties of cumulus clouds which cannot
be explained by simple entraining plume models. These characteristics include the lateral distribution of cloud properties, the
magnitude of the ratio of actual to adiabatic liquid water content (Warner [5]), the frequent appearance of dry holes in the
bases of clouds (Warner [4]), the weak dependence of maximum
liquid water content on cloud diameter in all but the smallest
clouds (Squires [1]), and the breakup of cloud updraft at higher
levels (Malkus [2]). In light of these observations and Squires'
theory, it seems surprising that many theoretical investigations
of cumulus dynamics continue to rely on lateral entrainment to
provide the necessary mixing. This weakness is especially apparent in cumulus parameterization schemes.

Subsequent to Squires' (1) initial analysis, little has been
done in the way of describing the individual unsaturated downdrafts. In the following section, the capacity of similarity
theory to adequately describe the dynamics of unsaturated downdrafts is explored.

2. ON THE USE OF SIMILARITY THEORY IN DESCRIBING PENETRATIVE
UNSATURATED DOWNDRAFTS

The description of the mean properties of fully turbulent dry
convective plumes and thermals using similarity theory has been
developed primarily by Schmidt (6), Batchelor (7) and Morton et
al. (8). The latter group also were among the first to carry out
detailed measurements of laboratory plumes and thermals in stratified fluids. The similarity theory is found to provide an excellent description of the laboratory phenomena.

The basic assumptions on which the similarity theories rely
are that (i) the radial profiles of vertical velocity and buoyancy are geometrically similar at all heights, (ii) the mean rate
of entrainment of environmental fluid is proportional at all
heights to a characteristic mean velocity, and (iii) local variations of density throughout the convective elements are small
compared to a reference density. It is further assumed that the

environment of the convective elements is stationary, and that the convective elements are steady in some coordinate system. The stationary environment is applicable to convection which is fundamentally local rather than global in character; in the latter case the entire fluid is presumed to be in motion, rendering use-less the concept of maintained or instantaneous point sources.

The fundamental premise upon which assumptions (i) and (ii) strongly rely is that no velocity or length scales may be formed from the parameters specifying the boundary and initial conditions of the fluid system. In that case, the length scales describing the size of turbulent eddies and the lateral variation of mean velocity and buoyancy may only depend on the distance at any time from the source. This will be precisely the case, for example, in a plume over a maintained point source of heat in a semi-infinite homogeneous fluid, provided that the Reynolds number is effectively infinite and that the buoyancy is small compared to the acceleration of gravity. Then the only relevant boundary condition is the maintained buoyancy flux, from which one cannot form a length scale. If the fluid is stably stratified, however, an external length scale can be formed from the boundary buoyancy flux and the Brünt-Väisälä frequency; this scale determines the maximum penetration height of the plume and makes questionable the similarity assumption. Morton et al. (8) have shown that in this case, the similarity description fails only near the top of the plume and still provides a useful description of the plume properties in its middle and lower sections.

The similarity approach, however, cannot be adequate for treating the dynamics of cumulus clouds since these merely repre-sent the ascending branches of a global instability and, as such, must possess horizontal scales which are related to the vertical scale of the unstable layer. Indeed, the visual appearance of cumulus clouds strongly suggests a fundamental relationship between vertical and horizontal scales and does not suggest the conical expansion of plumes over a point source of heat. It must be pointed out, however, that moist convective updrafts are more local in character than their dry counterparts since the one-way nature of the condensation insures that the dry environmental downdrafts will be relatively weak and broad compared to the cloudy updrafts.

The dynamics of evaporatively driven penetrative downdrafts are fundamentally distinct from those of moist convective updrafts in that mixing is _necessary_ to sustain the former, while it always works against the latter. Thus cumulus clouds are relatively broad so as to minimize the effects of lateral entrainment, while the scale of penetrative downdrafts is small enough that lateral entrainment can provide the liquid water necessary to drive the downdraft. Observations of cumulus clouds (Warner [4]; McCarthy

[9]) suggest that lateral entrainment is important only in clouds
with diameters $\lesssim$ 1 km, while the scale of the horizontal fluctuations of vertical velocity, liquid water content and buoyancy
within larger clouds is very much smaller than the lateral dimensions of the clouds themselves (e.g., Malkus [2]; Warner [4];
Warner and Squires [10]). As in the case of moist convection,
the one-way nature of the penetrative downdraft process insures
that downdrafts resulting purely from the cloud-top instability
will be relatively intense and isolated, while any upward return
circulation forced by the downdrafts will be broad and weak.
Deardorff (11) suggests that this is the case for penetrative
downdrafts in stratocumuli. The isolated character of penetrative
downdrafts may not apply near the tops of growing cumuli, where
the downdrafts are probably initiated by the static instability
in the upper portion of the cloud.

These ideas suggest that in the ideal case of an inert,
homogeneous cloud of great vertical extent, the lateral scale of
penetrative downdrafts is internally rather than externally
determined and, to the extent to which this is true, their dynamics may be described using the similarity approach. While it
must be admitted that clouds, expecially cumulus clouds, are far
from being homogeneous and inert, it appears that the similarity
approach may constitute a plausible means of isolating and high-
lighting the dynamics of the individual penetrative downdrafts.
The theory, which ideally pertains to the properties of unsaturated plumes and thermals in deep, inert clouds, provides a wealth
of physical insight regarding the dynamics of these motions. We
therefore proceed to develop one-dimensional equations describing
the radially averaged properties of evaporatively cooled unsaturated thermals.

3. PENETRATIVE THERMALS

Penetrative convection can occur when the environmental air overlying the cloud is sufficiently cool and dry. When such air is
mixed downward into the cloudy air, the cloud water evaporates
and cools the mixture to the point where it is negatively buoyant
with respect to the surrounding cloudy air; hence it accelerates
downward. Were it not for the effect of water vapor and liquid
water on the buoyancy of air, it could be easily seen that the
criterion for this instability is that the moist static energy
h of the overlying environmental air be less than that of the
cloudy air. In fact, the effect of liquid water and water vapor
is not negligible, as shown by Randall (12) who states the exact
instability criterion

$$\Delta h < \alpha L_v \Delta (q + \ell),$$

where

$$\alpha \equiv \frac{c_p T}{L_v} \left| \frac{1 + \frac{L_v}{c_p} \left(\frac{\partial q^*}{\partial T}\right)_p}{1 + (1 + \gamma) T \left(\frac{\partial q^*}{\partial T}\right)_p} \right| \quad ,$$

$$\gamma = 0.608$$

and q* is the saturation mixing ratio. Δh and $\Delta(q + 1)$ are the jumps across cloud top of moist static energy and total water, respectively. Since the latter is generally negative, the actual criterion for instability is more stringent than $\Delta h < 0$.

Depending on the degree of instability, the initiation of the downward convection may be characterized by the magnitude of the downward fluxes of negative buoyancy and total water deficit which result from the instability. We shall use these fluxes as initial or boundary conditions for penetrative convection under the assumption that the cloud top instability criterion is satisfied.

Idealized laboratory convection is generally produced by maintained localized sources of buoyancy or the instantaneous release of a small volume of buoyant fluid within a much larger volume of stationary fluid. In the former case a steady plume is generated while in the latter a discrete element results. There is, however, little to suggest which form would result in nature; fortunately the general properties of both are very similar. We proceed to develop equations governing the behavior of discrete thermals, since in this case the equations admit analytic solutions for some simple cloud structures. A full discussion of the treatment of plumes is given in Emanuel (13).

The derivation of conservation equations describing the radially averaged properties of a spherically symmetric thermal descending from an instantaneous point source closely follows the development of the thermal equations given by Morton et al. (8). The assumptions regarding the nature of the thermal are as follows:

(i) Radial profiles of vertical velocity, buoyancy and water vapor are geometrically similar at all heights.
(ii) The mean entrainment velocity is proportional to the radially averaged vertical velocity.
(iii) The total buoyancy is small compared to the acceleration of gravity.
(iv) The Froude number is small.
(v) Molecular viscosity is negligible, i.e., the plume is

fully turbulent.
   (vi) The environment is stationary or moving with uniform vertical velocity.

   Assumptions (i), (ii), (iii) and (vi) have been discussed in Section 2. The assumption that the Froude number is small asserts that, as a result of the small scale of the plume, aerodynamic effects are negligible compared to buoyant accelerations; this assumption together with (v) have been very well supported by laboratory experiments.

   An additional assumption also must be made with regard to the effects of phase transition:

   (vii) All entrained liquid water evaporates immediately provided the thermal is unsaturated.

   This assumption will be very nearly valid for liquid cloud droplets, but must fail for precipitation particles.

   In the present case we assume spherical geometry and a "top hat" profile of scalar thermal quantities. This assumption is made for convenience and, since the equations are radially integrated, only affects certain numerical   constants and does not alter the desired parameter dependences.

   The plume is taken to propagate downward from z = 0 so that the entrainment assumption takes the form

$$\bar{u} = \alpha w \quad ,$$

where $\bar{u}$ is a mean radial turbulent entrainment velocity, w the thermal's vertical velocity (taken to be negative) and α the entrainment constant. Following Morton et al. (8), the conservation equations for mass, momentum, heat and water are

$$\frac{d}{dt}R^3 = -3R^2\alpha w \qquad\qquad \text{mass,} \qquad\qquad (1)$$

$$\frac{d}{dt}R^3 w = R^3(B + gl_c) \qquad\qquad \text{momentum,} \qquad\qquad (2)$$

$$\frac{d}{dt}R^3 B = -N^2 R^3 w - Ml_c\frac{d}{dt}R^3 \qquad \text{heat,} \qquad\qquad (3)$$

$$\frac{d}{dt}R^3(q - L_c) = -R^3\frac{d}{dt}L_c \qquad \text{water deficit,} \qquad\qquad (4)$$

where R is the mean thermal radius, l_c is the cloud liquid water, q is the thermal's mixing ratio, L_c is the total cloud water content $(q_c + l_c)$, M is a latent heat constant $(= L_v g/c_p \overline{T}_v$, where $\overline{T}_v$ is a constant mean virtual temperature), B is the virtual temperature surplus of the thermal, defined:

$$B = g \left(\frac{T_{vp} - T_{vc}}{\overline{T}_v} \right) \quad ,$$

and N is the Brünt-Väisälä frequency, defined:

$$N^2 \equiv \frac{g}{\overline{T}_v} \left(\frac{dT_{vc}}{dz} - \Gamma_d \right) \quad ,$$

where Γ_d is the dry adiabatic lapse rate, T_{vc} is the cloud's virtual temperature, and T_{vp} is the virtual temperature of the thermal.

The independent variable may be transformed from time to height using

$$\frac{d}{dt} = \frac{dz}{dt} \frac{d}{dz} = w \frac{d}{dz} \quad .$$

For convenience in defining the boundary conditions, new dependent variables proportional to volume, kinetic energy, total water deficit and total heat deficit are defined as follows:

$$V \equiv R^3$$

$$K \equiv w^2$$

$$Q \equiv R^3 (q - L_c) \tag{5}$$

$$F \equiv R^3 B$$

In terms of the new variables, the conservation equations (1)-(4) are written

$$\frac{dV}{dz} = -3\alpha V^{\frac{2}{3}} \quad , \tag{6}$$

$$\frac{dK}{dz} - 6\alpha K V^{-\frac{1}{3}} = 2FV^{-1} + 2gl_c \quad , \tag{7}$$

$$\frac{dF}{dz} = -N^2 V - Ml_c \frac{dV}{dz} \quad , \tag{8}$$

$$\frac{dQ}{dz} = -V \frac{dL_c}{dz} \quad . \tag{9}$$

The boundary conditions defining a point source of heat and water deficit at $z = 0$ are

$$V = K = 0$$

$$F = F_0 \qquad , \qquad \text{at} \quad z = 0. \tag{10}$$

$$Q = Q_0$$

For simplicity, we treat the case of constant N, constant cloud water ℓ_c and constant lapse rate of cloud saturation mixing ratio. As in the case of dry thermals, the penetrative thermal equations have analytic solutions which can be arrived at by solving (6), (8) and (7) in that order. The water deficit equation (9) can also be solved given the solution of (6). For clarity, we express the solutions in terms of the radius, vertical velocity, heat and saturation deficit:

$$R = -\alpha z, \tag{11}$$

$$w^2 = -\frac{1}{2}\alpha^{-3} F_0 z^{-2} - \frac{2}{7}(M - g)\ell_c z - \frac{1}{16}N^2 z^2 , \tag{12}$$

$$B = -F_0 \alpha^{-3} z^{-3} - M\ell_c - \frac{1}{4}N^2 z , \tag{13}$$

$$(q_s - q) = Q_0 \alpha^{-3} z^{-3} - \ell_c$$
$$+ q_c \left[\exp\left(\frac{L_v}{R_v Tg}B\right) - 1 \right] - \frac{1}{4}\left(-\frac{dq_c}{dz}\right)z . \tag{14}$$

The last expression has been derived from (9) and the Clausius-Clapeyron equation; the subscripts zero denote the initial values of the quantities. The expression (14) clearly shows the various contributions to the moisture budget. The first term on the right is the contribution from the initial flux of water deficit and decays as z^{-3}, while the second term results from the entrained cloud liquid water. The third term represents the effect of the thermal's temperature deficit on the saturation mixing ratio, and the last term represents the effect of the increasing cloud mixing ratio along the path of the thermal. The depth at which saturation will occur, provided the vertical velocity is still negative and finite, can be obtained from (14). The third term on the right of (14) can usually be neglected in this calculation provided that saturation does not occur too close to the initial source. Provided that

$$\ell_c \gg 2^{-\frac{3}{2}}\alpha^{-\frac{3}{4}}(-Q_0)^{\frac{1}{4}}\left(-\frac{dq_c}{dz}\right)^{\frac{3}{4}} ,$$

analysis of the balance of the remaining terms reveals that the
level at which saturation occurs is approximately

$$-z_{sat} \approx \alpha^{-1}(-Q_0)^{\frac{1}{3}}\ell_c^{-\frac{1}{3}} \; . \tag{15}$$

A more precise analysis of (14) shows that saturation will never
occur if

$$-Q_0 \gtrsim 6.75\ell_c^4\alpha^3 \left(-\frac{dq_c}{dz}\right)^{-3} \; . \tag{16}$$

These conditions on the initial flux of water deficit are ultimately related to the cloud-top instability criterion.

Provided that the thermal remains unsaturated, the maximum
penetration depth and maximum vertical velocity attained by the
descending thermal may be assessed using (12). Analysis of the
dominant balance of terms in (12) shows that the first term on
the right may be neglected in computing the maximum penetration
depth and vertical velocity if

$$\ell_c \gg F_0^{\frac{1}{4}}\alpha^{-\frac{3}{4}}N^{\frac{3}{2}}(M - g)^{-1} \; .$$

The right-hand side of the above is $0(10^{-5})$ so that the condition
is easily satisfied even in small clouds. Neglecting the first
term on the right of (12), then, the maximum penetration depth
is

$$-z_{max} = \frac{32}{7}(M - g)N^{-2}\ell_c \tag{17}$$

and the maximum vertical velocity is

$$-w_{max} = \frac{4}{7}(M - g)N^{-1}\ell_c \; , \tag{18}$$

occurring at a height

$$-z = \frac{16}{7}(M - g)N^{-2}\ell_c \; . \tag{19}$$

ℓ_c (g kg^{-1})	$-z_{max}$ (km)	$-w_{max}$ (m s^{-1})
0	0.91	—
0.5	1.65	2.92
1	3.30	5.83
2	6.60	11.67
3	9.90	17.50
4	13.20	23.34
5	16.50	29.17

Table 1. Maximum penetration depth and vertical velocity of penetrative thermals as a function of cloud
liquid water content. Calculations are performed assuming the thermal remains unsaturated, and for
$N^2 = 5 \times 10^{-5}$ s^{-2}, $\alpha = 0.285$, $M = 82$ m^2 s^{-1} and
$F_0 = 10^5$ m^4 s^{-2}. The results are insensitive to both
α and F_0 except when $\ell_c = 0$.

Note that the maximum vertical velocity occurs at exactly half
the maximum penetration depth and that none of the above expressions depends on the initial conditions or the entrainment parameter. The maximum penetration depth and vertical velocity depend linearly on the liquid water concentration. Table 1 shows
these quantities as a function of liquid water concentration.
It is assumed that the thermal remains unsaturated.

It is evident that penetrative convection can reach great
depths at characteristic velocities similar to those associated
with typical convective updrafts. The thermal loses its negative
buoyancy when the surface-area-to-volume ratio becomes too small
for the entrainment and evaporation of liquid water to keep pace
with adiabatic warming.

In the case of laboratory dry thermals, the motion follows
a damped oscillation after the maximum height has been attained.
In the present example, however, a different behavior is implied.
Unlike the dry thermal, the penetrative thermal, were it to come
to rest at its maximum penetration, would be unstable in the same
sense as it was initially since the unsaturated neutrally buoyant
air within the thermal has a smaller θ_e than the surrounding dry
air (unless the thermal "undershoots" the cloud base). Since
the instability is always greatest for perturbations of small
horizontal cross section, it would appear that at some point in
its descent the thermal will break up into smaller entities.

In dry thermals or plumes, the maximum magnitudes of the
buoyancy and vertical velocity generally occur near the central
axis, and the radial profiles are observed to conform roughly to

Gaussian distributions. As the penetrative downdraft contains an unsaturated core and since most of the evaporation may occur away from the central axis, such a distribution may not be maintained and geometric similarity should break down at some point. The magnitude of the negative buoyancy should decrease and perhaps even reverse sign as the downdraft expands, leading to reversed momentum generation along the central axis and a breakup of the downdraft into smaller entities. This branching behavior should continue indefinitely, as long as there remains cloud water available for evaporation and as long as the equivalent potential temperature of the cloud decreases upward. The length of each branch will be proportional to $(M - g)N^{-2}\ell_c$, as expressed by (17). Thus an inert cloud whose top is unstable by the criterion developed by Randall (12) will continue to cool and dry through its entire depth by the action of the penetrative thermals until the cloud top criterion can no longer be satisfied, or until the cloud dissipates entirely. It should also be pointed out that since the cloud top may be unstable to penetrative disturbances even when it is statically stable, the cloud may continue to dissipate even after the bulk of it is in hydrostatic equilibrium with its environment. Thus the "equilibrium cloud" proposed by Telford (14) may still be unstable to penetrative disturbances, although a means of initiating those disturbances may be absent in that case.

4. CUMULUS CLOUDS AND PENETRATIVE DOWNDRAFTS

The pervasive and strong instability of inert clouds to deeply penetrating downdrafts initiated at their tops carries strong implications for the development of cumulus clouds. Not only are the summits of developing cumulus clouds unstable in the classical sense but, in general, they are also unstable to the penetrative downdraft. While the previously discussed similarity theory for such thermals is clearly inadequate for the detailed treatment of penetrative convection in cumulus clouds, which are highly turbulent and whose tops are globally rather than locally unstable, the theory does suggest that penetrative convection will occur under a wide variety of circumstances and with characteristic velocities comparable to those associated with convective ascent. Indeed, it would appear that the only moist convective motions immune to the influence of penetrative downdrafts are those within clouds whose tops do not meet the cloud-top instability criterion and those associated with quasi-steady convective updrafts which are so intense as to preclude the penetration of downdrafts from aloft. The existence of regions of essentially moist-adiabatic ascent within severe thunderstorms (Heymsfield et al. [15]) provides supporting evidence for these possibilities and lends further credence to the notion that entrainment from the sides of the cloud is comparatively insignificant.

   The idea, first suggested by Squires (1), that penetrative
downdrafts dominate the mixing process within cumuli has been
strongly reiterated by Telford (14), who stressed that direct
observations of cumulus clouds strongly support this premise in
preference to the lateral entrainment model. Some of these observations have been summarized in the Introduction. To those
we may now add the recent work of Paluch (16), who, with the use
of glider observations, shows that the properties of air deep
within growing cumulus clouds in Colorado are attributable primarily to the mixing of air from the subcloud layer with environmental air from near the cloud top, rather than to lateral mixing.
In her investigation, Paluch identifies two adiabatically invar-
iant scalar quantities which mix in a linear or nearly linear
fashion. These quantities are the total water content Q (when
precipitation is absent) and a modified equivalent potential
temperature θ_q. These are measured by a glider within the cloud
and plotted together with an environmental sounding of the same
quantities in a Q-θ_q coordinate system. The measurements generally fall along a straight line connecting points on the environmental curve representing subcloud and cloud-top air, respectively, suggesting that the mixture involves relatively little environmental air from middle levels. This is perhaps the first
persuasive study of the origin of air within extratropical cumulus
clouds.

   The objections to the lateral entrainment model raised by
the observations are supported by the fact that one-dimensional
steady-state models built on the lateral entrainment assumption
consistently predict excessive liquid water contents when the
simulated cloud top is made to conform with the observations, as
has been pointed out by Warner (5), who suggested that such simulations are essentially exercises in empirical curve fitting.
By contrast, a simple model proposed by Telford (14) assumes
that actual clouds are far closer to a state of hydrostatic equilibrium with their environment than they are to a state wherein
the cloud properties reflect moist adiabatic ascent. Telford
argues that the equilibrium is achieved by the mixing of environmental air at cloud top with cloudy air at each level within the
cloud. The equilibrium clouds computed using several thermodynamic soundings have liquid water distributions which are closer
to those of observed clouds than can be achieved with models
which rely on lateral entrainment.

   Finally, it may be noted that the dominant role of penetrative downdrafts in cumulus dynamics, implied by both observations
and the present work, casts some doubt on the validity of many
numerical simulations of small convective clouds performed to
date. The very small horizontal scale together with the substantial vertical extent of penetrative downdrafts suggest that
they can neither be resolved explicitly within most models, nor

can their existence be accounted for through the use of the type
of turbulence parameterizations currently employed. An important
exception might be the simulation of large convective storms with
stable ice anvils, above which θ_e is too high to permit the
formation of penetrative downdrafts, or whose quasi-steady updrafts are so strong as to prevent the downdrafts from penetrating
substantial depths into the clouds. Numerical simulations of
this type of cloud have been relatively successful (e.g., Klemp
and Wilhelmson [17]). For smaller clouds, some progress has been
made in representing penetrating downdrafts, most noteably by
Raymond (18) who treats moist convection as a two-scale process
that accounts for the penetrative downdrafts. His model is
successful in producing realistic distributions of liquid water,
velocity and turbulence, especially when these quantities are
compared to those produced by lateral entrainment models. These
results, together with the evidence presented here, suggest that
the failure to account for the presence of penetrative downdrafts
in most cumulus clouds may lead to a serious misrepresentation
of their dynamics.

NOTE

1. Portions of this article are reproduced from an article of the
same title published in the Journal of the Atmospheric Sciences,
38, pp. 1541-1557.

REFERENCES

1. Squires, P., 1958: "Penetrative downdraughts in cumuli."
   Tellus, 10, pp. 381-389.
2. Malkus, J.S., 1954: "Some results of a trade cumulus cloud
   investigation." J. Meteor., 11, pp. 220-237.
3. Stommel, H., 1947: "Entrainment of air into a cumulus cloud."
   J. Meteor., 4, pp. 91-94.
4. Warner, J., 1955: "The water content of cumuliform clouds."
   Tellus, 7, pp. 449-457.
5. Warner, J., 1970: "On steady-state one-dimensional models of
   cumulus convection." J. Atmos. Sci., 27, pp. 1035-1040.
6. Schmidt, W., 1941: "Turbulent propagation of a stream of
   heated air." Z. Angew. Math. Mech., 21, 265 and 351.
7. Batchelor, G.K., 1954: "Heat convection and buoyancy effects
   in fluids." Quart. J. Roy. Meteor. Soc., 80, pp. 339-358.
8. Morton, B.R., Taylor, G., and Turner, J.S., 1956: "Turbulent
   gravitational convection from maintained and instantaneous
   sources." Proc. Roy. Soc. London, A234, pp. 1-23.
9. McCarthy, J., 1974: "Field verification of the relationship
   between entrainment rate and cumulus cloud diameter." J. Atmos.
   Sci., 31, pp. 1028-1039.

10. Warner, J., and Squires, P., 1958: "Liquid water content and the adiabatic model of cumulus development." Tellus, 10, pp. 390-394.

11. Deardorff, J.W., 1980: "Cloud top entrainment instability." J. Atmos. Sci., 37, pp. 131-147.

12. Randall, D.A., 1980: "Conditional instability of the first kind upside-down." J. Atmos. Sci., 37, pp. 125-130.

13. Emanuel, K.A., 1981: "A similarity theory for unsaturated downdrafts within clouds." J. Atmos. Sci., 38, pp. 1541-1557.

14. Telford, J.W., 1975: "Turbulence, entrainment and mixing in cloud dynamics." Pure Appl. Geophys., 113, pp. 1067-1084.

15. Heymsfield, A.J., Johnson, P.N., and Dye, J.E., 1978: "Observations of moist adiabatic ascent in northeast Colorado cumulus congestus clouds." J. Atmos. Sci., 35, pp. 1689-1703.

16. Paluch, I.R., 1979: "The entrainment mechanism in Colorado cumuli." J. Atmos. Sci., 36, pp. 2462-2478.

17. Klemp, J.B., and Wilhelmson, R.B., 1978: "Similations of right- and left-moving storms through storm splitting." J. Atmos. Sci., 35, pp. 1097-1110.

18. Raymond, D.J., 1979: "A two-scale model of moist, non-precipitating convection." J. Atmos. Sci., 36, pp. 816-831.

CONVECTIVE OVERTURNING AND THE SATURATION POINT

Alan K. Betts

West Pawlet, Vermont   05775

A unified approach to the thermodynamics of cloudy air, cloud-clear
air mixing processes, atmospheric thermodynamic equilibrium
structure and instability is formulated, using a new concept:
the Saturation Point. This permits the representation of mixing
processes and virtual potential temperature isopleths for clear
and cloudy air on a thermodynamic diagram, and their comparison
with the atmospheric stratification. Illustrative examples will
be given for evaporative mixing instability and convective equilibrium structure for stratocumulus, cumulus, and cumulonimbus
convection.

1.   INTRODUCTION

This paper discusses a new unified approach to the thermodynamics of cloudy air (but not the formation of precipitation)
and the relationship of the vertical structure of the convective
atmosphere to cloud mixing processes and convective overturning.
A key new concept, the Saturation Point (see Section 2) (which
will be abbreviated SP) and a corresponding Saturation Level (SL)
will be introduced to represent the properties of clear and cloudy
air (but not precipitation particles). For unsaturated air this
"SL" is the familiar lifting condensation level (LCL). In general,
however, we may use the Saturation Point concept to understand
mixing processes between cloud and environment, and to interrelate
updraft and downdraft structure, atmospheric stratification and
stability. Extensive use will be made of the tephigram to illustrate these concepts. Other thermodynamic diagrams could easily
be substituted. The use of the SP also permits the representation
of virtual potential temperature of both clear and cloudy air on

E. M. Agee and T. Asai (eds.), Cloud Dynamics, 117–133.
Copyright © 1982 by D. Reidel Publishing Company.

on a tephigram which facilitates the visual understanding of cloud
parcel buoyancy as well as atmospheric instability and equilibrium
structure. These are significant advances in the use of the thermodynamic diagram. This paper, which is abbreviated from (1) draws
on many strands of research: most notably (2) (3) (4) for the
thermodynamics of moist processes; (5) (6) (7) (8) and (9) for
the modelling of stratocumulus, (10) for her analysis of cloud-
top mixing, and (11) for the concept of convective equilibrium
structure.

2.   SATURATION POINT FORMULATION OF MOIST THERMODYNAMICS

2.1 Conservative variables with cloud water

In this paper, we distinguish between cloud water, and precipitation water with a fall-speed significant compared with vertical
air motions. Cloud water is treated as a parcel property. The
difference between the pseudoadiabat and the reversible adiabat
will be neglected (we shall simply refer to the moist adiabat),
as will freezing processes. To this approximation, the variables
q_T, Θ_E, Θ_L (total water, equivalent potential temperature, liquid water potential temperature) are conserved in adiabatic motions
in non-precipitating cloud systems, and are defined by integrating
the three approximate equations (3).

$$0 = dq_T = dq_S + dl \tag{1a}$$

$$0 = c_p \frac{d\Theta_{ES}}{\Theta_{ES}} = c_p \frac{d\Theta}{\Theta} = \frac{L dq_S}{T} \tag{1b}$$

$$0 = c_p \frac{d\Theta_L}{\Theta_L} = c_p \frac{d\Theta}{\Theta} - \frac{L dl}{T} \tag{1c}$$

The chief approximation is the use of dry air values for the specific heat and the gas constant (in the definition of Θ). Clearly
only two of these equations are independent. Integration gives

$$q_T + q_S + l \tag{2a}$$

$$\ln\left(\frac{\Theta_{ES}}{\Theta}\right) = \int_0^{q_S} L dq_S / c_p T \tag{2b}$$

$$\ln\left(\frac{\Theta_L}{\Theta}\right) = -\int_0^1 L dl / c_p T \tag{2c}$$

The integrals are both evaluated along the moist adiabatic through
a cloudy parcel (Θ, q_S, p), but in <u>opposite</u> directions since

Θ_{ES} is that value of Θ where $q_S = 0$, and Θ_L that value of Θ where $1 = 0$ (see Fig. 1). Just as Θ_E is the highest value of Θ attainable by condensing all the vapor q_S, so is Θ_L the lowest Θ, attainable by evaporating all the parcel cloud water.

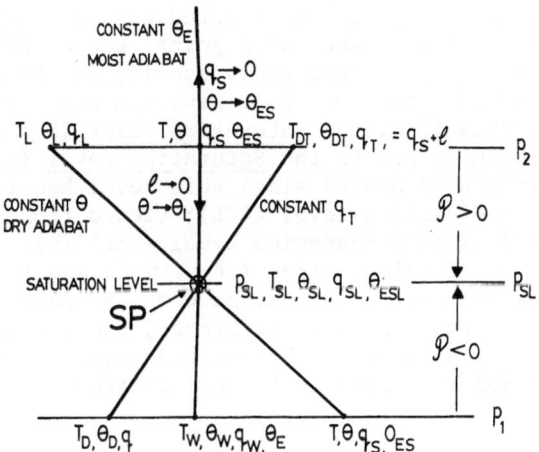

Figure 1. Sketch thermodynamic diagram (tephigram), showing the relationship of Saturation Point, SP, (T_{SL}, P_{SL}) to the conserved parcel properties $(\Theta_{SL}, \Theta_{ESL}, q_{SL})$ which are independent of parcel pressure, and to the parcel properties at other pressure levels, such as (T, T_w, T_p) at p_1, for unsaturated air.

Because the latent heat L and parcel temperature T vary along the moist adiabat, the integrals in Eq. (2) must be evaluated numerically, or using empirical functions, or approximated. For most observational purposes, it is sufficiently accurate (error$\sim$ 0.2K at warm temperatures) to approximate the integrals as (see (12) for Θ_E)

$$\ln(\Theta_{ES}/\Theta) = 2 \cdot 67 q_S/T \tag{3a}$$

$$\ln(\Theta_L/\Theta) = 2 \cdot 371/T \tag{3b}$$

These formulae are useful because they now contain only parcel parameters at one pressure level. The coefficients differ because the integrations are in opposite directions along the moist adiabat (Fig.1 arrows). For unsaturated air   is given by

$$\ln(\Theta_E/\Theta) = 2 \cdot 67 \ q/T_{SL}$$

Where T_{SL} is the temperature at the saturation level, which can be readily computed from parcel parameters. Θ_L for unsaturated air reduces to Θ , and q_T to vapour mixing ratio q .

2.2 The Saturation Point

The three lines in Fig. 1 representing the conservation of three variables ((Θ, Θ_E, q) for unsaturated air and (Θ_L, Θ_{ES}, q_T) for cloudy air: (3), (4)) intersect at a point, where the temperature and pressure (T_{SL}, p_{SL}) completely specify the three conserved   parameters, which we shall give symbols: Θ_{SL}, Θ_{ESL}, q_{SL}. We shall call this point the <u>Saturation Point</u> (SP) and the pressure level at which it occurs the <u>Saturation Level</u> (SL). Viewed from below (the unsaturated side) this level has long been called the Lifting Condensation Level (LCL); viewed from above it has been called the Sinking Evaporation Level (SEL) (4). The more general terminology is of value, since (T_{SL}, p_{SL}) and the three derived conserved parameters do not change with the reference level of an observation. In this frame of reference, the thermodynamic state of an air parcel is specified by the Saturation Point properties (T_{SL}, p_{SL}) and its pressure difference from the Saturation Level.

$$P = p_{SL} - p \qquad\qquad\qquad (4)$$

On a thermodynamic diagram, air parcels that have a given (T, p) may have SL's at <u>any</u> level, p_{SL} , depending on their total moisture content. Parcels with total moisture content $q_T > q_S$ (T,P)will be saturated and cloudy with P>0 , while if parcel q< q$ T,p), it will be unsaturated with P<0 . Positive P is directly related to liquid water content; negative P to subsaturation. We shall call P the parcel <u>saturation pressure difference</u> and the parcel level or data pressure level to distinguish it from the Saturation Level, p_{SL} . The procedure we shall use is to analyze parcel processes, including cloud-environment mixing and diabatic processes in terms of their Saturation Point. This is a shift to a (T_{SL}, p_{SL}, P) coordinate system for the three independent parcel variables from (T, q, p).

2.3 Mixing diagrams and the Saturation Point

The parcel conserved parameters (Θ_{SL}, Θ_{ESL}, q_{SL}) are conserved in adabatic motion and approximately in isobaric mixing processes (3), (7). Since the SP characterizes parcel conserved properties, we need only consider their SP's in mixing two parcels. This is a great simplification. For example, if we mix equal masses of air parcel with different SP's, the SP of the mixture, therefore will have   properties given by the simple average of their conserved parameters (to slight approximation). Fig. 2 and Table 1 show a specific example of mixing between two parcels with saturation points (T_{SL}, p_{SL}) of (20C, 900mb) and (5C, 700mb).

The SP's of <u>all</u> mixtures lie on the dashed <u>mixing line</u>, which
can easily be constructed on a thermodynamic diagram by computing
Θ_{SL}, q_{SL} for different mixtures and finding their associated SP.

   The distinction between the <u>saturation level</u> of a parcel P_{SL}
and the <u>data pressure level</u>, p, where a parcel may find itself in
the atmosphere, is of crucial importance. The tephigram is being
used to represent <u>both</u>. For example, suppose we consider a mixing
process at <u>800mb</u> between cloudy air which has risen adiabatically
from cloud-base at 900mb where it had the SP C shown. Its SP has
stayed at C, while $P = p_{SL} - p$ has become positive and its liquid
water has increased. This cloudy air mixes isobarically at 800mb
with air that has an SP at E. All the mixtures of C and E have
SP's on the dashed line. The mixtures that are unsaturated have a
saturation level $p_{SL} < 800$mb, corresponding to $P < 0$, while the mixtures that are cloudy are those which have $p_{SL} > 800$mb. The 1:1
mixture in Table 1, for example, with SL at 804mb, is only just
cloudy at 800mb; if it sinks to pressures above 804mb, it becomes
unsaturated. Its (T, T_w, T_p) or (T_L, T, T_{pT}) at any other pressure
level are found by drawing dry and moist adiabats and constant
lines through its SP as shown.

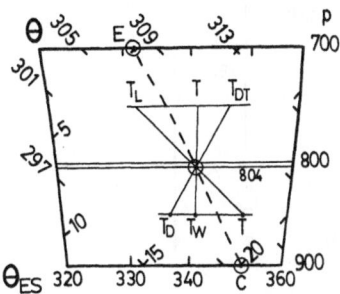

Figure 2. Tephigram showing mixing between parcels with
SP's at C and E (circled). All mixtures have SP's on
the dashed mixing line. The 1:1 mixture has SP (T_{SL}, p_{SL})
at (13.5C, 804mb) shown. This gives its conserved parameters: its parcel properties at other pressure levels
are given by the construction shown.

Table 1: Mixing of Two Parcels C and E in Various Mass
Ratios

C	900	20.0	302.1	16.6	351.3
E	700	5.0	308.0	7.9	332.0
Mixture					
1:3	754	9.6	306.5	10.0	336.9
1:1	804	13.5	305.1	12.2	341.7
3.1	853	17.0	303.6	14.4	346.5

The dashed line, representing the SL's and SP's of all
mixtures (of C-and E) is <u>unchanged</u>, whatever the pressure level
of mixing; although for every mixture, P changes as p changes.
It is clear that it is easy to find with this diagram the temperature, humidity, liquid water content, etc., of any mixture at any
pressure level. Thus, comparisons of cloud-environment mixtures
with sounding properties can be made.

2.4 Slope of the mixing line

This analysis of the mixing of two parcels in terms of the
mixing line between their SP's is particularly useful for convective
systems where there are essentially two different "source regions"
of air with distinct properties. Among the examples to be explored
later are stratocumulus and the trade cumulus layer where warm dry
air is subsiding into a well-mixed layer. An important parameter
for stability questions is the slope of the mixing line in relation
to the dry and moist adiabats, and the slope of the θ_v isopleths
(see section 4).

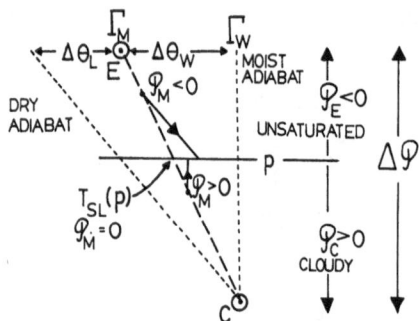

Figure 3. Schematic mixing diagram for two parcels, one
cloudy, one unsaturated (environmental), showing mixing
line (heavy dashes) with a gradient between dry and moist
adiabats, and the minimum temperature from cloud-clear
air mixing at pressure level p.

To the linear approximation the mixing line gradient is a
function only of the SP's of the two "source regions." Fig. 3
defines this gradient $\Gamma_M = \partial\theta/\partial p$ for the mixing line in relation
to the slope of the moist adiabat $\Gamma_W = (\partial\theta/\partial p)_{\theta_{ES}}$.

$$\Gamma_M = \frac{\Delta\theta_L}{\Delta P} = (\frac{\Delta\theta_L}{\Delta\theta_L + \Delta\theta_W})\Gamma_W \qquad (5)$$

Fig. 3 shows a mixing process typical of shallow cumulus, where
Γ_M lies between wet and dry adiabats.

3. MIXING IN CUMULUS CLOUDS

The mixing line analysis (Fig. 3) can be used to formulate many cumulus mixing processes. Two schematic examples are presented.

3.1 Minimum temperature from cloud-clear air evaporative mixing

Fig. 3 elegantly solves the well-known problem of the equilibrium temperature of cloud-environment mixtures. For the mixture with SP where the mixing line crosses any pressure level p, $p=p_{SL}$ and $P_M=0$. This is by definition the mixture that is just saturated with no liquid water at the pressure level p, with temperature T_{SL} (P). The temperature of mixtures that have SL's away from p are found by drawing dry adiabats (for unsaturated mixtures $P_M<P_o$) or moist adiabats (for cloudy mixtures $P_M>0$) from the mixing line back to p, as shown in Fig. 3. It is obvious that provided the mixing line lies between dry and moist adiabats, then the just saturated mixture has the minimum temperature, T_{SL} (p). The Θ depression of T_{SL} (p) below that of the environment is given by

$$\Delta\Theta = -\Delta\Theta_L \; P_E/(P_C - P_E) \approx (1 - \sigma)\Delta\Theta_L \tag{6}$$

where σ is the fraction of environment in this just saturated mixture. The water budget for this mixture gives σ (1).

3.2 Cumulus mixing and downdraft equilibrium level

A comparison of the mixing line (Fig. 3) with the environmental stratification is clearly of great significance. Certain mixtures will be colder than the environment and will tend to form downdrafts, others will be warmer and might be expected to characterize updrafts.

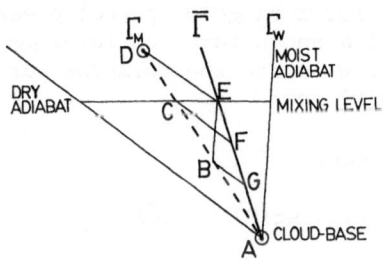

Figure 4. Construction of thermal equilibrium level of mixtures given the mixing line and an environmental sounding.

Fig. 4 shows a schematic tephigram with dry and moist adiabats, an environmental stratification $\overline{\Gamma}$, and a mixing line between cloudy air (which has risen undiluted from cloud-base at A) and environ-mental air at the mixing level (which is unsaturated with SP at D). DE, CF, BG are dry adiabats, and EB is a moist adiabat. All the cloud-environment mixtures have SP's on AD, which may be cate-gorized as follows: Mixtures with SP's
i) between D and C are unsaturated, cooler than the environmental Θ (E) and may sink to equilibrium between E and F (ignoring the virtual temperature correction).
ii) at C, is just saturated, has a minimum temperature and an equi-librium level F.
iii) between C and B are saturated and cloudy, cooler than the environment Θ (E) and may sink, first moist adiabatically to their SP, and then along a dry adiabat to thermal equilibrium between F and G. G is the lowest equilibrium level for downdrafts produced by mixing.
iv) at B, this parcel has the same potential temperature as the environment at E, and the same Θ_{ES} .
v) between B and A, are cloudy and warmer than the environment and are likely to ascend a moist adiabat.

4.   VIRTUAL POTENTIAL TEMPERATURE

      The virtual temperature correction is important for instability and buoyancy equilibrium analyses. Isopleths of virtual potential temperature Θ_v , for unsaturated and cloudy air can be drawn on buoyancy mixing diagrams (8): $(\Theta_{SL}$, $q_{SL})$ diagrams in our notation.

4.1 Tephigram plot of Θ_v

      Θ_v overlays can also be drawn for a tephigram for any parcel pressure level p (and a stability analysis identical to that of (8) performed.) The value of Θ_v is read at the SP (T_{SL}, p_{SL}) not (T,p). We shall find that qualitatively a single overlay depicts the shape of the Θ_v isopleths for a range of parcel pressure P .
The distinction between parcel p and saturation level p_{SL} is again crucial, and become clearer if we write the formulae for Θ_{vu} , Θ_{vc} (for unsaturated and cloudy air) as

$$\Theta_{vu} \ (p_{SL}) = \Theta_{SL} \ (1 + 0.61 q_{SL}) \qquad\qquad (7a)$$

$$\Theta_{vc} \ (p_{SL}, \ p) = \Theta(p) \ (1 + 0.61 \ q_S(p) - 1(P) \qquad\qquad (7b)$$

For unsaturated air, Θ_{vu} is a unique function of (T_{SL}, p_{SL}) at the SP, so isopleths of Θ_{vu} can be drawn on a thermodynamic diagram, with the convention that Θ_{vu} is read at the SP. For cloudy air, Θ_{vc} depends on Θ , q_S which are defined at p , and 1 which increases as $P=p_{SL}-p$ increases. Specifically, this means that the Θ_v isopleths appropriate to a given parcel pressure level p , <u>kink</u> at P .

Fig. 5 shows a section of a low level tephigram with isopleths (heavy dashes) overlaid for a parcel level of 850mb. If we plot the SP of an air parcel which is at 850 mb, the dashed lines give its Θ_v . For SL's below 850mb (the "<u>unsaturated region</u>" with respect to the given data level of 850mb) Θ_{vu} increases along the dry adiabat ED (for example) as the saturation mixing ratio ($q_S=q_{SL}$) rises. For pressures above p , SP's on the moist adiabat BC (the "<u>cloudy region</u>" with respect to 850mb) have the same Θ,q_S at point B, but increasing cloud liquid water 1 . Θ_{vc} thus decreases as   increases on BC (Eq 7b). At the parcel level p, cloudy and environmental parcels which have SP's at C, E respectively have the <u>same buoyancy</u>, although the environmental parcel is actually warmer.

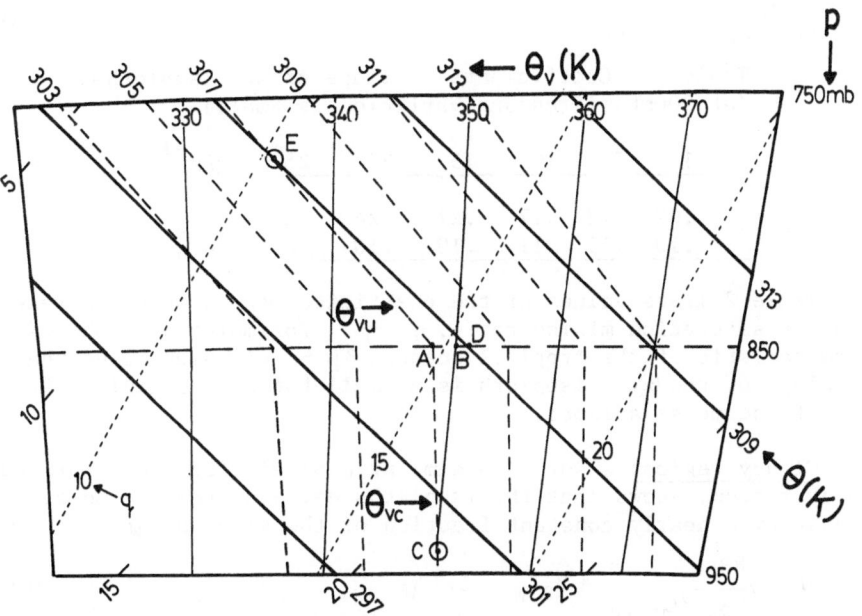

Figure 5. Section of tephigram showing isopleths of (heavy dashed, in degrees K) for a parcel level of 850mb. The light dashed lines are saturation mixing ratio.

If an air parcel moves from 850mb dry or moist adiabatically to another pressure level, p, its <u>SP does not change</u>, but the isopleths now kink at the new p level. If the parcel is <u>cloudy</u>, its Θ (and Θ_{vc}) will be different, but if it is unsaturated, neither Θ nor Θ_{vu} changes. In fact, an overlay of simply the

dashed lines on Fig. 5 (with q_S labeled and Θ_V unlabeled) is of
practical use because it shows the slope of the isopleths in both
cloudy and unsaturated regions (with respect to a parcel pressuure
level p), since the isopleths depend predominantly on the
field.

5.2 Relation of Θ_V isopleths to dry and moist adiabats

The slope of the Θ_V isopleths and their deviation from the
dry and moist adiabats is of importance to stability analyses.
It is informative to express these as fractions of the slope of
the moist adiabat, Γ_W . Simple linearized expressions are derived
in (1).

<u>Unsaturated region</u>: $P<0$. The gradient of the isopleth can
be written

$$\left(\frac{\partial \Theta_{SL}}{\partial P_{SL}}\right)_{\Theta vu} = \beta_1 \Gamma_W \qquad (8)$$

Table 2: Coefficients in Slope of Θ_V Isopleths
(at 900mb) (Pressure Variation is Small)

1	5	10	15	20	25	gKg^{-1}
.07	.12	.17	.21	.24	.27	
.10	.11	.11	.10	.10	.09	

Table 2 lists values of the coefficient β_1 , which increases
with the saturation mixing ratio, q_S . For moist atmospheres,
characteristic of the tropical oceans, $\beta_1 \simeq 0.2$ which means that
the slope of the Θ_{vu} isopleth is a small but not negligible fraction of the moist adiabat.

<u>Cloudy region</u>: $P>0$. A similar analysis for the saturated,
cloudy region, shows that the isopleths deviate from the moist
adiabat by a nearly constant fraction of the slope of Γ_W

$$\Gamma_W - \left(\frac{\partial \Theta_{SL}}{\partial P_{SL}}\right)_{\Theta_{VC},P} = \beta_2 \Gamma_W \approx 0.1 \Gamma_W \qquad (9)$$

In this case the coefficient β_2 is nearly constant (Table 2).
The slope of the Θ_{VC} isopleth gives the criterion for cloud-top
entrainment instability for a stratocumulus layer (see Section 5).
This slope is also of significance to cloud parcel buoyancy. Although an ascending cloud parcel Θ may follow the moist adiabat
Γ_W , its <u>buoyancy</u> is effectively following a Θ_{VC} isopleth of
only $0.9 \Gamma_W$, a considerable reduction in buoyancy (see Section
5), until cloud water is converted to precipitation water and falls
out. (9) can be written in terms of the Θ_{ES} gradients on the

isopleth and the dry adiabat (1).

$$\left(\frac{\partial_{ESL}}{\partial_{P_{SL}}}\right)_{\Theta_{VC},P} \approx 0.1 \ \left(\frac{\partial\Theta_{ES}}{\partial_{P}}\right)_{\Theta} \tag{10}$$

Since the Θ_V correction affects parcel buoyancy, several conventional tephigram definitions (such as level of free convection, convective available potential energy) are affected significantly (1).

5. ILLUSTRATIVE EXAMPLES

5.1 Stratocumulus instability on a tephigram

Several authors have discussed the instability of a stratocumulus layer which results from cloud-top mixing if the Θ_E of air above the inversion is low enough. The analysis of (5) has been extended by (8), (9) to include virtual temperature effect due to liquid water and show that the criterion for instability requires a decrease of Θ_E across the capping inversion. The tephigram mixing diagram analysis shows this same criterion in terms of $\Delta\Theta_E$ ΔP (between cloud and clear air). The critical criterion for instability then become very simply a comparison of the slope $\Delta\Theta_E/\Delta P$ and that of the Θ_{VC} isopleths in the cloudy region (8). Fig. 6 shows the construction.

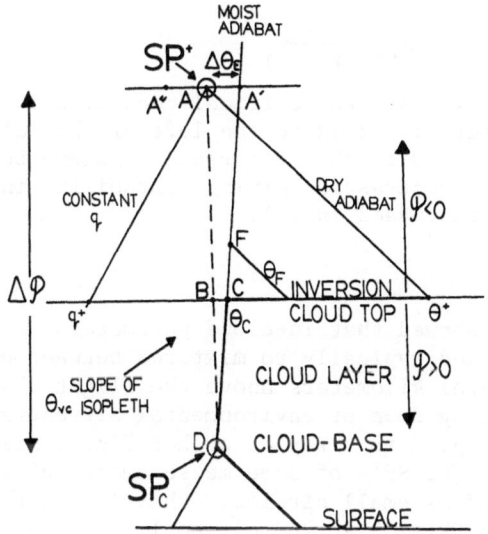

Figure 6. Schematic tephigram for stratocumulus cloud-top entrainment instability. Saturation points SP for ascending cloudy air and warm dry air just above

inversion (SP), with the mixing line between them
(heavy dashed line). The critical mixing line for
instability is drawn which is parallel to the Θ_{VC} isopleth (short dashes).

The dry and moist adiabats drawn as heavy solid lines correspond
to properties of air in the ascending branch of the stratocumulus
circulation. Cloudy air reaches the inversion at cloud-top with
Θ_C shown, but retains an SP at cloud-base. It mixes at cloud-top
with warm dry air (Θ^+ , q^+) above the inversion with an SP$^+$ as
shown. The heavy dashed mixing line (AD) shown is parallel to
the Θ_{VC} isopleth, which means that all cloudy mixtures with SP's
on BD have exactly the same buoyancy as the ascending unmixed
cloudy air. We can see that AD must be the critical mixing line
dividing stability and instability of cloudy mixtures by considering
the points A' and A". Cloudy mixtures with A' air would be stable
and not sink into the cloud, while cloudy mixtures with A" air are
unstable because they have a Θ_{VC} cooler than the unmixed cloudy
air. These mixtures can sink freely into the cloud layer as negatively buoyant, saturated downdrafts until they run out of cloud
water at their saturation level.

While cloudy mixtures with A" air are unstable because they
have a Θ_{VC} cooler than the unmixed cloudy air. Thus, the critical
condition for instability in terms of jumps across the inversion
can be written (using linearized gradients):

$$\frac{\Delta\Theta_E}{\Delta P} > (\frac{\partial\Theta_{ESL}}{\partial P_{SL}})_{\Theta_{VC},p} \sim 0.1 \; (\frac{\partial\Theta_{ES}}{\partial_p})_\Theta \tag{11}$$

using Eq. (10). In this form the critical condition is easily
visualized on a tephigram: it is to the left of the moist adiabat
through cloud-base by 1/10 of the difference in slope between dry
and moist adiabats. A contrasting mixing instability in the severe
storm environment is discussed in (1).

5.2 Cloud-top mixing

A study by (10) showed that in-cloud parameters measured by
a sailplane corresponded typically to mixtures between cloud-base
air and air form several kilometers above the flight observation
level, suggesting mixing down of environmental air entrained
through cloud-top. Fig. 7 is a replot of her Fig. 4 mixing diagram on a tephigram. The SP's of some measurements of in-cloud
properties are plotted as small circles: they lie on the mixing
line between air from 8km (SP at 282mb) and 1.5km (just above the
surface, with SP at the observed cloud-base of 3.8km, 645mb). The
flight level of the in-cloud observations was 5.2km. The mixing
line is also drawn between cloud-base air and air from 4.65km,
corresponding approximately to "lateral entrainment" of air into

the cloud between cloud-base and the observation level. (10) concluded that it is clear that this mixing process does not characterize the measured in-cloud properties.

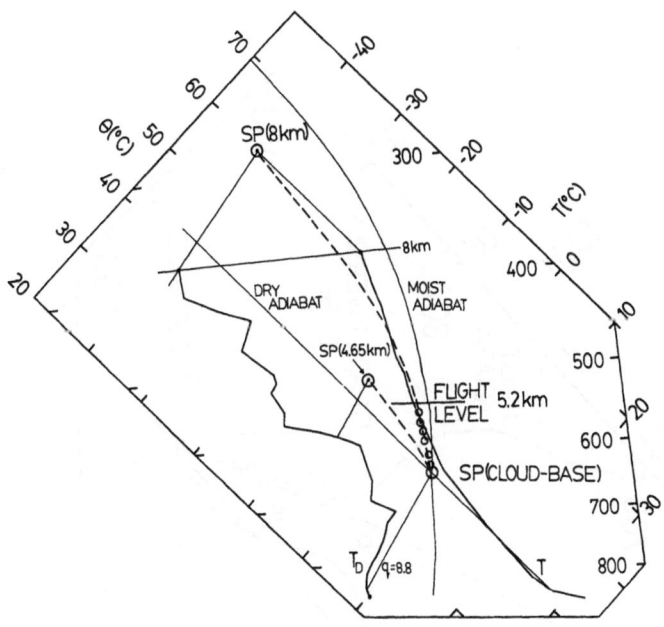

Fig. 7. 1730Z sounding at Potter on 25 July, 1976, showing
        temperature and dewpoint; two mixing lines between SP's
        for cloud-base air and air from 8km and 4.65km. SP's
        for in-cloud observations are shown (open circles).
        Sounding from Paluch, 1979)

5.3 Trade cumulus equilibrium structure

     The trade cumulus layer involves the mixing of a moist boun-
dary layer with overlying drier, potentially warmer air above the
trade inversion. The comparison with the stratocumulus layer is
instructive. Fig. 8 presents the average of three days' soundings
during BOMEX, a trade-wind experiment, during an undisturbed period. The dashed line is the mixing line between cloud-base air
and the subsiding air just above the trade inversion. Several
facts are readily apparent:
i) The instability criterion for the breakup of a stratocumulus
layer is easily satisfied: θ_E of the subsiding dry air at inversion top is well to the left of both moist adiabat through cloud-
base, and the θ_{VC} isopleth through cloud base (not shown).

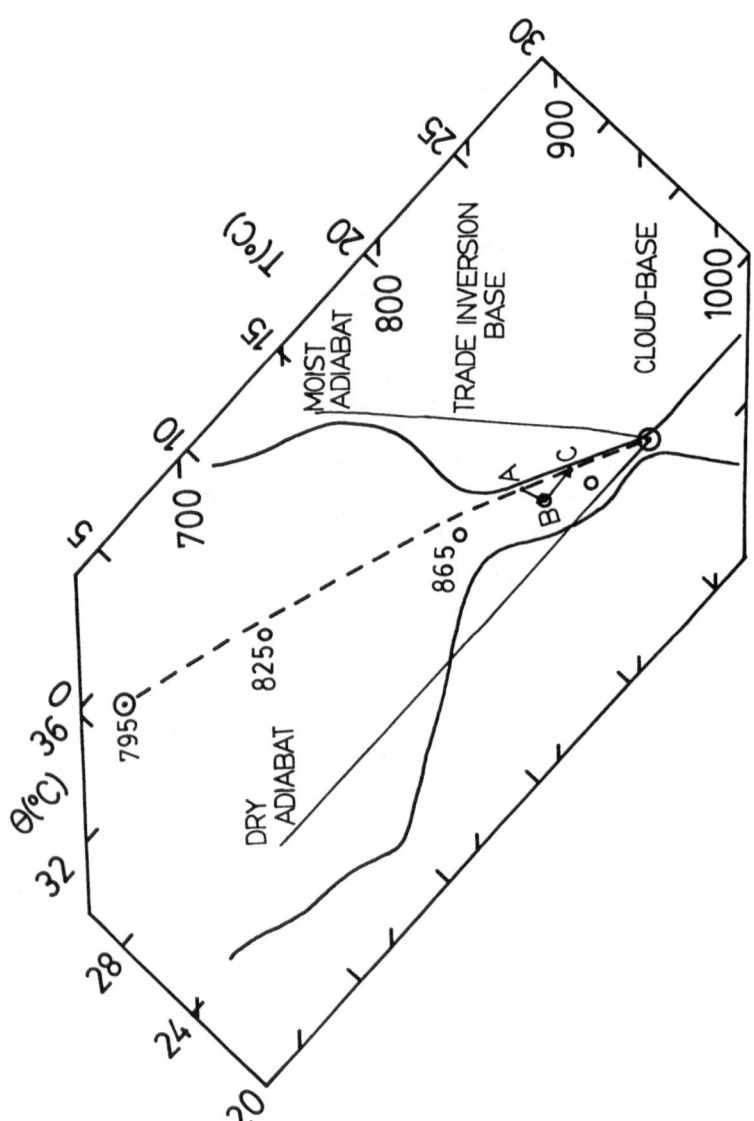

Figure 8. Three-day average sounding for undisturbed tradewind convection (BOMEX 22-24 June, 1969). Dashed curve is mixing line between inversion top and subcloud air SP's. Open circles are SP's of environment.

ii) Above the base of the trade inversion, just saturated mixtures
are very cold and will readily sink in downdrafts to near the base
of the inversion (3).
iii) The lapse-rate in the conditionally unstable cumulus layer,
between cloud-base and inversion base, parallels remarkably closely
the mixing curve, which does not favor downdrafts in this layer.
This suggests that the lower cumulus layer lapse rate is controlled
by the mixing of dry air from the inversion.
iv) Within the cumulus layer and trade inversion. layer, saturation
points of the environment lie to the left of the mixing line. This
equilibrium structure cannot be explained by mixing alone: radiative cooling must be involved (1), indicated schematically by the
arrows.

5.4 Deep convective equilibrium structure

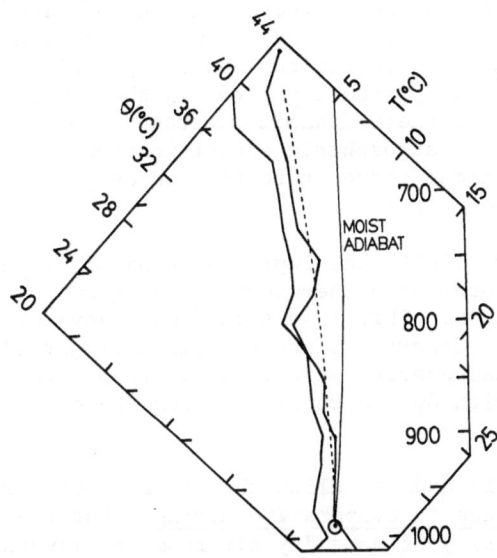

Fig. 9. Sounding for GATE ship <u>Dallas</u>, for 1213Z, Julian Day 248,
1974. The Θ_{VC} isopleth (short dashes) and moist adiabat
are drawn through a low level SP (at 990mb).

A number of GATE soundings in highly disturbed convective
episodes show nearly saturated soundings with lapse-rates through
a deep layer that are between dry and moist adiabats and are,
therefore, apparently highly unstable. Fig. 9 shows an example:
1213Z on Julian Day 248 (1974) from the ship <u>Dallas</u>. The sounding
(during a major disturbance) is nearly saturated from 1000-650mb
and the moist adiabat through the low level SP (cloud-base) suggests

a large convective available potential energy. However, the Θ_{VC} isopleth through the cloud-base SP at 990mb to 625mb (the sounding Θ_{ES} minimum) shows that without fallout of cloud water, updrafts are almost neutrally buoyant. This is consistent with the weak drafts observed in GATE. Furthermore, it suggests that in major disturbances, the atmosphere approaches an equilibrium structure that is near neutral to the convective process. Above 625mb, for the environment again increases: this is consistent with the fact that above this level precipitation and freezing (freezing level is about 575mb) significantly increase parcel buoyancy. Other examples are shown in (1) for the hurricane eyewall and con- vection over land in Venezuela.

CONCLUSIONS

    i) The use of the <u>Saturation Point</u> (SP) consolidates and simplifies the description of the moist thermodynamics of cloudy air. It compactly represents the conserved parameters and enables the representation of the mixing process between air parcels (whe- ther clear or cloudy) originating from any level in the atmosphere, in terms of a <u>mixing line</u> on a thermodynamic diagram. The relation- ship of the mixing line to the atmospheric stratification is made readily visible, and the minimum temperature·from an evaporative mixing process can easily be computed.

    ii) The use of the SP permits the representation of <u>virtual</u> <u>potential temperature</u> isopleths on a thermodynamic diagram for both unsaturated and cloudy domains. This is a significant advance in the use of the thermodynamic diagram, since questions of parcel buoyancy, available potential energy, parcel equilibrium level, level of free convection (with Θ_V correction) become readily vis- ible.

    iii) Taken together, i) and ii) allow the diagrammatic repre- sentation of <u>instabilities due to evaporative mixing</u>. For the stratocumulus layer, the mixing of some dry air from the inversion into the cloud layer produces unstable cloudy downdrafts in the moist adiabatic cloud layer, provided Θ_E of the entrained clear air is low enough.

    iv) The use of the mixing line and the Θ_V isopleths suggests explanations for the <u>equilibrium structure</u> of the convective atmos- phere. A specific illustration shows the equilibrium thermal struc- ture of the lower trade cumulus layer lies on the mixing line be- tween air from the subcloud layer and from the inversion top. An- other example for the highly disturbed tropical atmosphere over the ocean (GATE) shows that the apparently conditonally unstable thermal structure is near neutral in Θ_V until liquid water is precipitated.

Acknowledgements: This work was supported by the Atmospheric Science Division (GARP program) of the National Science Foundation under grant ATM-7915788. I am grateful to E.J. Zipser and I.R. Paluch for discussions.

References

1. Betts, A.K., 1982: Saturation Point analysis of moist convective overturning. J. Atmos. Sci. 39, (submitted)

2. Rossby C.G., 1932: Thermodynamics applied to air mass analysis. Massachusetts Institute of Technology, Meteor. Papers Vol. 1, No. 3, pp. 7-24.

3. Betts, A.K., 1973: Non-precipitating convection and its parameterization. Quart. J. Roy. Meteor. Soc. 99, pp. 178-196.

4. Betts, A.K., 1978: Convection in the tropics. Quart. J. Roy. Meteor. Soc. Supplement, 1978: "Meteorology over the tropical oceans," pp. 105-132.

5. Lilly, D.K., 1968: Models of cloud-topped mixed layers under a strong inversion. Quart. J. Roy. Meteor. Soc. 94, pp. 292-309.

6. Schubert, W.H., 1976: Experiments with Lilly's cloud-topped mixed layer model. J. Atmos. Sci. 33, pp. 436-446.

7. Deardorff, J.W., 1976: On the entrainment rate of a strato-cumulus topped mixed layer. Quart. J. Roy. Meteor. Soc. 102 pp. 563-582.

8. Deardorff, J.W., 1980: Cloud-top entrainment instability. J. Atmos. Sci. 37, pp. 131-147.

9. Randall, D.A., 1980: Conditional instability of the first kind upside-down. J. Atmos. Sci. 37, pp. 125-130.

10. Paluch, I.R., 1979: The entrainment mechanism in Colorado cumuli. J. Atmos. Sci. 36, pp. 2467-2478.

11. Ludlam, F.H., 1966: Cumulus and cumulonimbus convection. Tellus 18, pp. 687-698.

12. Bolton, D., 1980: The computation of equivalent potential temperature. Mon. Wea. Rev. 108, pp. 1046-1053.

13. Zipser, E.J., and M.A. Lemone, 1980: Cumulonimbus vertical velocity events in GATE. Part 2: Synthesis and model core structure. J. Atmos. Sci. 37, pp. 2458-2469.

ATTEMPT TO DIRECTLY SIMULATE CLOUD-RADIATION INTERACTION
IN THE CASE OF SMALL CUMULI

Ph. VEYRE, Ecole Nationale de la Météorologie
2, Avenue Rapp, 75007 PARIS
G. SOMMERIA, Laboratoire de Météorologie Dynamique
24, rue Lhomond 75231 PARIS Cedex 05
and Y. FOUQUART, Laboratoire d'Optique Atmosphérique
Université de LILLE

ABSTRACT

The object of this paper is a numerical study of the interaction between cloud dynamics and infrared radiative effects,
in the case of small cumuli. For this purpose, an infrared
radiation model has been developed for use in a three-
dimensional planetary boundary layer model. It takes into
account the CO_2 , the water vapor including its continumum
in the 8-14 μ window as well as the liquid water inhomogeneities.
A comparison has been made between two numerical simulations
which differ only by the presence or absence of liquid water
effect on the radiative field. This comparison is made during
80 minutes in the case of small trade-wind cumuli.

Main results are presented and discussed. They show a strong
increase in cloud activity and of all related turbulence
characteristics, due to the local cloud top radiative cooling.
This cannot be explained by a mean radiative effect on the
cloud layer and is mainly due to the local interaction between
the turbulent cloud circulation and the radiative field.

1. INTRODUCTION

The object of this paper is an attempt to study the interaction
between cloud dynamics and radiative effects at cloud scale.

The importance of cloud-radiation interaction is recognized for
medium-range weather forecast and climate studies. What is needed
in those cases is a statistical relationship between cloud cover

E. M. Agee and T. Asai (eds.), Cloud Dynamics, 135–147.
Copyright © 1982 by D. Reidel Publishing Company.

and radiative fluxes. One of the ways to obtain them is through
detailed studies of the physical processes involved. Understanding
cloud-radiation interaction is also important in micro-
meteorology for example in order to forecast fog or low-level
stratus cloud cover. In addition cloud-radiation probably has
a triggering effect in the development of deep convective
systems (cf. Gray and Jacobson, 1977 - Stephens and
Wilson, 1980).

If we restrict ourselves to small-scale processes located within
the planetary boundary layer, one interesting example is given
by the diurnal cycle of convective activity over the ocean,
which shows a maximum around sunrise. Several integral type
models such as those by Augstein (1980) and Albrecht (1981)
try to explain these features by including some schematic
radiative effects. A study with a three-dimensional model
however allows a more complete description of the physical
processes involved. It has been done by Deardorff (1980) in
the strato-cumulus case and we attempted it here in the case
of small cumuli.

Main questions to be answered are :

- What is the relative importance of clear air and cloud
radiative effects in the development of one individual
cumulus cloud ?

- What are the relative magnitude of the various heat sources
in a cloudy boundary layer : temperature vertical advection,
condensation and evaporation, radiative flux divergence ?

- Which changes are produced by cloud radiative effects in the
boundary layer convective activity : cloud cover, turbulent
fluxes and energies ?

- Is there a way to parameterize the effect of cloud-radiation
interaction for a layer of small cumuli ?

This study uses a three-dimensional boundary layer model for
which an adequate radiative scheme has been developed (including
only infra-red fluxes).

2. MAIN FEATURES OF THE DYNAMICAL MODEL

The dynamical model is a slightly improved version of the one
developed by Sommeria (1976) in collaboration with
J.W. Deardorff.

The main purpose of the model is to describe the detailed
evolution of the turbulent meteorological field within a
three-dimensional grid network covering a small domain of the
atmosphere (a cube 2 x 2 x 2 km with a grid interval of
50 meters in the present case). The basic variables include
the three components of velocity, the potential temperature,
the specific humidity, the specific content of cloud water and
the pressure deviation from the mean field. The usual set
of meteorological equations for moist air is used, with
the anelastic approximation for density : an equation of state,
the continuity equation, a Navier Stokes equation with a Coriolis
and a buoyancy term including water loading, a thermodynamic
equation including latent heat release and radiative effects,
evolution equations for water vapor and cloud water.

Subgrid scale processes are parameterized with a simplified
second order closure, taking into account sub-grid condensation.
The grid scale is chosen so that most of the turbulent processes
are explicitely computed and sub-grid scale turbulence represents
only a small fraction of them from the energetical point of view.

Lateral boundary conditions are periodic, and large-scale forcing
can be applied, such as a prescribed subsidence or advection.

3. MAIN FEATURES OF THE INFRA-RED RADIATIVE SCHEME

The objective was to develop a radiative scheme which includes
liquid water effects but remains economical in computer time
and storage requirement. This lead to the following approximations, which can be justified :

- Visible fluxes are not taken into account.

- Lateral radiative exchanges are neglected in comparison with
vertical exchanges : computations are done column by column.

- The radiative scheme uses emissivities integrated over the
whole spectral range, derived from Sasamori, and including carbon
dioxide and water vapor e-type absorption. It has been developed
in collaboration with Fouquart (Université de Lille) (cf.Veyre et
al, 1980).

- Clear airfluxes are computed once per level and use horizontally averaged variables.

- Liquid water is considered as a grey body with an absorption
coefficient of 120 m^2/kg. It is only included in the exchange
terms with the ground and with space, which is justified by the
fact that liquid water is mostly active within the atmospheric
window.

The result of the computation, done at each time-step, is that radiative cooling at a given point is function of the mean temperature and moisture profiles and of the liquid water amount below and above it.

4. MAIN RESULTS OF COMPUTATIONS

Comparison runs have been done under typical undisturbed tropical conditions, similar to those observed during the 1972 NCAR Puerto Rico experiment, and which have been already used to validate the dynamical model against experimental data (Sommeria and Le Mone, 1978). Initial profiles of temperature, potential temperature and humidity are presented here. (fig.1)

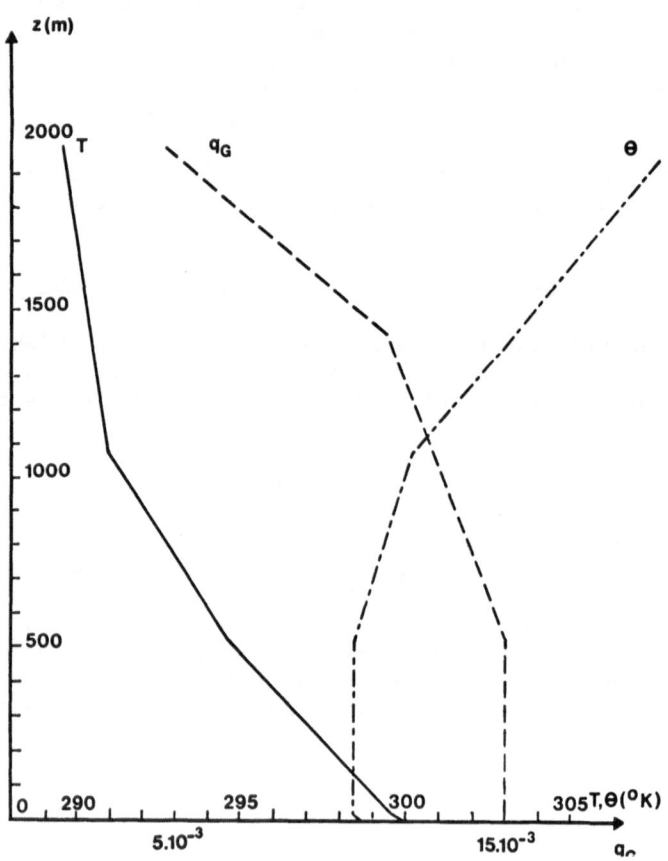

Figure 1. Initial profiles of temperature (T), potential temperature (θ) and specific humidity (q_G).

Three layers are traditionally distinguished : a mixed layer
extending from the ground up to 500 meters, a conditionally
unstable layer where clouds are forming, between 500 and
1000 meters, a stable layer where strongest clouds are dissipated
(passive layer between 1000 and 1400 meters).

After a 2.2 hours simulation which enables the model to reach a
statistically steady state with turbulent eddies and clouds, two
parallel simulations are carried along during 1.25 hours. The
first one includes only clear air radiative fluxes and the second
one the complete radiative scheme with cloud effects. This
allows to isolate cloud radiative effects.

Various types of comparisons are made with the two runs, with
respect to cloud cover, turbulent features and the various terms
of heat budget.

The temporal evolution of the total cloudy volume and of the
mean liquid water content (fig.2) show fluctuations with a
pseudo-period of around 0.5 hours which approximately corresponds
to the time needed for a cloud parcel to go from the bottom

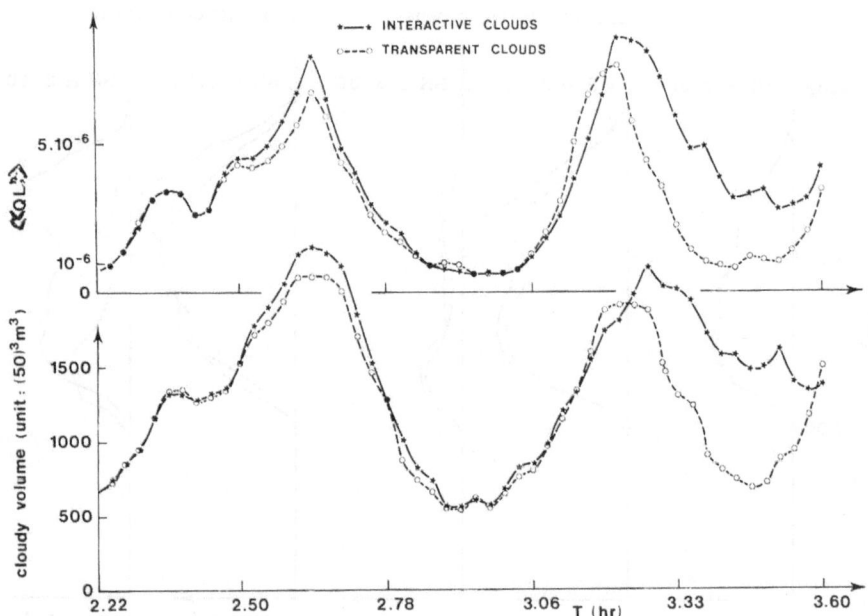

Figure 2.   Temporal evolution of the cloudy volume and of the mean
            liquid water content for the total integration volume.

to the top of the cloud layer. The magnitude of these
fluctuations (25 to 35 %) is strongly related to the small
size of the domain which samples only one or a few clouds at
a given time. A significant comparison would then require a
temporal average over a sufficiently long period, at least of
the order of one pseudo-period.

Comparison of the cloud volume and mean liquid water content
curves (fig.2) show a slow decoupling of the two runs with
larger values corresponding to the complete radiative scheme.
After one hour of parallel simulations, the two runs are clearly
decoupled and the comparison indicates a strong influence of the
cloud radiative effect.

The same remark applies to the vertical profiles of liquid water
content, which are averaged here over 200 time steps, that is
1000 seconds (fig.3). The interactive model shows slightly less
temporal oscillations which indicates a kind of damping of the
cloud oscillations by the radiative effect. If we compare
liquid water contents averaged over the last half hour, the
relative increase due to cloud radiative interaction is of the
order of 50 %.

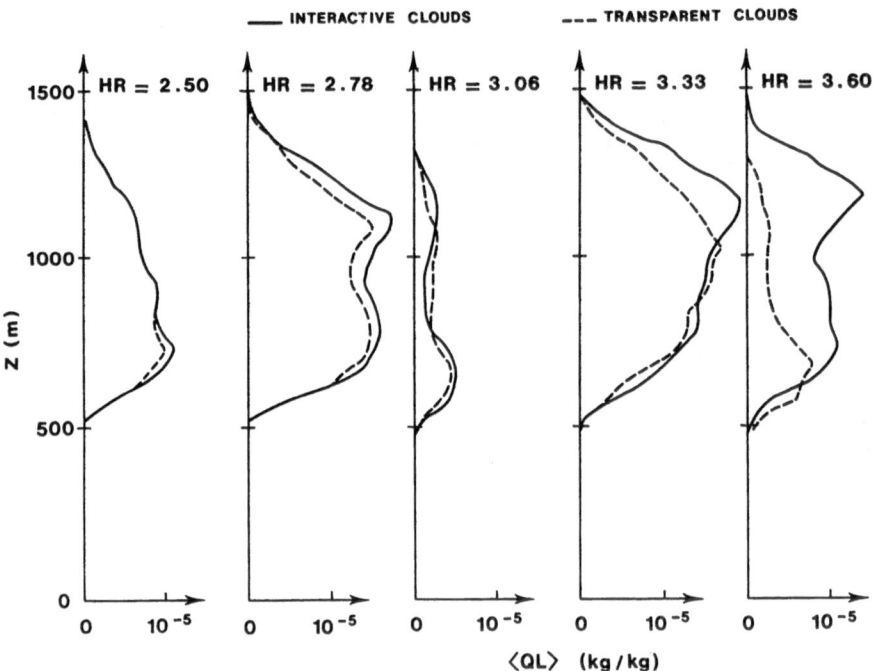

Figure 3.   Temporal evolution of the vertical profile of liquid
            water content (horizontally averaged   over 1000 time
            steps).

Most turbulent characteristics of the cloud layer are also
increased by the cloud-radiative interaction. This is the case
for the turbulent kinetic energy and for one of its components,
the variance of vertical velocity (horizontally averaged and
time-averaged over 1000 s). (fig.4). The maximum of this
variance around 1300 m is located near the maximum of the
vertical velocity of cloud elements. It partly includes the
effect of internal gravity waves which develop in the stable
layer and are not sufficiently damped by the model.

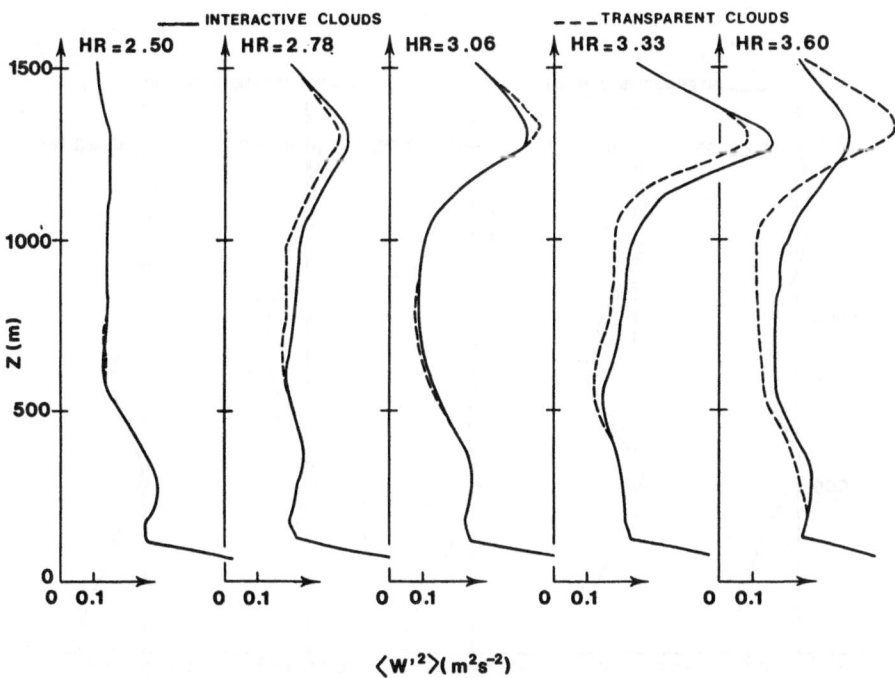

Figure 4. Id.as   fig.3 for the variance of vertical velocity.

The increase of turbulent kinetic energy in the interactive
case is linked to an increase in the vertical flux of virtual
potential temperature, which is one of the main sources
of kinetic energy (fig.5). During the last half hour this
quantity is approximately doubled inside the cloud layer but
not significantly modified within the mixed layer. The virtual
heat flux at the top of the mixed layer, generally negative,
becomes slightly larger and even sometimes positive. This
absence of a negative peak may be due to the temporal averaging
which has a tendency to smooth extreme values.

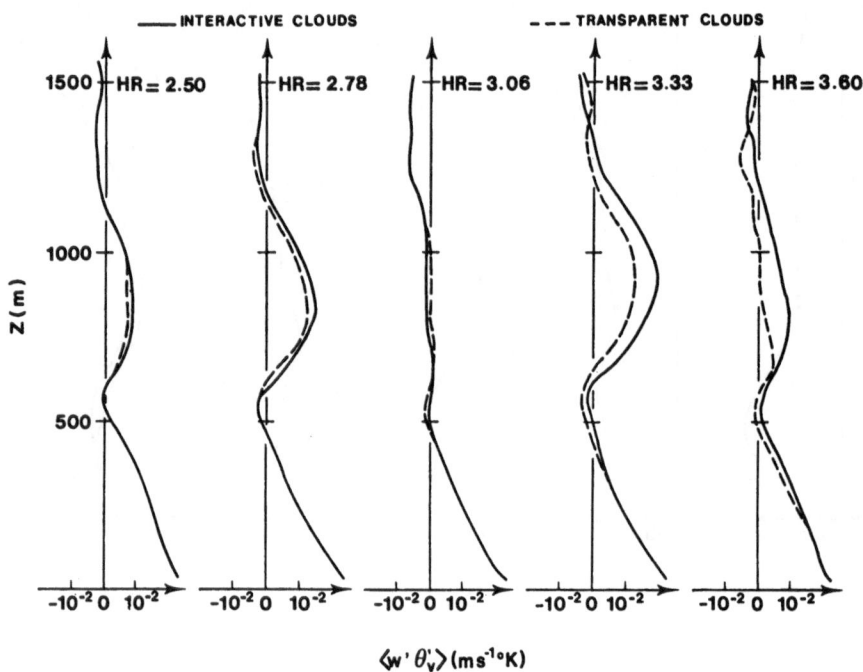

Figure 5.   Id. as fig.3 for vertical turbulent flux of virtual
            potential temperature.

Vertical fluxes of conservative variables, the total specific
water content q_W (fig.6) and the liquid water potential
temperature θ_L (fig.7), show a strong increase in the cloud
layer in the interactive case, along with a smaller intermittency.

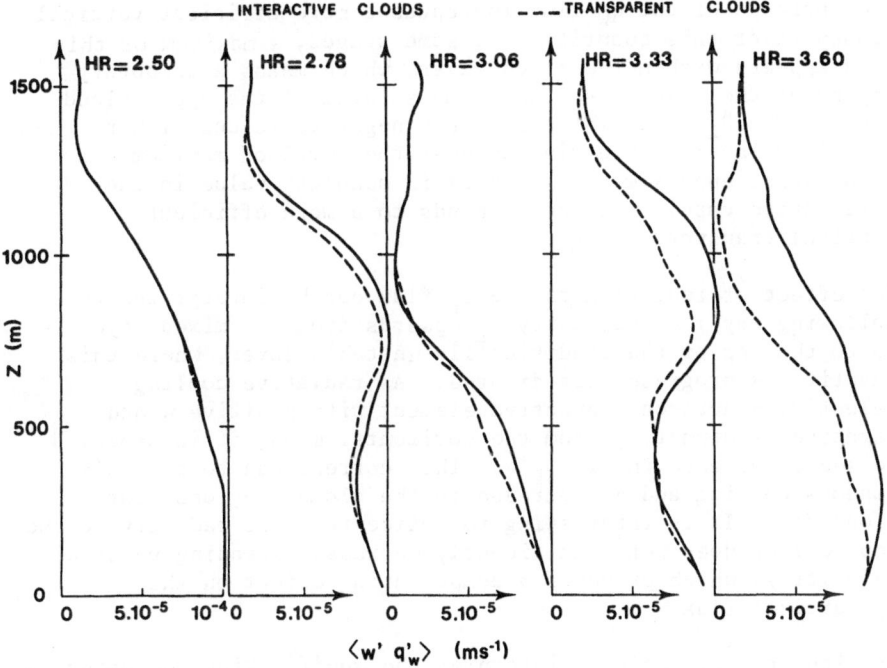

Figure 6

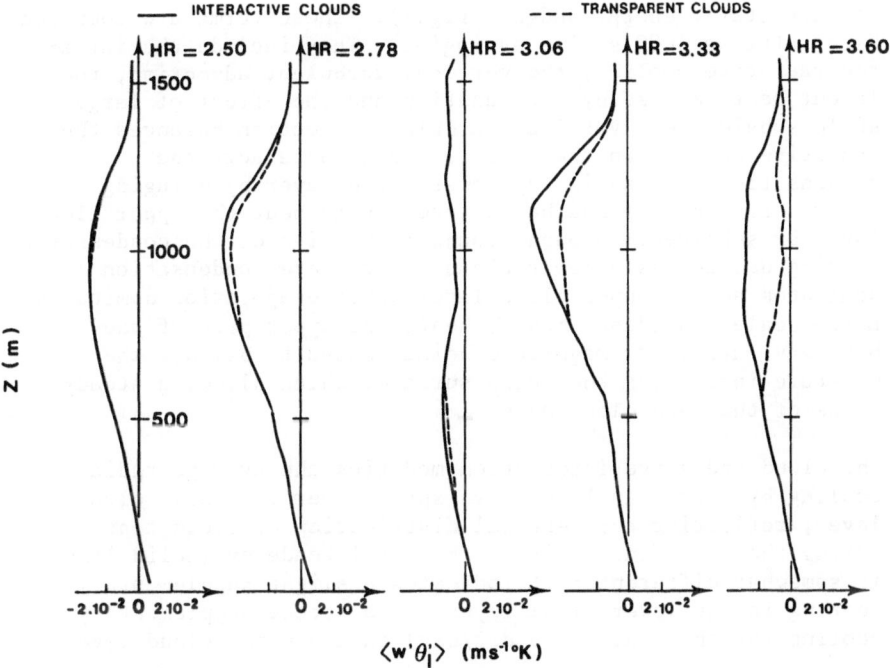

Figure 7

The increase of the q_W flux indicates a more efficient vertical transport of this quantity. At some stages, a maximum of this flux appears within the cloud layer, which means a temporary drying of the mixed layer and a moistening of the upper cloud layer. The θ_L flux, which presents negative values within the cloud layer with a minimum near the level of maximum cloudiness, approximately doubles in absolute value in the interactive case. This corresponds to a more efficient vertical transport of θ_L.

The effect of radiation on the θ_L flux can be interpreted the following way : clouds carry θ_L upwards from the mixed layer up to the top of the conditionally unstable layer, where this quantity is progressively diluted. As radiative cooling selectively acts on convective element with positive w and negative θ_L departure from the horizontal mean, it is expected to see a decrease in $<w' \theta_L'>$. This corresponds to a simul-taneous cooling and q_L increase in the cloudy regions near cloud top. It is interesting to notice that the radiative cloud top cooling does not significantly decrease ascending vertical velocities, which prevents a compensating effect on the θ_L vertical flux.

Previous remarks help to interpret the modifications occuring in the various terms of the potential temperature budget (or dry static energy budget (fig.8). These terms are compared here, after a 5000 s time averaging. They include the infra-red radiative cooling, the vertical turbulent advection, the latent heat release by condensation and the effect of large scale subsidence. Vertical turbulent advection balances the radiative cooling in the mixed layer, counteracts the condensation heating in the lower cloud layer by bringing lower θ from below, and has a tendency to heat the upper cloud layer by subsidence around clouds. The sign of the condensation heating delineates a lower cloud layer where condensation dominates and an upper cloud layer where evaporation dominates. Large scale subsidence mostly heats the upper part of the boundary layer, its magnitude being chosen to balance the moisture input from the ocean surface, which allows a steady value of the mean cloudiness.

The cloud radiative interaction modifies the average radiative cooling by less than 1 deg/day, spread over the whole cloud layer, reflecting the vertical distribution of cloud tops during the averaging period. The total tendency (solid line) is somewhat different : it indicates a slight supplementary heating in the lower cloud layer and a strong supplementary cooling (of the order of 5 deg/day) in the upper cloud layer.

This means a destabilization of the cloud layer and the building up of a stronger inversion at the top of the cloud layer.

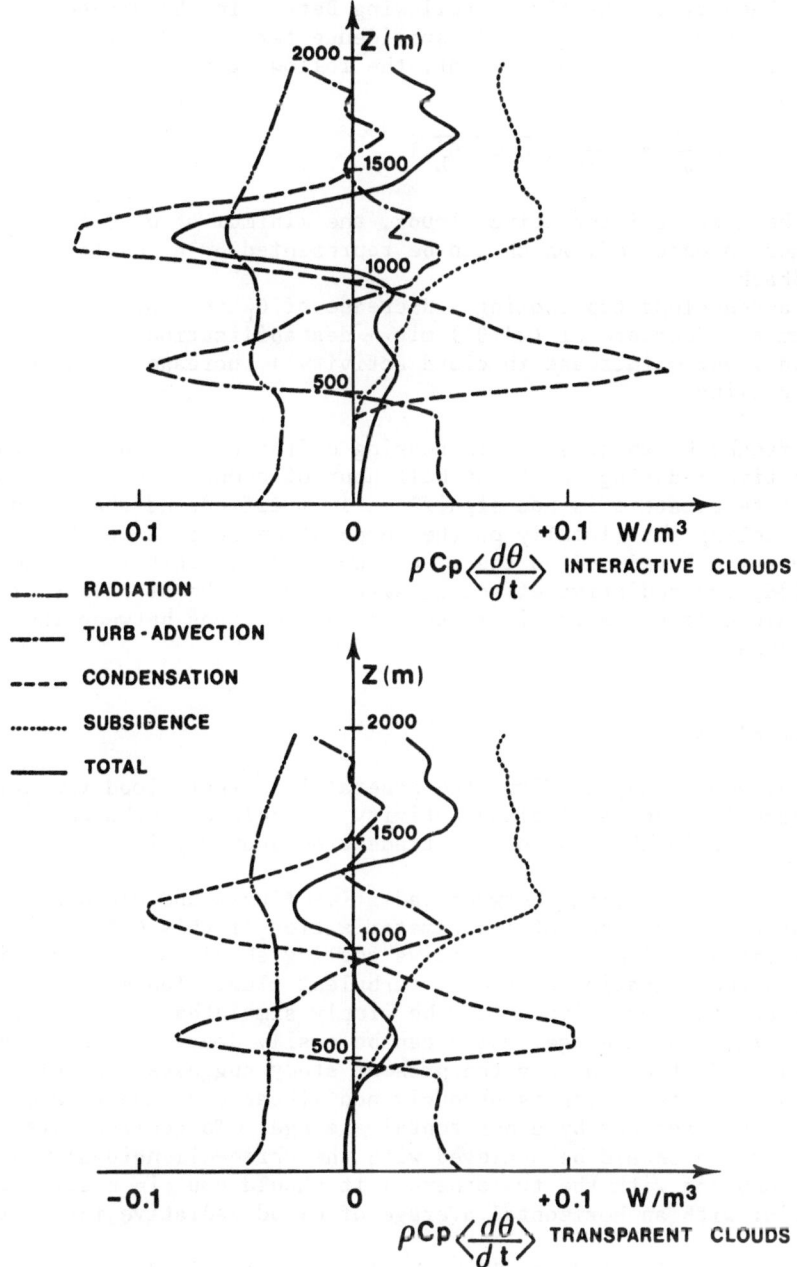

Figure 8.   Source terms for dry static energy $C_p\theta$ (averaged over 5000 s).

The changes in the temperature profile can be deduced from
the shape of the θ_L flux. Following Betts, in the absence
of radiative and large-scale subsidence terms, and for a
stationary cloud water content, the following equation is
verified :

$$\bar{\rho} \frac{\partial \bar{\theta}}{\partial t} = - \frac{\partial}{\partial z} (\bar{\rho} \; \overline{w' \; \theta_L'})$$

In the case of interactive clouds, the minimum of $\overline{w' \; \theta_L'}$
is more pronounced, which can be represented by a positive
feedback :
Infra-red cloud top cooling → decrease of θ_L in cloud
columns → decrease of (w' θ') min → destabilization of
cloud layer → increase in cloud activity → increase in cloud
top cooling.

The feedback explains the increasing difference of the two runs
with time : during the first half hour of comparison, the cloud
layer temperature is not significantly modified and the cloud
top cooling acts locally on the temperature that is on the
saturation level and on the q_L content. After this first pseudo-
period, the radiative effect is also felt on the stability of
the cloud layer, which increases the differences between the
two runs.

CONCLUSIONS

The present study confirms the general idea that cloud radiative
interaction increases cloud activity and related turbulent
quantities in the case of the trade-wind boundary layer.

The positive feedback between radiative fluxes and cloud activity
shows the importance of a parameterization of this effect. In
one-dimensional models, radiative cooling should be included in
the parameterization of the θ_L turbulent flux. The main question
is : can the radiative effect be simply simulated by considering
its horizontal average, which can be easily done in a one-dimensional
model ? On the contrary the present study suggests that cloud-
radiation interaction is strongly non linear and cannot be easily
taken into account by a horizontal average. To confirm this idea
another run should be achieved with the three-dimensional model,
and compared with the two others ; it should contain clear air
cooling with an horizontal average of cloud radiative interaction.

Results of this study could probably be generalized for any case
of oceanic boundary layer where lower boundary conditions are
steady. Conclusion would however certainly be different over a
solid ground where boundary conditions and surface fluxes are
rapidly changing.

REFERENCES

ALBRECHT, B.A. (1981) - Parameterization of Trade-Cumulus Cloud
          Amount. J. Atmos.Sci, 38 pp. 97-105.

AUGSTEIN, E. and WENDEL,M. (1980) - Modelling of the Time-
          Dependent Atmospheric Tradewind Boundary Layer with
          Non-Precipitating Cumulus Clouds.
          Cont. to Atm. Phys., 53 pp. 509-538.

DEARDORFF, J.W. (1980) - Stratocumulus-Capped Mixed Layers
          Derived From a Three-Dimensional Model
          Boundary-Layer Meteorol., 18 pp. 495-527.

GRAY, W.M. and R.W. JACOBSON (1977) - Diurnal variations of
          Deep Cumulus Convection. Month. Wea. Rev., 105
          pp. 1171-1188.

SASAMORI, T. (1968) - The Radiative Cooling Calculations for
          Application to General Circulation Experiments.
          J. Appl. Meteo., 7 pp. 721-729.

SOMMERIA, G. (1976) - Three-Dimensional Simulation of Turbulent
          Processes in an Undisturbed Trade-Wind Boundary Layer
          J. Atmos. Sci., 33 pp. 216-241.

SOMMERIA, G. and LE MONE, M.A. (1978) - Direct Testing of a
          Three Dimensional Model of the Planetary Boundary
          Layer Against Experimental Data. J. Atmos. Sci., 35
          pp. 25-39.

STEPHENS, G.L. and K.J. WILSON (1980) - The Response of a Deep
          Cumulus Convection Model to Changes in Radiative
          Heating. J. Atm. Sci., 37 pp. 421-434.

VEYRE, P., SOMMERIA, G. and FOUQUART, Y. (1980) - Modélisation
          de l'Effet des Hétérogénéistes du Champ Radiatif
          Infra-Rouge sur la Dynamique des Nuages.
          J. Rech. Atmos., 14 pp. 89-108.

ON THE PREFERRED MODE OF CUMULUS CONVECTION IN A CONDITIONALLY
UNSTABLE ATMOSPHERE

Tomio Asai

Ocean Research Institute, University of Tokyo

Numerical experiments are made to determine a preferred mode of
cumulus convection in a conditionally unstable atmosphere and to
investigate its dynamical properties. The preferred mode of
cumulus convection is regarded as the steady convection cell
attained eventually after a random potential temperature disturbance is imposed initially. It is shown that the preferred
size of the convection cell and the preferred cloud coverage
depend on mean vertical velocity, static stability and relative
humidity in the atmospheric layer in which the convections are
imbedded. Inspection of each term of energy equations indicates
that the preferred convection is of the mode for which the
potential energy of the layer is at the lowest and consequently
the mean temperature lapse-rate is a minimum.

1. INTRODUCTION

It was pointed out first by Bjerknes (1938) and Petterssen
(1939) that the cumulus convection characterized by the adiabatic
ascent of saturated air through a dry-adiabatically descending
environment in a conditionally unstable atmosphere could never
grow unless the ratio of the area of the ascending cloud region
to that of the descending cloudless region is below a critical
value which depends on the static stability of the atmospheric
layer. Extending the theory introduced by Bjerknes and Petterssen
to the atmospheric layer with a net vertical mass transport,
Cressman (1946) investigated the effect of mean vertical motion
on cumulus cloud mass and showed that the cloud mass can increase
as mean ascending motion is larger while it can decrease as mean
descending motion is larger.

E. M. Agee and T. Asai (eds.), Cloud Dynamics, 149–162.
Copyright © 1982 by D. Reidel Publishing Company.

The linear hydrodynamic stability analysis of a conditionally unstable atmosphere has been performed by Haque (1952), Syono (1 953) and some others. First of all Haque considered an infinite area which allows the static stability to change sign with the sense of the vertical motion in order to approximate the effects of moist updraft and dry downdraft in a conditionally unstable atmosphere. Then he found that a critical finite size of saturated ascending motion region exists to unstabilize the perturbation with an infinite size of dry descending motion region. The growth rate of unstable perturbation, however, increases as the size of the ascending region decreases without limitation. Lilly (1960) considered a slab symmetry convection with a finite domain of the descending motion associated with an ascending motion in a conditionally unstable atmosphere. Then he showed that a smaller size of the descending domain tends to stabilize the convection and the growth rate becomes zero when the ratio of the ascending area to the descending area is equal to the ratio of the static stability for the ascending area to that for the descending area. This is coincident with Bjerknes'result based on the slice method. Taking into account viscosity and conductivity in the perturbation analysis mentioned above, Kuo (1961) obtained a definite size of the ascending area at which a critical Rayleigh number of the ascending area is a minimum. Still he could not obtain a finite area of the descending motion to minimize the critical Rayleigh number because the descending area is infinitely large as the critical Rayleigh number is smaller. Extending Kuo's analysis to the supercritical convection and the finite-amplitude convection, Yamasaki (1972,1974) showed that the preferred size of an ascending area for which the growth rate is a maximum is uniquely determined while the growth rate increases with an increase in descending area similarly to Kuo's. This comes from the fact that the descending motion always acts as a stabilizing factor in the pseudo-adiabatic process in a conditionally unstable layer which is completely different from the convection in an absolutely unstable layer.

Kuo (1965) thus introduced a hypothesis that the preferred mode of cumulus convection is the one for which a net percentage rate of production of available potential energy due to release of latent heat is a maximum and he suggested that a definite descending area as well as an ascending one is determined as the preferred convection employing the selection hypothesis. Asai and Kasahara (1967) proposed a simplified cumulus convection model which consists of two circular concentric air columns, the inside column corresponding to the updraft cloud region and the outside concentric annular column to the downward motion region. Then it was found that the ratio of the updraft area to the downward motion area is uniquely determined for a given static stability by imposing the assumption that the cumulus convection transports heat upwards most efficiently. However, they were not able to determine a horizontal scale of the preferred convection. Revising the cumulus convection model by taking into account the vertical eddy exchange of momentum, Asai (1967) obtained a steady convection with a finite horizontal scale transporting heat upwards most efficiently which was regarded as the preferred mode of cumulus convection. Furthermore for the moist atmospheric layer heated below and cooled above Asai (1968) adopted the selection hypothesis that the preferred steady cumulus convection is the one which realizes at a minimum value of the temperature lapserate. Including the perturbation pressure term of the equation of vertical motion which is ignored in Asai and Kasahara (1967) as well as in most other one-dimensional cloud models, Holton (1973) demonstrated that the dynamic pressure perturbation has a profound effect on the growth of larger clouds and its inclusion in a one-dimensional model makes it possible to predict the cloud's size at which the maximum cloud growth rate will occur. On the other hand Schlesinger and Young (1970) examined the transient behavior of the moist convection model devised by Asai (1967) and showed that final appoach to a steady convection cell discussed by Asai is dependent upon the relative importance of entrainment of heat and momentum.

Kitade (1972) dealt with the perturbation in a conditionally unstable layer with a mean vertical motion. He showed that the cumulus convection in the layer with mean upward motion has a smaller lowest Rayleigh number, larger preferred updraft area and smaller compensating downward motion area than does the cumulus convection in the layer with mean downward motion. Thus a definite horizontal scale of the preferred cumulus convection was found for the layer with mean upward motion. His study, however, is restricted to the marginal stability.

Asai and Nakasuji (1977) made an attempt to seek a preferred cumulus convection in a conditionally unstable atmospheric layer and to see how the preferred mode depends on mean vertical motion and stratification in the layer based on numerical experiments. In their study the preferred mode of cumulus convection is regarded as the steady convection cell which consists of an ascending saturated and a descending unsaturated regions attained eventually after small random temperature disturbances are imposed initially. Inspecting of each term of energy equations, they found that the preferred convection is of the mode for which the potential energy of the layer is the lowest and consequently the mean temperature lapes-rate is a minimum. Asai and Nakasuji (1982) extended their previous work by introducing water vapor into the model in an explicit form without assuming the ascending motion always saturated with water vapor while the descending motion always unsaturated. Then they confirmed the conclusions obtained in their previous paper(1977).

The present report is to summarize the results obtained by Asai and Nakasuji (1977 and 1982).

2. MODEL AND FORMULATION OF PROBLEM

2.1 Basic equations

Consider a horizontally uniform atmospheric layer of the depth, h, which has a conditionally unstable stratification. Convective motions are restricted in the layer between two horizontal planes fixed at z=0 and z=h, respectively. Horizontally averaged vertical motion does not necessarily vanish and the air can go through these two boundaries. The convective motion is confined to the vertical (x,z) plane, where the x and y are the horizontal and vertical coordinates, respectively. A pseudo-adiabatic process is assumed, i.e., condensed water falls out of the system immediately. Then the conservation equations of momentum, heat energy, water vapor and mass for the shallow convection under the Boussinesq approximation can be obtained as follows (Ogura and Phillips,1962).

$$\frac{\partial u}{\partial t} + u\frac{\partial u}{\partial x} + w\frac{\partial u}{\partial z} = -C_p\theta_m\frac{\partial \pi}{\partial x} + \nu\nabla^2 u, \tag{2.1}$$

$$\frac{\partial w}{\partial t} + u\frac{\partial w}{\partial x} + w\frac{\partial w}{\partial z} = -C_p\theta_m\frac{\partial \pi}{\partial z} + \frac{g}{\theta_m}\theta + \nu\nabla^2 w, \tag{2.2}$$

$$\frac{\partial \theta}{\partial t} + u\frac{\partial \theta}{\partial x} + w\frac{\partial \theta}{\partial z} + w\frac{\partial \theta_0}{\partial z} = \frac{L}{C_p\pi_0}C + \nu\nabla^2\theta, \tag{2.3}$$

$$\frac{\partial q}{\partial t} + u\frac{\partial q}{\partial x} + w\frac{\partial q}{\partial z} + w\frac{\partial q_0}{\partial z} = -C+\nu\nabla^2 q, \tag{2.4}$$

$$\frac{\partial u}{\partial x} + \frac{\partial w}{\partial z} = 0, \tag{2.5}$$

where

$$C = \begin{cases} \frac{\partial}{\partial t}(q_0+q-q_s) & \text{for } q_0+ q>q_s, \\ 0 & \text{for } q_0+ q\leqq q_s, \end{cases} \tag{2.6}$$

where u and w are the velocity components in the x and z directions, respectively. The non-dimensional variable $\pi \equiv (p/p_{00})^{(C_p-C_v)/C_p}$ is used instead of the pressure p in the equations of motion. Here p_{00} is a reference pressure, C_p and C_v are the specific heats of air at constant pressure and constant specific volume, respectively. p_0, θ_0, q_0, π_0, p, θ, q and π are the initial basic values of pressure, potential temperature, specific humidity, pressure equivalent and their departures from the respective initial basic values. θ_m, L, C and q_s are the constant mean potential temperature, the latent heat of condensation, the condensation rate and the saturation specific humidity. g is the acceleration of gravity and ν is the coefficient of eddy diffusion which is assumed constant.

2.2 Boundary conditions

Both the upper and the lower boundaries are fixed and smooth
for the convective motion while a uniform mean vertical motion is
allowed to go through both the boundaries. The potential temperature and the specific humidity are assumed constant at the lower
boundary, while constant fluxes of heat and water vapor are
assumed at the upper boundary. The symmetrical conditions are
adopted with respect to the lateral boundaries x=0 and x=d. In
summary,

$$\frac{\partial u}{\partial z} = \theta = q = 0, \; w = \bar{w} \quad \text{at } z = 0$$

$$\frac{\partial u}{\partial z} = \frac{\partial \theta}{\partial z} = \frac{\partial q}{\partial z} = 0, \; w = \bar{w} \quad \text{at } z = h \qquad \Big\} \quad (2.7)$$

$$u = \frac{\partial w}{\partial x} = \frac{\partial \theta}{\partial x} = \frac{\partial q}{\partial x} = 0 \quad \text{at } x = 0 \text{ and } x = d,$$

where $\bar{w}$ is a uniform mean vertical velocity and d is the horizontal width of the domain.

2.3 Initial conditions

The intial basic field is set up as follows.

$$u = 0, \qquad\qquad (2.8)$$
$$w = \bar{w}, \qquad\qquad (2.9)$$
$$T_0 = T_{00} - \gamma z, \qquad\qquad (2.10)$$
$$p_0 = p_{00}(T_0/T_{00})^{g/R\gamma}, \qquad\qquad (2.11)$$
$$q_0 = \frac{0.622 r}{p_0} \times 6.11 \exp[17.27(1 - \frac{237.3}{T_0 - 35.85})], \qquad (2.12)$$

where T_0 is the initial basic temperature. T_{00} and p_{00} are the
temperature and pressure at the lower boundary, respectively.
γ is the constant temperature lapse-rate, R the gas constant of
dry air and r the constant relative humidity. θ_0 and π_0 are
obtained from T_0 and p_0 following their definitions.

The following potential temperature disturbance is superimposed on the initial basic field.

$$\theta (x,z) = a R (x,z), \qquad\qquad (2.13)$$

or

$$\theta (x,z) = A(x) \sin^2(\pi z/h), \qquad\qquad (2.14)$$

$$A(x) = \begin{cases} a \cos^2(\pi x/2b) & \text{for} \quad 0 \leq x \leq b, \\ 0 & \text{for} \quad b < x \leq d, \end{cases}$$

where a is an amplitude of the initial potential temperature disturbance, b is its horizontal scale and R (x,z) is the two-
dimensional random function which ranges from -0.5 to 0.5. (2.
13) is adopted for determination of the preferred mode of convection while (2.14) for the other cases.

3. COMPUTATIONAL PROCEDURE

Introducing a stream function ψ and eliminating π from (2.1) and (2.2), we obtain a complete set of equations for ψ, $\eta(= -\nabla^2 \psi)$, θ and q. These differential equations are approximated by a set of finite-difference equations and solved numerically. The condensation rate C is computed by the same method as Asai (1965). We first estimate θ (t + Δt), q (t + Δt) and q_s (t + Δt) by assuming C=0 in (2.3) and (2.4). These tentatively estimated values of θ, q and q_s are denoted by θ^*, q^* and q_s^* respectively. When $q_0 + q^* < q_s^*$, the air is not saturated with water vapor during the time interval from t to t + Δt. Thus C=0, $\theta(t + \Delta t)$ = θ^*, $q(t + \Delta t) = q^*$ and $q_s(t + \Delta t) = q_s^*$. When $q_0 + q^* \geq q_s^*$, the air has become saturated during the period of time Δt from t to t + Δt. We can adopt the following approximation,

$$C = (q_0 + q^* - q_s^*)/[1 + \frac{17.27 \times 23.3 L q_s^*}{C_p(T_0 + \theta^* \pi_0 - 35.85)^2}]\Delta t,$$

$$\theta = \theta^* + \frac{L}{C_p \pi_0} C\Delta t,$$

$$q = q^* - C\Delta t,$$

$$q_s = q_0 + q,$$

because the restriction $q_0 + q \leq q_s$ must be satisfied.

Test computations were performed by varying the horizontal grid size Δx, the number of subdivided layers n and the time increment Δt. When $\Delta x \leq$ 400m, $n \geq 8$ and $\Delta t \leq$ 30 sec, sufficiently accurate solutions were obtained. In the following numerical experiments we adopt n = 8 and Δt = 20 sec and 80m $\leq \Delta x \leq$ 340m. More detailed computational procedure can be referred to Asai and Nakasuji (1977, 1982).

4. ENERGY EQUATIONS

The equations of the convective kinetic energy, the convective potential energy and the potential energy can be derived from (2.1),(2.2)and(2.3) with the aid of (2.5) and (2.7) as follows, respectively.

$$\frac{\partial}{\partial t} <K_c> = \frac{g}{\theta m} <w'\theta'> - \nu <\eta'^2> - \frac{1}{h}[\bar{w} \ \bar{K}_c]_0^h, \tag{4.1}$$

$$\frac{\partial}{\partial t}<P_c> = -<w'\theta'(\frac{\partial \theta_0}{\partial z} + \frac{\partial \bar{\theta}}{\partial z})> + \frac{L}{C_p} <\frac{C\theta'}{\pi_0}> - \nu <(\nabla \theta')^2> - \frac{1}{h}[\bar{ } \ \bar{P}_c]_0^h, \tag{4.2}$$

$$\frac{\partial}{\partial t} <P> = -<w\theta> + <zw\frac{\partial \theta_0}{\partial z}> - \frac{L}{C_p} <\frac{Cz}{\pi_0}> + \frac{\nu}{h}[\bar{\theta}]_0^h - \frac{1}{h} [\bar{w} \ \bar{P}]_0^h \tag{4.3}$$

where K_c, P_c and P are the convective kinetic energy, the convective potential energy and the potential energy which are defined respectively as

$$K_c \equiv \frac{1}{2}(u'^2 + w'^2), P_c \equiv \frac{1}{2}\theta'^2 \text{ and } P \equiv -\theta z$$

Here $<A>$, $\bar{A}$ and A' denote the average over the whole domain, the horizontal average and the deviation from the horizontal average of A, respectively. $[A]_0^h$ denotes the value of A at z=h subtracted the one at z=o.

The first terms on the right-hand side of (4.1), (4.2) and (4.3) express the conversion rates of the respective energies due to vertical heat transport. The second term of (4.3) indicates the conversion rate of the potential energy due to the mean vertical motion. The second term of (4.2) and the third term of (4.3) represent the conversion rates of the convective potential energy and the potential energy due to the latent heat release, respectively. The last two terms of (4.1), (4.2) and (4.3) express the diffusional dissipation of the energies and the vertical flux divergence of the energies through the upper and lower boundaries due to the mean vertical motion, respectively.

Asai and Kasahara (1967) and Asai (1967) adopted an assumption that the preferred convection transports heat upward most efficiently. Kasahara and Asai (1967) further introduced a hypothesis that the preferred convection maximizes the production rate of the kinetic energy. Asai (1968) conjectured that the preferred cumulus convection realizes at a minimum value of the temperature lapse-rate in a moist layer heated below. On the other hand Kuo (1965) enunciated that the preferred convection maximizes the production rate of the convective potential energy due to latent heat release. The selection hypotheses adopted so far are based on some portion of the energy equations. In the following section we will examine dependence of magnitudes of each term of the energy equations on the horizontal scale of convection and then compare them with those for the preferred convection which is eventually attained from initial random disturbances.

5. RESULTS

5.1 Preferred mode

A random potential temperature disturbance denoted by (2.13) is imposed to set up convective motion in the computational domain with a horizontal scale wide enough to determine a possible preferred convection. An amplitude of the initial disturbance is taken to be as small as possible to avoid a serious influence of the imposed initial disturbance on properties of the final steady convection (e.g., Ogura, 1971). We adopt a=0.001K for the cases with mean ascending currents or initially saturated basic field without mean descending currents. However, a=1K is adopted for

the cases with an initially saturated basic field accompanying
mean descending currents and a=10K which may fail to provide
meaningful results for the cases with an initially unsaturated
basic field without mean ascending currents, since larger amplitudes of the initial disturbance are required to set up convection.
As a horizontal scale, d, of the domain wide enough to contain
many convection cells, we adopt 30km $\leq$ d $\leq$ 60km. We set T_{00} =
20°C, P_{00} = 900 mb, h = 1 km, ν = 100 m^{2}s^{-1} for which Rayleigh
parameter Ra defined as $g(\gamma_d-\gamma_m)h^4/(T_{00}\nu^2)$ is about 1.8×10^4.
Here γ_d is the dry adiabatic lapse-rate and γ_m is the moist
adiabatic lapse-rate.

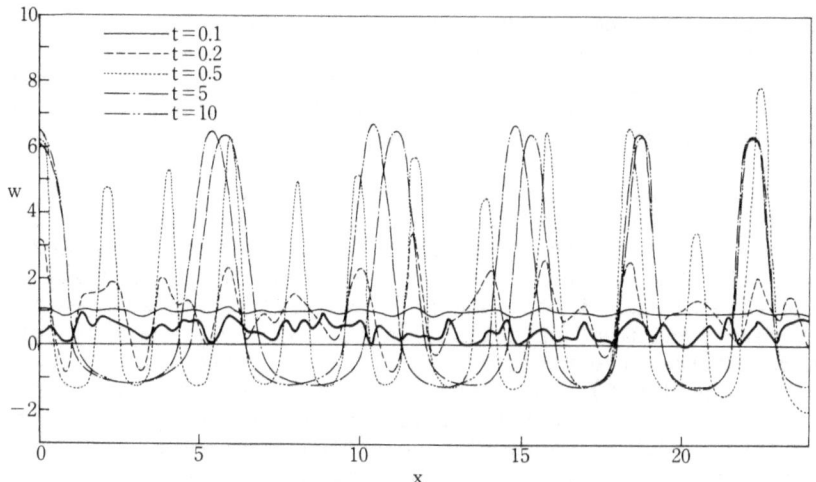

Figure 1. Horizontal distributions of vertical velocity
          W at the mid-level for different times t
          for Ra=10^4, δ=0.5 and $\overline{W}$=1. The thick solid
          line indicates the initial potential tempera-
          ture disturbance at the mid-level in units
          of 10^{-3}. x, t and W are non-dimensionalized
          by h, h$^2/\nu$ and ν/h, respectively.

     An example is shown in Fig.1 to demonstrate evolution of con-
vective motions. Fig.1 shows the horizontal distributions of vertical
velocity at the mid-level at different time steps for the case of
Ra=10^4, δ=0.5, and $\overline{W}$=10cm s^{-1}, where $\delta\equiv(\gamma-\gamma_m)/(\gamma_d-\gamma_m)$. Convective
motions of irregular and smaller sizes observed at an early stage
tend to reduce their variety and shift toward larger sizes as time
elapses and prevalence of a definite size of convection is finally
shown in the steady state. Similar features could be observed
for different types of random function R and different values of
a imposed as initial disturbances. Thus we can determine a
preferred mode of convection having a definite size of cell and

a definite area ratio of an ascending region to the entire region
which is prevalent in a quasi-steady state attained after a random
disturbance is given initially.

5.2 Dependence of the preferred mode on parameters

Fig. 2 shows variations of the horizontal scale, ℓ/h, nondimensionalized by the depth of the convection layer and the
cloud coverage, σ, of the preferred cumulus convection cell with
a value of $\bar{W}$ for $\gamma = 7K \ km^{-1}$ and $r = 1$. This value of γ corresponds to $\delta = 0.5$. The horizontal scale ℓ is defined as a spacing
between neighboring maximum and minimum of vertical velocity at
the mid-level, while σ is defined as a ratio of the saturated
area to the entire one in each cell at the mid-level of the cloud
layer. The preferred scale of convection cell decreases from
5 to 2.7 and the preferred cloud coverage increases from 0.08 to
0.2 when the mean ascending current increases from 0 to 10cm s^{-1} .
On the other hand the preferred scale of convection cell increases
rapidly and the preferred cloud coverage tends to decrease to
zero as the mean descending current increases. In other words
cumulus clouds will be scattered and then disappear as the mean
descending motion becomes dominant. In any case a preferred mode
of cumulus convection in a conditionally unstable layer may appear
to be much flatter than that of dry convection in an absolutely
unstable layer.

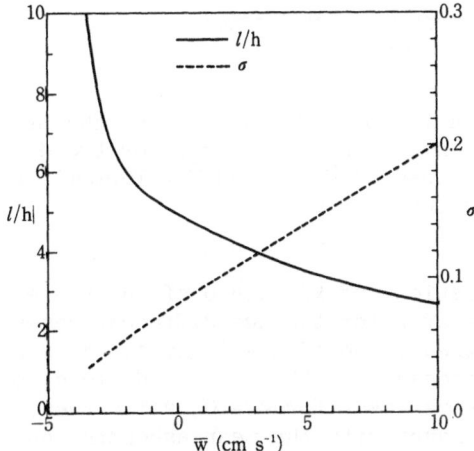

Figure 2.   Variations of the horizontal scale ℓ/h and
            the cloud coverage σ of the preferred cumulus
            convection cell with $\bar{W}$ for $\gamma = 7K \ km^{-1}$ and
            $r = 1$.

Fig. 3 shows variations of ℓ/h and σ of the preferred convection cell with a value of δ for two different values of $\bar{W} = 0$
and $\bar{W} = 10$ cm s^{-1} at the fixed value of $r = 1$. As γ decreases,
that is, the stratification is less unstable, ℓ/h increases and σ
decreases for $\bar{W} = 0$, while ℓ/h decreases and σ increases as γ decreases for $\bar{W} = 10$ cm s^{-1}. It should be noted here that the preferred mode of cumulus convection depends heavily upon the mean
vertical motion in less unstable atmospheric layer.

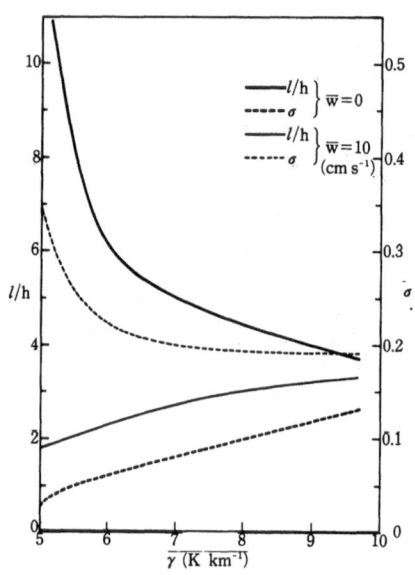

Figure 3.   Variations of ℓ/h and σ of the preferred
            convection cell with γ for two different
            values of $\bar{W} = 0$ and $\bar{W} = 10$ cm s^{-1} at the fixed
            value of $r = 1$.

Fig. 4 shows variations of ℓ/h and σ of the preferred convection cell with a value of r for the two different cases $\bar{W} = 0$ and
$\bar{W} = 10$ cm s^{-1} at the fixed value of $\gamma = 9$K km^{-1}. As r decreases,
the preferred size increases and the preferred cloud coverage
decreases in a convective layer for $\bar{W} = 0$, while these variations
are very small in the layer with the mean ascending current. As
r decreases, a thin cloud layer is confined in an upper layer and
eventually cumulus convection vanishes for $r \lesssim 0.8$.

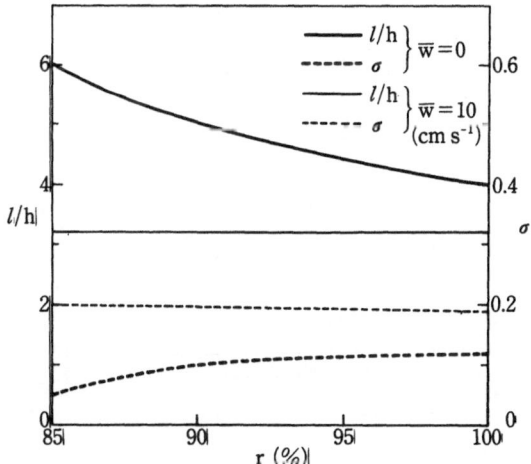

Figure 4. Variations of ℓ/h and σ of the preferred
convection cell with a value of r for two
different cases of $\bar{W} = 0$ and $\bar{W} = 10$ cms^{-1} at
the fixed value of $\gamma = 9K$ km^{-1}.

5.3 Examination of selection hypotheses of the preferred mode

Some hypotheses on selection of a preferred mode of cumulus
convection proposed so far are fairly well coincident with the
result obtained above. It is difficult, however, for them to be
compared quantitatively with each other, because the models employed are different from each other. Here we examine variations
of magnitude of each term of the energy equations (4.1), (4.2)
and (4.3) with a horizontal size of convection cell. In order to
do this we obtain a steady convection in a domain in which only
one cell of convection is observed for an initial disturbance
given by (2.14). We can have a number of different sizes of the
steady convection which may differ from the preferred convection
by varying the initial disturbance and the corresponding domain.

Fig. 5 shows variation of magnitude of each term of the
potential energy equation (4.3) with a horizontal scale of con-
vection cell ℓ/h for $\bar{W} = 10$cm s^{-1}, $\gamma = 7K$ km^{-1} and r = 1. Numbers
of occurrence of convection cells in a steady state attained
eventually after the random potential temperature disturbance is
imposed initially are also shown for each cell size in Fig. 5.
The potential energy (solid line) becomes minimum at the preferred
scale $\ell/h = 2.7$, and it is seen from the dotted line that the
mean temperature lapse-rate $\langle\gamma\rangle$ defined as $\gamma - [\pi_0\bar{\theta}]_0^h/h$ also becomes minimum at the preferred scale. The horizontal scale of
convection cell which maximizes the total vertical heat transport
(dashed line) is larger than the preferred scale.

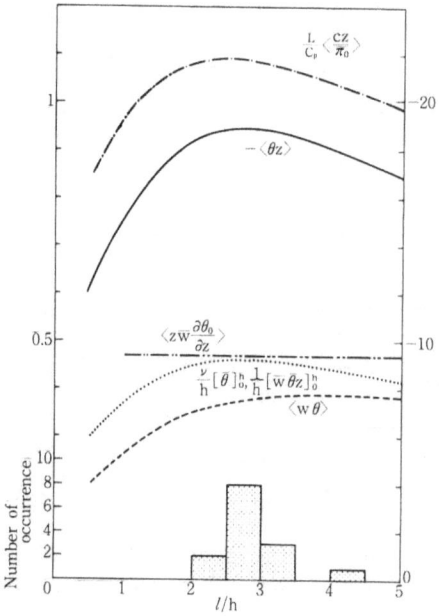

Figure 5.   Variations of magnitude of each term of the
            potential energy equation (4.3) and number
            of occurrence with a horizontal scale of
            convection cell, ℓ/h for $\overline{W}$ = 10cm s^{-1}, γ =
            7K km^{-1} and r = 1.   The scale for the
            potential energy is indicated on the right
            and for the other terms of (4.3) on the left.

All terms of the kinetic energy equation (4.1) and the convective
potential energy equation (4.2) have no maximum in the same range
of the horizontal scale of convection cell as in Fig. 5, though
they are not shown here.   Thus the selection hypothesis that the
preferred convection maximizes any term of them is not supported.
The same results are obtained for the other cases.

The "maximum upward heat transport" hypothesis is based on
the idea that convection realizes so as to release the gravitational instability of a fluid layer most efficiently.   As for
the cumulus convection, however, the release of latent heat as
well as the upward heat transport plays an important role in
reducing the potential energy.   Thus the potential energy is
reduced most efficiently by the convection for which the sum of all
the potential energy consuming terms is a maximum.   It is thus
concluded that a preferred convection in the steady state is the
one which realizes at a minimum level of the potential energy.

6. CONCLUSIONS

An attempt is performed to determine a preferred mode of cumulus convection in a conditionally unstable atmospheric layer by a numerical experiment. When a random temperature disturbance is imposed initially in the atmospheric layer within a sufficiently wide domain, the reslting convective motion tends to be occupied by a number of cumulus convection cells characterized by a definite cell size and a definite cloud coverage as time elapses. Thus a preferred mode of cumulus convection is determined as one which has a maximum frequency of occurrence in the steady state eventually attained. The preferred scale of covection cell decreases and the preferred cloud coverage increases as the mean vertical velocity increases. The preferred scale increases and the preferred cloud coverage decreases as the initial basic temperature lapse-rate γ decreases in the atmospheric layer without a mean ascending current, while in the layer with a mean ascending current this tendency of the variation is opposite so that the preferred scale decreases and the preferred cloud coverage increases. As the initial basic relative humidity decreases, the preferred scale increases and the preferred cloud coverage decreases in an atmospheric layer without a mean ascending current, while the preferred convection cell slightly changes its mode in the layer with a mean ascending current.

The preferred cumulus convection thus obtained is compared with different steady cumulus convections in terms of the energy equations to examine some of the selection hypothesis proposed so far. The examination indicates that the preferred convection realizes so as to minimize the potential energy and hence the mean temperature lapse-rate in the convective layer.

REFERENCES

1. Asai, T. : 1965, J. Meteor. Soc. Japan 43, pp. 1-15.
2. Asai, T. : 1967, J. Meteor. Soc. Japan 45, pp. 251-260.
3. Asai, T. : 1968, J. Meteor. Soc. Japan 46, pp. 301-307.
4. Asai, T. and Kasahara, A. : 1967, J. Atmos. Sci. 24, pp. 487-496
5. Asai, T. and Nakasuji, I. : 1977, J. Meteor. Soc. Japan 55,
     pp. 151-167.
6. Asai, T. and Nakasuji, I. : 1982, J. Meteor. Soc. Japan 60,
     pp.
7. Bjerknes, J. : 1938, Quart. J. Roy. Meteor. Soc. 64, pp. 325-
     330.
8. Cressman, G.P. : 1946, J. Meteor. 3, pp. 85-88.
9. Haque, S.M.A. : 1952, Quart. J. Roy. Meteor. Soc. 78, pp. 394-
     406.
10. Holton, J.R. : 1973, Mon. Wea. Rev. 101, pp. 201-205.
11. Kasahara, A. and Asai, T. : 1967, J. Meteor. Soc. Japan 45,

        pp. 280-291.
12. Kitade, T. : 1972, J. Meteor. Soc. Japan 50, pp.243-258.
13. Kuo, H.L. : 1961, Tellus 13, pp. 441-459.
14. Kuo, H.L. : 1965, Tellus 17, pp. 413-433.
15. Lilly, D.K. : 1960, Mon. Wea. Rev. 88, pp. 1-17.
16. Ogura, Y. : 1971, J. Atmos. Sci. 28, pp. 709-717.
17. Ogura, Y. and Phillips, N.A. : 1962, J. Atmos. Sci. 19, pp. 173-179.
18. Petterssen, S. : 1939, Geofys. Publ. 12, pp. 5-23.
19. Schlesinger, R.E. and Young, J.A. : 1970, Mon. Wea. Rev. 98, pp. 375-384.
20. Syono, S. : 1953, Tellus 5, pp. 179-195.
21. Yamasaki, M. : 1972, J. Meteor. Soc. Japan 50, pp. 465-482.
22. Yamasaki. M. : 1974, J. Meteor. Soc. Japan 52, pp. 365-379.

TOWARD A UNIFIED THEORY OF ATMOSPHERIC CONVECTIVE INSTABILITY

Hampton N. Shirer

Department of Meteorology, The Pennsylvania State
University, University Park, PA   16802

ABSTRACT

   Convection in the planetary boundary layer can develop via
three mechanisms, two dynamic and one thermal. By use of a nonlinear three-dimensional truncated spectral model of shallow
moist Boussinesq convection, we show that two of these mechanisms,
parallel instability and thermal forcing, are linked because only
one convective mode exists when either or both mechanisms are
operating.

   In addition, we clarify the roles that the basic environmental wind field and latent heating play in the initiation and
development of the nonlinear convective state. The wind field
causes the convection to form two-dimensional rolls, to align in
a preferred way with respect to the wind, and to propagate with
a speed related to the wind component perpendicular to the roll
axis. Latent heating causes the critical value of the environmental lapse rate (Rayleigh number) to decrease in accordance
with the slice method stability criterion, and, when only the
upper part of the upward branch is moist and all of the downward
branch is dry, causes a finite-amplitude convective solution to
exist for values of the Rayleigh number that are less than the
critical one obtained from a linear analysis.

1.   INTRODUCTION

   Convection organized into bands is a commonly observed
feature of the atmospheric boundary layer, particularly in the
strongly shearing winds and unstable air behind cold fronts.
These rolls typically are dominated by single wavenumbers in the
horizontal and vertical (15), and they are quasi-two-dimensional

163

E. M. Agee and T. Asai (eds.), Cloud Dynamics, 163–177.
Copyright © 1982 by D. Reidel Publishing Company.

features. Moreover, they usually appear in an environment that
has a strongly curved wind profile (5,6).

At least three mechanisms, two dynamic and one thermal,
have been proposed to explain the development of these rolls
in the atmosphere (1). The wind component orthogonal to the
rolls can be destabilizing if it contains an inflection point,
while the component parallel to the rolls can be destabilizing
only in a rotating system (3,7). Adverse potential temperature
gradients lead to thermal instability of Rayleigh-Bénard type,
and the orthogonal component of the basic wind is stabilizing
if its profile contains no inflection point (5,6). In addition,
the thermal mechanism is modified by latent heating effects;
this conditional instability causes the necessary critical environmental lapse rate to be less than dry adiabatic.

In order to gain more insight into some of the linear and
nonlinear aspects of convection in the atmospheric boundary
layer, we study a truncated spectral model of three-dimensional,
shallow, moist, Boussinesq convection in the presence of a
height-dependent, horizontal basic wind field. Although we study
the characteristics of only the first convective branch, we
generalize and clarify the earlier results given in (12,13) by
expanding the model from two to three dimensions, by allowing
the domain bottom to be below cloud base so that only some of
the upward moving air cools moist adiabatically, and by adding
the Coriolis parameter so that the parallel and thermal instability mechanisms can be studied simultaneously. The model
is of Lorenz form (8), in which the motion field is represented
by single harmonics in the lateral, longitudinal and vertical
directions, but the thermal field is represented by single
harmonics in the horizontal and two harmonics in the vertical.
Thus, the nonlinear interactions present in the model represent
only vertical thermal advection. Although the present model
does not contain any secondary branches, the results presented
here suggest how these effects might be studied in a larger
model. Apparently, this truncation is too restrictive in the
vertical for the inflection point instability to exist.

In the results presented below we clarify the roles of the
basic shearing wind field, of the Coriolis parameter f, and of
the latent heating. The presence of a basic wind causes the nonlinear solution to represent physically a two-dimensional solution
so that the wind field organizes the convection into bands. The
parallel component of the wind reduces the critical value R_c of
the Rayleigh number R when f is included. Thus, the parallel
and thermal instabilities actually give rise to one branching
mode. When $f \neq 0$, the orientation angles are a function of the
Reynolds number Re given by the magnitude of the basic wind at
the domain top. The amount of variation depends sensitively on

the magnitude of the dimensional combination $z_T^4/\nu\kappa$, in which z_T is the domain height, ν is the eddy viscosity and κ is the eddy thermometric conductivity.

Latent heating effects also cause the value of R_c to be decreased from that of the dry, motionless case. Here the value of R_c corresponds to the value of the environmental lapse rate given by the slice method criterion for instability. When the areas of clear and cloudy air are unequal, a nonlinear convective solution exists for values of $R < R_c$, and a sudden transition to convection would be possible. The critical value R_n of R at which this subcritical solution first exists also depends sensitively on the combination $z_T^4/\nu\kappa$.

2. THE SPECTRAL MODEL

As in (12,13) we use the shallow Boussinesq equations to model three-dimensional moist convection. We note that introduction of latent heating does not conflict with the other scaling assumptions used to derive the shallow Boussinesq system (10). As before, the convective state is considered to be a perturbation superimposed on a hydrostatic conductive state in which the basic temperature field varies linearly with height.

We include latent heating by assuming that only upward motion above cloud base h is moist adiabatic, while upward motion below h and all downward motion is dry adiabatic. Thus, we assume that entrainment of dry warm air from above the domain top is sufficient to ensure that the entire downward branch is dry. The advantage of this approach is that we do not have to introduce a specific humidity variable. Here h is given by the lifting condensation level obtained from the surface temperature-dew-point depression.

We include an arbitrary horizontal basic flow $V(z)$ in the reference state, but following (12) we do not require it to be a solution to the Boussinesq set because $V(z)$ is of much larger spatial and temporal scales that those of the convection. Thus, we investigate how $V(z)$ through its Fourier coefficients initiates and alters the convective solution without allowing the convective solution to affect $V(z)$.

The Coriolis force is neglected normally in Boussinesq convection because the convective time scale is on the order of one hour (2). However, it is shown in (7) that f must be included in order for parallel instability to be possible. Thus, we will add f to the Boussinesq system, but we will require that the **Coriolis term be an order of magnitude smaller than the**

the viscous term $\nu\nabla^2 v'$. Although Lilly (7) notes that these two terms must be the same order of magnitude for parallel instability to operate in an Ekman flow, we find that this is not the case in agreement with (3).

2.1 Nondimensional forms

The Boussinesq system is obtained by expressing each dependent variable velocity v, temperature T, pressure p and density ρ as the sum of a height-dependent basic state and a perturbation (12). Because the convective cell is assumed to fill the domain $0 \le x \le L_x$, $0 \le y \le L_y$, $0 \le z \le z_T$, we choose the nondimensionalization so that the domain corresponds to $0 \le x^* \le 2\pi$, $0 \le y^* \le 2\pi$, $0 \le z^* \le \pi$. We use scaling relations similar to those in (12,13).

If we define the constants γ_e, γ_d, γ_m to be the environmental, dry adiabatic and moist adiabatic lapse rates and set $a=2z_T/L_x$, $b=L_x/L_y$, $P=\nu/\kappa$ and

$$f^* = f z_T^2 / \nu \pi^2 \qquad (2.1)$$

$$H = g(\gamma_d - \gamma_m) z_T^4 / (\nu \kappa T_0 \pi^4) \qquad (2.2)$$

$$R = g(\gamma_e - \gamma_d) z_T^4 / (\nu \kappa T_0 \pi^4) \qquad (2.3)$$

then we may write the nondimensional Boussinesq system as

$$\frac{\partial v^*}{\partial t^*} + (v^* + V_H^*) \cdot \nabla^* v^* + w^* \frac{\partial V_H^*}{\partial z^*} + \frac{f^* P}{a} k \times v^* + \nabla_H^* p^*$$

$$+ (\frac{1}{a^2} \frac{\partial p^*}{\partial z^*} - \frac{PT^*}{a^2}) k - \frac{P}{a} \tilde{\nabla}^2 v^* = 0 \qquad (2.4)$$

$$\frac{\partial T^*}{\partial t^*} + (v^* + V_H^*) \cdot \nabla^* T^* - R w^* - H w_h^* - \frac{1}{a} \tilde{\nabla}^2 T^* = 0 \quad (2.5)$$

$$\nabla^* \cdot v^* = 0 \qquad (2.6)$$

in which the asterisk denotes a nondimensional variable, p^* is the nondimensional form of p'/ρ_0, and ∇^* and $\tilde{\nabla}^2$ are appropriate forms of the gradient and Laplacian operators. In (2.5), $H w_h^*$ represents the latent heating, where $w_h^* = w^*$ when both $w^* > 0$ and $h^* < z^* < \pi$, and $w_h^* = 0$ otherwise. We note that as $b \to 0$, (2.4)-(2.6) becomes a two-dimensional model. Also, because we will substitute sinusoidal forms into (2.4)-(2.6) to obtain our

spectral model, the requirement that the Coriolis term be small
compared to the viscous term holds provided that $f* << (a^2+a^2b^2+1)$,
in which we have used $\tilde{\nabla}^2 u* \approx u* \times |a^2+a^2b^2 + 1|$. Thus, for example
we require that $f* < 0.2$ for $a^2 \approx 0.5$ and $b^2 \approx 1$. From (2.1) we see
that this requirement puts a limit on the magnitude of $z_T^4/\nu\kappa P$.

2.2   The spectral expansion

Because the domain is cyclically continuous, we may use
sinusoidal basis functions for the dependent variables. We
only require that the vertical motion and temperature perturbations vanish at $z*=0$ and $z*=\pi$. The truncation we choose is of
the same form as that used by Lorenz (8): the motion variables
$u*$, $v*$, and $w*$ are described by one harmonic in each of the
three directions, but the temperature perturbation $T*$ is described
by one in the horizontal and two in the vertical. This wavenumber
choice adequately represents the first convective branch in
Rayleigh-Bénard convection (11), and we assume that it is the
appropriate one for study of the boundary layer circulation.
We note that a two-dimensional truncation of this type led to
good orientation angle predictions for the developing roll when
a height-dependent basic wind was included in the problem (12).

In order that we allow for all possible phasing among the
variables, we must use a 21-coefficient truncation: 4 coefficients for each of $u*$, $v*$, $w*$, $p*$ and 5 for $T*$. Thus we have

$$w* = [(C_1\cos\phi x* + C_2\sin\phi x*)\cos\delta y*$$

$$+ (C_3\cos\phi x* + C_4\sin\phi x*)\sin\delta y*]\sin z* \qquad (2.7)$$

$$T* = [(E_1\cos\phi x* + E_2\sin\phi x*)\cos\delta y*$$

$$+ (E_3\cos\phi x* + E_4\sin\phi x*)\sin\delta y*]\sin z* + E_5\sin 2z* \qquad (2.8)$$

and the expansions for $u*$, $v*$ and $p*$ are similar to (2.7) except
that we use Fourier components A_i for $u*$, B_i for $v*$ and D_i for
$p*$. Here ϕ and δ are integers and the spectral coefficients are
functions of only time $t*$. All possible rectangular planforms
are obtained by varying the value of b.

Because we use the function w_h* in our approximation of the
latent heating effects, we use the Fourier series for the upward
branch ($w* > 0$) of $w*$. This can be obtained from (2.7) and from
it we find that the latent heating leads to three contributions:
uniform heating throughout the domain, heating at the same wavenumber as that of $w*$, but having only half the magnitude, and
heating at larger wavenumbers. Only the first two contributions
remain in the spectral equations.

The number of spectral equations can be reduced from 21 to 13 by using (2.6) to replace C_i and D_i with functions of A_i and B_i. The resulting thirteen-coefficient model is too complicated to display conveniently here. However, the Fourier coefficients arising from the basic wind field and from w_h* join the other nondimensional numbers as important parameters in the spectral system. These Fourier coefficients are

$$n_1 = (1 - h*/\pi)/2 + (\sin 2h*)/(4\pi) \tag{2.9}$$

$$n_2 = -2(\sin^2 h*)/\pi^2 \tag{2.10}$$

$$\Lambda_1 = 2\pi^{-1} \int_0^\pi U*(z*)\cos^2 z* dz* \tag{2.11}$$

$$\Lambda_2 = 2\pi^{-1} \int_0^\pi V*(z*)\cos^2 z* dz* \tag{2.12}$$

$$\Lambda_3 = 2\pi^{-1} \int_0^\pi U*(z*)\sin^2 z* dz* \tag{2.13}$$

$$\Lambda_4 = 2\pi^{-1} \int_0^\pi V*(z*)\sin^2 z* dz* \tag{2.14}$$

Physically, Hn_1 represents heating at the same wavenumber as that of $w*$ itself and n_1 is related to the cloud cross-sectional area in the domain; Hn_2 measures the effects of uniform heating and n_2 is related to the asymmetry of the temperature variations in the upward and downward branches of the circulation.

The forcing provided by the basic wind field is measured often in terms of a Reynolds number Re, here defined as

$$Re = |V_H*(\pi)| \pi P^{-1} \tag{2.15}$$

If we know the functional form of the basic wind field $V(z*)$, then we can relate the Fourier coefficients Λ_i to Re. In anticipation of later orientation angle results, we introduce an angle θ between the $x*$-axis of our coordinate system $(x*, y*, z*)$ and east. Following (12), we find that the preferred angle θ will vary with the wind field and will be given by the minimum value of the critical Rayleigh number. For example, we may set

$$\Lambda_1 = (a_2\cos\theta + a_4\sin\theta)P \, Re \tag{2.16}$$

with similar forms for $\Lambda_2 - \Lambda_4$.

3.   THE CONVECTIVE STATE

The properties of the nonlinear developing convection are affected in fundamentally different but significant ways by the wind field and asymmetric latent heating. We found in our study of a two-dimensional model (12) that the wind plays two important roles. First, the convective solution is temporally periodic, corresponding to rolls propagating downwind at a speed given by the Fourier coefficient of the wind component perpendicular to the roll. Second, the rolls have a preferred orientation, given by the vanishing of the Fourier coefficient of the wind shear perpendicular to the roll. Also, we found in (12,13) that the primary role of the latent heating was to decrease the value of the critical Rayleigh number, so that convection developed in a conditionally unstable atmosphere. We extend these results here, and we discuss them in the following two subsections.

3.1 Effects of the wind field

We find that the wind plays a third significant role when the problem is cast in three dimensions. To determine this, we consider the stability properties of the trivial, or basic conductive, solution to our thirteen-coefficient model. We find that the characteristic polynominal, whose roots are the eigenvalues of the system linearized about the trivial solution, factors into a linear expression and two sixth-degree polynomials. From earlier results (12), we expect that the convective solution will be temporally periodic, and we can detect such a branching periodic solution by finding a Hopf bifurcation. A Hopf bifurcation occurs at a value R_h of R when only the real parts of a conjugate pair of eigenvalues vanish as the value of R is increased past R_h (4,9).

To find the stability of the basic state, we can consider each sixth-degree polynomial separately. We find that the critical values R_h are given by the roots of two cubic polynomials in R_h. The coefficients of them are complicated functions of the other parameters. These cubic polynomials will always admit of one real root each, so that at least two convective solutions, branches 1 and 2, bifurcate from the conductive state.

By minimizing the value of R_h, we can find the preferred orientation θ of the coordinate system (x*, y*, z*) (and therefore the convective cell) with respect to east for one of the branches. The angle θ varies depending on whether branch 1 or 2 is chosen, but in such a way that the angle between the diagonal and east is the same for each. Only one preferred orientation actually exists, because the other branch will be aligned at an angle 2β from the first one, where β is given by

$$\cos\beta = \delta b / (\phi^2 + \delta^2 b^2)^{1/2} \tag{3.1}$$

The diagonals of the two cells are displayed schematically in
Fig. 1 for the case $\beta = 20°$. In either case, the value of R_h
for the second branch must be greater than that of the first.
The maximum difference between the values of the critical Rayleigh
numbers occurs when the separation is given by $2\beta = 90°$.

Moreover, it can be shown that the value of the critical
Rayleigh number R_h only depends on an aspect ratio A defined by

$$A^2 = a^2(\phi^2 + \delta^2 b^2) = 4z_T^2(\phi^2/L_x^2 + \delta^2/L_y^2) \tag{3.2}$$

The aspect ratio A is therefore a measure of the height of the
cell divided by the length of its diagonal. In the two-dimensional
limit $L_x \to \infty$ or $L_y \to \infty$, A approaches the usual aspect ratio of
a roll, which is the height divided by the width. The important
thing to note here is that only preferred values of A and θ can
be determined by finding the minimum value of R_h so that no
expected value of b or separation angle 2β can be found from the
linear analysis. Thus, b and β are free parameters in the nonlinear thirteen-coefficient model, and so a larger system must
be used to find preferred values for them.

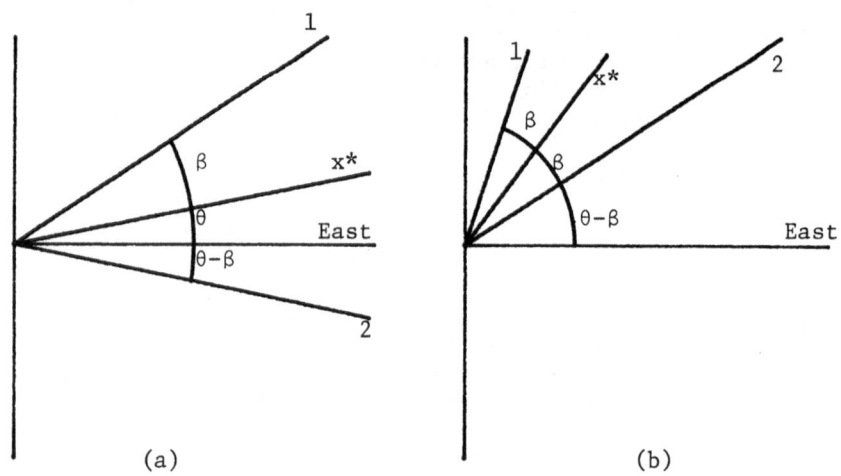

(a)   (b)

Fig. 1. Orientation of the two branching cells whose diagonals
are denoted by the numbers 1 and 2. The angle θ has been
determined in (a) so that 1 has the preferred orientation
and in (b) so that 2 has the preferred orientation. The
observed cell will always have the same orientation whether
branch 1 or branch 2 is considered.

A nonlinear analysis of the model gives the planform of the two branching convective solutions. All amplitudes are nonzero, and they must be related by one of the forms in

$$A_1 = \mp A_4, \quad A_2 = \pm A_3, \quad B_1 = \mp B_4, \quad B_2 = \pm B_3, \quad E_1 = \pm E_4, \quad E_2 = \mp E_3 \tag{3.3}$$

The relations (3.3) can be substituted into the spectral expansion to find the form of the nonlinear branching solution. For example, we find that the branches have the form

$$u^*(x^*, y^*, z^*, t^*) = [A_1(t^*)\sin(\phi x^* \mp \delta y^*)$$
$$+ A_2(t^*)\cos(\phi x^* \mp \delta y^*)]\cos z^* \tag{3.4}$$

Both possibilities in (3.4) represent flows of two-dimensional rolls, the axes of which are aligned along either the lines $\phi x^* - \delta y^* + 2n\pi$, or the lines $\phi x^* + \delta y^* + 2n\pi$. But these are rolls separated by the angle 2β, the angle we defined in (3.1) for the separation of the two convective branches (Fig. 1). Moreover, we see that the aspect ratio A in (3.2) can be viewed in the usual way as the height of the roll divided by the width.

We conclude that wind in the three-dimensional flow organizes the convective solution into rolls, a physical argument for which was given in (5). More importantly, earlier results of two-dimensional studies apply to the three-dimensional case, and we may study additional effects introduced by the Coriolis parameter or asymmetric latent heating by using one of the choices in (3.3) to reduce the thirteen-coefficient model further, to a seven-coefficient one. Thus, individual seven-coefficient models governing each roll can be studied separately because these rolls are not linked in the original model. In fact, the forms of the two seven-component models are similar enough that we need study the characteristics of only one of them. Secondary branching from one roll to the other, which arises from the term $\underset{\sim}{v} \cdot \nabla \underset{\sim}{v}$ in the equation of motion (2.4) (12,13), can occur only in a larger model than the one considered here.

Now we must find the preferred orientation angle and aspect ratio of the roll by minimizing the value of R_h. We consider here only branch 1 because as noted earlier the results for branch 2 are the same. We note that now the orientation angle α of the roll is given by $\alpha = \theta + \beta$ because when the x*-axis is directed toward the east and $\theta = 0$, then the roll is still aligned at an angle β with respect to east (Fig. 1). In general, analytic solutions for R_h cannot be obtained when $f^* \neq 0$, but numerical investigation shows that both α and A vary with the Reynolds number Re, which via (2.15) is a measure of the wind speed at the top of the domain.

In Fig. 2 we show the variations of α and A as a function of Re for an Ekman profile when $f*=0.1$, $z_T=1$ km, $P=1$, and $\nu=\kappa=100$ m^2s^{-1}. Here α is the angle between the roll axis and the wind direction at the domain top. From Fig. 2 we see that the values of α and A do not vary appreciably for realistic values of Re; but as the value of $f*$ increases, however, much larger variations of α and A can occur, and we trace this sensitivity to the magnitude of the dimensional combination $z_T^4/\nu\kappa$ used in the definitions (2.1)-(2.3) of $f*P$, H, and R. We avoid this problem if we require $f*<0.2$ as discussed in Sec. 2.1. The expected value of α for $f*=0$ was given in (12) to be 18.4°; we see from Fig. 2 that when $f*\neq0$ the orientation angle is about 16°, which is an alignment that is closer to the wind direction at the top of the boundary layer than that found when $f*$ is neglected. Moreover, we find that $A^2=0.5$ when $f*=0$, but that $A^2<0.5$ when $f*\neq0$ so that rotation leads to broader rolls when embedded in a basic wind field.

In the present model, the $f*=0$ orientation angle formula given in (12) corresponds to $\Lambda_2=\Lambda_4$; this is the alignment for which the Fourier coefficients of the shear perpendicular to

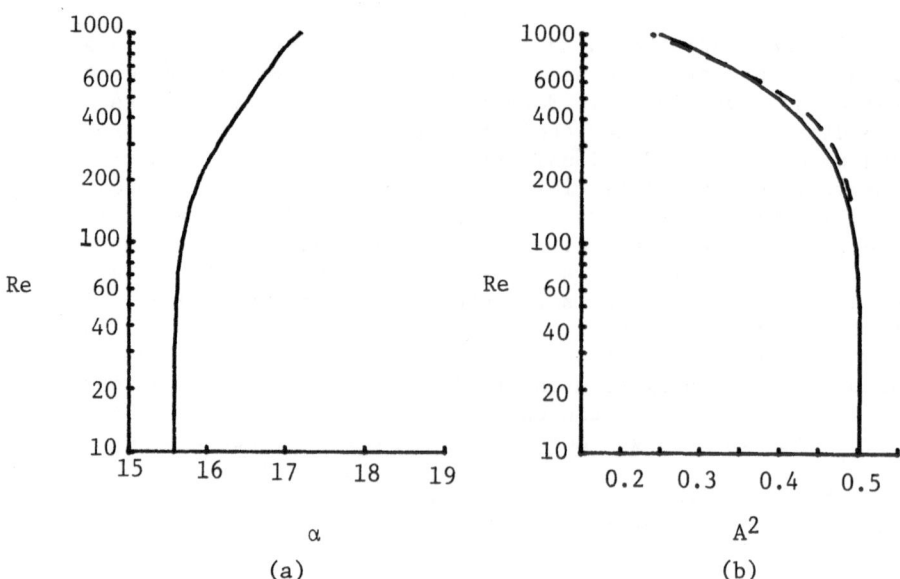

(a)   (b)

Fig. 2.   Preferred orientation angle $\alpha=\theta+\beta$ and aspect ratio A of the branch 1 roll embedded in an Ekman layer. When $f*=0$, the obtained values are $\alpha=18.4°$ and $A^2=0.5$ for all values of Re. The dashed curve in (b) is the approximate form derived from (3.6) under the assumption that $f*\approx0.1$, $P=1$ and $\alpha=18.4°$.

the roll vanishes. For this case when P=1, we can obtain an analytic solution for R_h, and we know from Fig. 2 that this solution is a good approximation to the actual one. When $\Lambda_2 = \Lambda'_4$, we have

$$R_h = (A^2+1)^3/A^2 + f^{*2}/A^2 - (\Lambda_1 - \Lambda_3)^2 f^{*2}/[4(A^2+1)^3] - Hn_1$$

(3.5)

The first two terms of (3.5) are the usual ones obtained in an analysis of the rotating Rayleigh-Bénard problem (16), and we see that rotation alone inhibits convective development. The third term in (3.5) represents a destabilizing effect that is related to the Fourier components of the wind parallel to the roll axis. Moreover, if $\Lambda_1 \neq \Lambda_3$, then the Fourier component of the shear parallel to the roll axis is nonzero, and so wind shear parallel to the roll is destabilizing only in the presence of rotation. This is the parallel instability mechanism of Lilly (7), and we conclude that it is linked to the thermal instability via (3.5). Thus, we may view the thermal instability as being modified by parallel instability, at least for small values of f*, and we find that only one convective mode develops from the conductive state. Of the three proposed mechanisms for roll development in the atmosphere, we conclude that at most two (inflection point instability and thermal forcing) can lead to separate developing modes.

We may find an approximation to the preferred aspect ratio A by differentiating (3.5) with respect to A^2 and setting the result to zero. When f*=0, we found in (12) that $A^2=0.5$ for all values of Re; when $f^* \neq 0$ but $f^* \approx 0.1$, we find here that

$$(\Lambda_1 - \Lambda_3)^2 \cong 4(A^2+1)^6 (1-2A^2)/(3f^{*2}A^4)$$

(3.6)

$$R_h \cong (A^2+1)^3 (5A^2-1)/(3A^4)$$

(3.7)

With the aid of (2.16) and a form similar to it, we can relate Re to $(\Lambda_1 - \Lambda_3)$ in (3.6). From (3.6), we see that $A^2 < 0.5$ as expected from Fig. 2 and that as Re → 0, $A^2 → 0.5$, the windfree limit. Because the motion and thermal equations become decoupled in the seven-coefficient model when R=0, we must require $R_h > 0$; from (3.6) and (3.7), we conclude that $0.2 < A^2 < 0.5$ in this case, and with the aid of (3.6) we find an allowable limit for the magnitude of Re.

3.2 Effects of asymmetric latent heating

As found in our earlier studies (12,13) we see from (3.5) that latent heating reduces the value of the critical Rayleigh number R_h for convective development. With the aid of (2.2) and (2.3) we see that $R+Hn_1 > 0$ corresponds to the condition

$\gamma_e > n_1\gamma_m + (1-n_1)\gamma_d$. This is the slice method criterion for
conditional instability if we interpret n_1 as the cloud cross-
sectional area in the domain. The limit $n_1 \to 1$ of moist air
corresponds to $\gamma_e > \gamma_m$ and the limit $n_1 \to 0$ of dry air corresponds
to $\gamma_e > \gamma_d$ as expected. Therefore, one important role of latent
heating is to reduce the critical value of the Rayleigh number
for convection. From (3.5) we see that this conditional instability is linked with the thermal and parallel instabilities.

The uniform heating term Hn_2 introduces an important new
qualitative effect into the nonlinear temporally-periodic solution. We are not able to find analytically the functional form
of this time-dependent solution, but we can find the amplitude
of it by moving the coordinate system (x^*,y^*,z^*) at the phase
speed of the propagating roll in the direction perpendicular to
the roll. We know this phase speed and direction from our
general calculation of the minimum value of the critical Rayleigh
number R_h.

The algebraic expressions for the amplitudes of the convective solution are very complicated, but the form of the solution
is given in Fig. 3. Here we show a single Fourier coefficient
as a function of R to emphasize the form of the branching diagram. Asymmetric latent heating $Hn_2 \neq 0$ leads to transcritical
bifurcation, in which the parabola of the usual trident-shaped

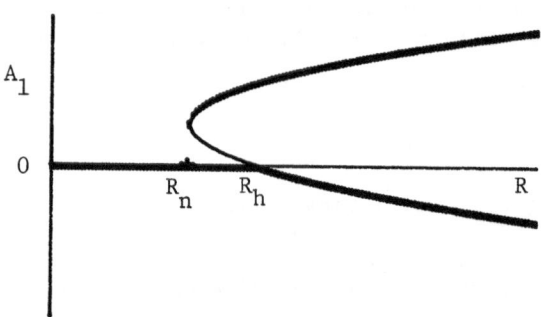

Fig. 3. Form of the branching temporally-periodic roll solution
when cloud base h* obeys $0 < h^* < \pi$ so that $n_2 \neq 0$. Stable solutions are denoted by the heavier lines. One stable branch exists
for $R > R_h$ but a second one exists for $R > R_n$. This transcritical
bifurcation leads to the possibility of a sudden transition to a
convective state at values of R that are less than the critical
bifurcation value R_h.

branching diagram (13), which occurs when $Hn_2=0$, is shifted in
the vertical leading to a stable supercritical branch and an
unstable subcritical one. But the subcritical branch curves
back toward larger values of R at a value R_n, and in so doing
the branch becomes stable (Fig. 3). Then a convective solution
can exist for values of R between R_n and R_h as well as the
expected range $R > R_h$. Thus, when asymmetric latent heating
occurs in the roll circulation, we find that a sudden transition
to a convective state can occur and hysteresis is possible because
the value of R at which a convective state develops may be
different than the value of R at which a convective state decays.
The introduction of subcritical convection is the second important
qualitative role of the latent heating, but we see from (2.10)
that cloud base h* must be between 0 and π for this effect to be
realized.

We can find an expression for R_n only in the case $\Lambda_2 = \Lambda_4$
and P=1 for which we could obtain (3.5) for R_h. Thus we find
that

$$R_n = R_h - H^2n_2^2[4(A^2+1)^6 + (\Lambda_1-\Lambda_3)^2f*^2A^2]/[8(A^2+1)^5R_h]$$

$$(3.8)$$

in which R_h is given by (3.5). This expression makes sense
provided that $R_h > 0$ for which we noted previously that the
thermal and motion equations remain linked. We see from (3.8)
that when there is asymmetric latent heating $Hn_2 \neq 0$, then the
parallel shear term $\Lambda_1-\Lambda_3$ contributes to the reduction of the
value of R_n, as well as R_h. Furthermore, for the preferred
values (3.6) and (3.7) of $(\Lambda_1-\Lambda_3)$, or Re, and R_h that were found
by minimizing the value of R_h with respect to A^2, we find that
(for $f* \approx 0.1$)

$$R_n \cong (A^2+1)^3(5A^2-1)/(3A^4)$$
$$- H^2n_2^2A^2(4-5A^2)/[2(A^2+1)^2(5A^2-1)] \qquad (3.9)$$

4.   SUMMARY

We have presented here preliminary results of the branching
and stability properties of the first branch in a three-dimensional truncated spectral model of shallow moist convection.
With this model we have identified some of the fundamental
qualitative effects introduced by the large-scale basic wind
field and by latent heating in the upward branch of the circulation between cloud base and cloud, or domain, top.

We find that when the convection is embedded in a height-dependent basic wind field, then only roll solutions are possible. Thus, the wind aligns the convective elements into lines; this is consistent with the observation that cloud streets are found in strongly shearing boundary layers. Breakup of the rolls into cells, therefore, is a secondary branching effect. Moreover, the wind field causes the rolls to be aligned in a favored way, and in many cases the orientation angle formula given in (12) gives good predictions of the orientation angle results obtained when the Coriolis parameter $f \neq 0$.

We find that the thermal and parallel instabilities are linked so that only one convective mode develops when both instability mechanisms are operating. We found that these two mechanisms are unified by determining how the critical Rayleigh number R_h is modified when the environment is not at rest. We identified the instability due to the wind shear as the parallel instability because it involved the Fourier component of the wind parallel to the roll axis and because it existed only when rotation $f \neq 0$ was present.

Latent heating plays two important roles. First, it leads to a reduction of the values of the critical Rayleigh number that corresponds to the slice method of conditional instability. Second, the nonlinear branching diagram has both a supercritical branch and a subcritical one that curves back toward larger values of R at a value R_n. Thus, a sudden transition to a finite-amplitude convective state is possible when the latent heating is asymmetrically distributed in the two branches of the roll circulation. This occurs when part of the upward motion is moist, part is dry and all the downward motion is dry as might be found when cloud base is between the bottom and top of the domain.

ACKNOWLEDGMENTS

The research in this paper was partially supported by the National Science Foundation through Grants ATM 78-02699 and ATM 79-08354, the National Aeronautics and Space Administration through Grant NAS 8-33794, and the Max-Planck Society by the granting of a Max-Planck Postdoctoral Fellowship for Foreigners to me. Most of the results reported here were obtained while I was visiting the Department of Atmospheric Processes of the Max-Planck-Institut für Meteorologie, Hamburg, West Germany at the invitation of Professor Dr. H. Hinzpeter. I gratefully acknowledge the many constructive criticisms of my work given me by members of his group.

REFERENCES

1.   Asai, T. and Nakasuji, I.:   1973, J. Meteo. Soc. Japan, 51, pp. 29-42.

2.   Dutton, J. and Fichtl, G.:   1969, J. Atmos. Sci., 26, pp. 241-254.

3.   Gammelsrød, T.:   1975, J. Geophys. Res., 80, pp. 5069-5075.

4.   Iooss, G. and Joseph, D.:   1980, "Elementary Stability and Bifurcation Theory", Springer-Verlag, 286 pp.

5.   Kuettner, J.:   1959, Tellus, 11, pp. 267-294.

6.   Kuettner, J.:   1971, Tellus, 23, pp. 404-425.

7.   Lilly, D.:   1966, J. Atmos. Sci., 23, pp. 481-494.

8.   Lorenz, E.:   1963, J. Atmos. Sci., 20, pp. 130-141.

9.   Marsden, J. and McCracken, M.:   1976, "The Hopf Bifurcation and Its Applications", Applied Mathematical Sciences, 19, Springer-Verlag, 408 pp.

10. Ogura, Y. and Phillips, N.:   1962, J. Atmos. Sci., 19, pp. 173-179.

11. Saltzman, B.:   1962, J. Atmos. Sci., 19, pp. 329-341.

12. Shirer, H.:   1980, J. Atmos. Sci., 37, pp. 1586-1602.

13. Shirer, H. and Dutton, J.:   1979, J. Atmos. Sci., 36, pp. 1705-1721.

14. Shirer, H. and Wells, R.:   1982, J. Atmos. Sci., 39, to appear.

15. Sommeria, G. and LeMone, M.:   1978, J. Atmos. Sci., 35, pp. 25-39.

16. Veronis, G.:   1959, J. Fluid Mech., 5, pp. 401-435.

CLOUD BANDS IN THE ATMOSPHERE

Wen-yih Sun

Department of Geosciences
Purdue University, West Lafayette, Indiana 47907 USA

ABSTRACT

   The cloud bands, which occur in a convective planetary boundary layer and/or above, can be triggered by either inflectional
instability, or internal gravity waves or thermal convection.
Theoretical studies and observations on cloud bands are discussed
here.

1. INTRODUCTION

   It is well known that clouds frequently form into organized
bands in the atmosphere. Figure 1 shows one Apollo 6 picture of
widespread convective bands developing in a southernly flow inland off the coastline of Georgia at 1500 GMT (10:00 EST) on April
4, 1968. Many of these bands extend over 100 km in length, and
the average spacing is about 2 to 2.5 km. They align parallel to
the wind with the cloud tops about 1 to 2 km. These shallow cloud
bands are also frequently observed over such warm water surfaces
as the Great Lakes, the Gulf stream and Gulf of Mexico during
winter. They are also observed over the tropical ocean where
Malkus and Riehl [1] discovered that the average distance between cloud bands is about 4 km; the orientation of the cloud
bands is closely parallel to the low-level flow. The tops of
these clouds are usually limited by the base of the trade wind
inversion or by a relatively dry layer in the lower atmosphere.

   Near the trough of an easterly wave, Malkus and Riehl [1] also
discovered the other type of cloud bands which consist of deep
clouds with tops extending up to at least 6 km. The observed
precipitation indicates that latent heat is important to the
formation of these cloud bands, which have a horizontal wavelength

179

E. M. Agee and T. Asai (eds.), Cloud Dynamics, 179–191.
Copyright © 1982 by D. Reidel Publishing Company.

from about 30 to 100 km. The orientation of deep cloud bands is
more variable. It will be convenient to define that a longitudinal mode is parallel to the direction of the wind shear in the
cloud layer, while a transverse mode is perpendicular to it.
Figure 2 indicates that both longitudinal and transverse modes of
clouds coexist to the east of a moderately disturbed wave trough
over the tropical ocean. The rows composed of medium or small
clouds are parallel to the wind shear and hence belong to the
longitudinal mode. The distance between two rows is about 25 km.
We can also find that large clouds line up at right angle to the
wind. The horizontal interval is about 100 km. These transverse
mode bands are referred to as a "cross-wind mode" by Malkus and
Riehl. The longitudinal rows and transverse rows coexist and
look just like a chessboard on the east side of the trough. However, to the west of the trough, there only exists a longitudinal
mode formed by congestus cumuli. The spatial interval between
two bands is about 30 km.

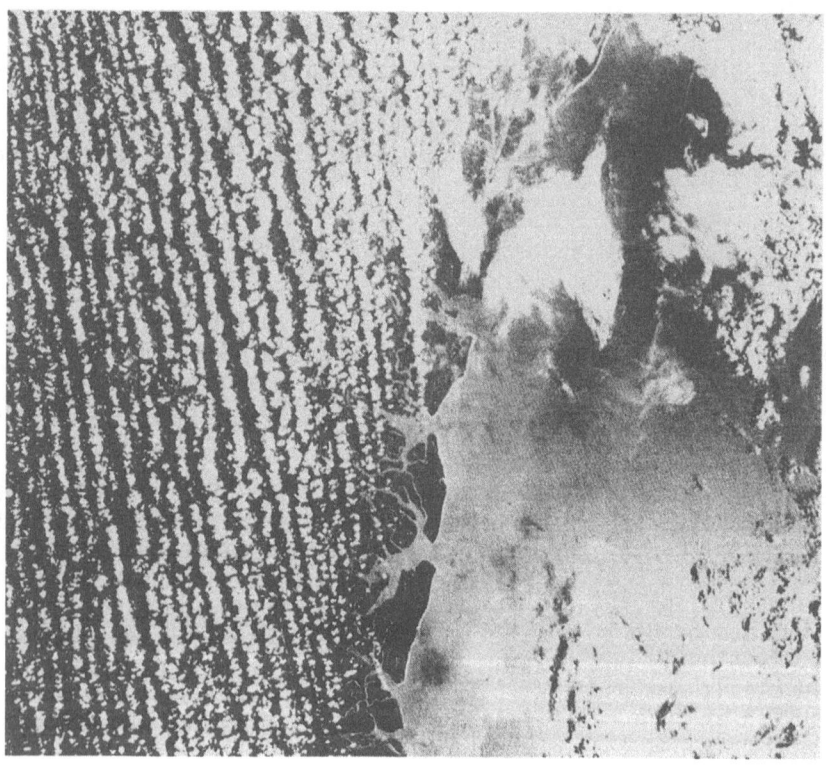

Figure 1. Satellite view of stratocumulus cloud streets over
Georgia.

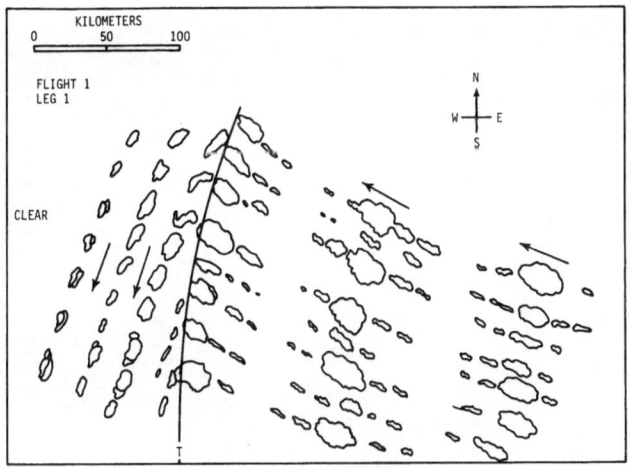

Figure 2. Cloud bands near a wave trough in the easterlies. The direction of the wind is indicated by an arrow (after Malkus and Riehl, [1]).

A completely different type of cloud bands are also observed over the tropical continents. Figure 3 is a satellite picture which shows the cloud bands over South America from the equator to 15°S. The cloud elements organize into lines parallel to the coast and extend deeply inland with a horizontal wavelength of a few hundred kilometers, while a clear sky is noted off the coast. It seems that the bands are triggered by the interaction between the sea breeze circulation and the variation of the stratification in the planetary boundary layer (Sun and Orlanski [2] and [3]).

This paper is limited to discuss the theories and observations relative to the cloud bands shown in Figures 1 and 2, since the mechanism for those bands in Figure 3 is quite different and more complicated.

2. SHALLOW CLOUD STREETS

The shallow cloud bands shown in Figure 1 seem reproducible by models based on inflectional instability in the Ekman spiral (e.g., Faller [4], Lilly [5], Brown [6]) or thermal instability with wind shear (Kuo [7], Asai [8], Kuettner [9]). The details are as in the following.

2a. Inflectional instability

The Ekman flow in a neutral atmosphere can be expressed as

Figure 3. GEOS East Satellite picture at 2130 GMT 5. September 1974, showing three cloud bands aligned parallel to the coast of South America. Their horizontal wavelength is of a few hundred kilometers. Their is a clear sky off the coast.

$$u = V_g \left(1 - e^{-z/D} \cos \frac{z}{D}\right) \cos \varepsilon + V_g \, e^{-z/D} \sin \frac{z}{D} \sin \varepsilon \qquad (2.1)$$

$$v = -V_g \left(1 - e^{-z/D} \cos \frac{z}{D}\right) \sin \varepsilon + V_g \, e^{-z/D} \sin \frac{z}{D} \cos \varepsilon \qquad (2.2)$$

where V_g, the geostrophic velocity; D, the characteristic depth of the boundary layer is defined as

$$D = \sqrt{\frac{2\nu}{f}} \qquad\qquad\qquad (2.3)$$

ν, the coefficient of kinematic viscosity; f, the Coriolis parameter; ε, the angle of the x-axis to the left of the direction of the geostrophic flow. The original famous Ekman spiral is

shown by the dashed lines in Figure 4. The y-component velocity
profiles with respect   to the nondimensional height z/D for
different angle ε are presented in Figure 5, which shows a less
shear but stronger inflection when x-axis is parallel to the geostrophic wind (ε = 0°).

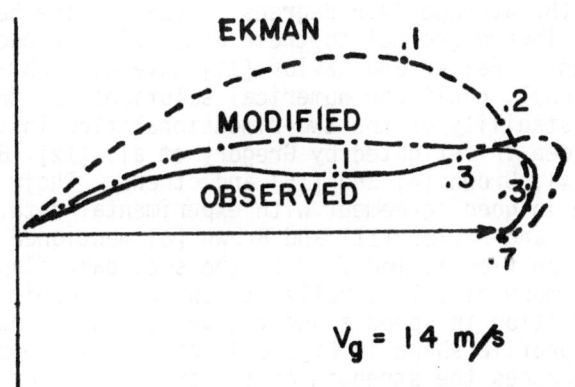

Figure 4. Comparison of observed hodograph with modified mean
hodograph and original Ekman spiral. The numbers give the height
in km. (after Brown [6])

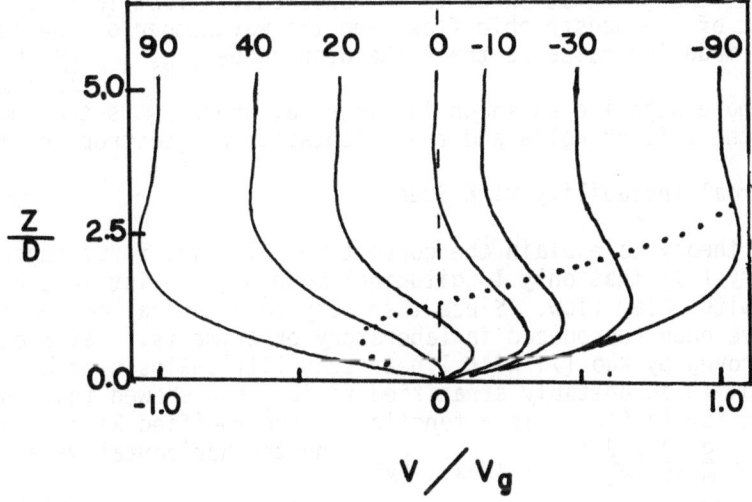

Figure 5. Two-dimensional velocity profiles taken in (y,z) vertical planes for different angle of ε.

The instability of the inflection point in the Ekman layer
has been well reproduced in a laboratory experiment by Faller [10]
and [4]. He found that the thin layer of dye forms a series of
light and dark spiral bands near the bottom of the boundary layer
when the Reynold number, $RE = V_gD/\nu$, reaches 125. The average
spacing is $L = 11D$, the angles of orientation of the bands are
mostly in the range 10-17 degrees to the left of the geostrophic
wind, with the average 14.5 degrees. Usually, the bands move
very slowly inward (normal to their axes) with respect to the
rotating tank. Faller and Kaylor [11] have also obtained the
two-dimensional, nonlinear numerical solutions for those rolls.
The linear stability of the two-dimensional flow in the Ekman
layer has been investigated by Gregory et al. [12], Brown [13],
Barcilon [14], Brown [6] and [15] and others. Their theoretical
results are in good agreement with experimental data. Furthermore, Faller and Kaylor [11] and Brown [6] mentioned that their
original Ekman flow is modified by the secondary flow such that
the flow is more closely parallel to the geostrophic flow with
a large variation in speed along the geostrophic wind. The modified wind profile shown in Figure 4 not only increases the shear
but also enhances the strength of the inflection for wind component
along the geostrophic wind. Hence, it may also produce transverse
rolls as discussed by Asai [16]. Since the modified Ekman profile
may still be unstable, it remains to be investigated by using a
three-dimensional numerical model.

When the stratification is taken into account, Brown [15]
found that the orientation of those bands shifts from left to
the right of the geostrophic flow, and the wavenumber of the most
unstable mode increases as the Richardson number, $Ri = \frac{g}{\Theta} \frac{\partial \Theta}{\partial z} \left(\frac{\partial V}{\partial z}\right)^{-2}$,
becomes more negative as shown in Figure 6, where ε_c is the angle
between the axis of rolls and the orientation of geostrophic flow.

2b.   Thermal Instability With Shear

The theory to explain the convective rolls was first suggested
by Jeffrey [17] that only longitudinal mode might exist in a convection with shear flow. Since then many longitudinal convective
bands have been reproduced in laboratory experiments. Later on
it was proved by Kuo [7] with linear stability analysis of a
shear flow in an unstably stratified fluid. Kuo showed that the
amplification factor σ_r is a function of the modified Richardson
number $\bar{J} = \frac{g}{\Theta} \frac{\partial \Theta}{\partial z} \left(\frac{dU}{dz}\right)^{-2} (1 + k_x^{-2} k_y^2)$ and the horizontal wavenumber, where k_x and k_y are the wavenumbers which are parallel and
perpendicular to the mean flow U, respectively. Those longitudinal rolls propagate with the mean wind of flows. The horizontal
spacing is about 2.8 times of the depth of fluid for a free-free
boundary condition. The numerical results of the linearized

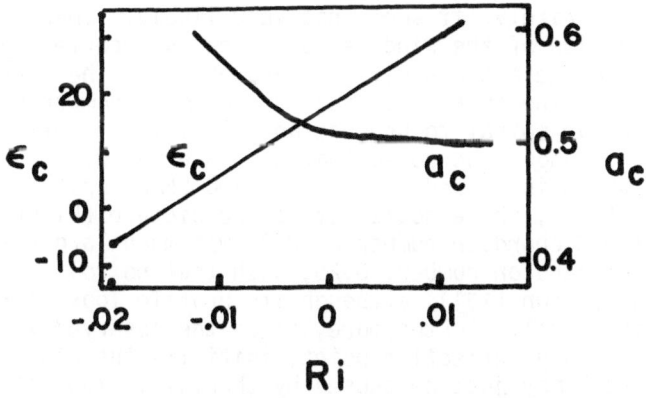

Figure 6. The behavior of wavenumber (a_c) and orientation angle (ε_c) at maximum growth rates vs Ri (Brown [15]).

equations obtained by Asai [8] not only confirmed Kuo's analytical results but also showed that energy conversion between the kinetic energy of the basic flow and perturbation tends to intensify the shear of the basic flow for a transverse mode but it reverses for a longitudinal mode. In an unstably stratified fluid, the longi- tudinal rolls are more unstable than the transverse rolls not only in the plane Couette flow but also in many other two-dimensional shear flow (Gage and Riehl, [18]; Asai, [16]; Kuettner, [9]). It is also found that the longitudinal rolls remain as the preferred mode when the rigid top of the unstable layer is replaced by a deep stably stratified fluid (Sun, [19]).

2c. Comparison With Observations

Observations of Kuettner [9] and LeMone [20] and others show that the longitudinal band consistently occurs in a moderately strong wind and slightly unstable condition. The ratio of the horizontal wavelength to the depth of the mixed layer is from 2 to 6. The bands are parallel to the wind in the mixed layer according to Kuettner [9] and others but they orient from 10^0 to 20^0 to the left of the geostrophic wind according to LeMone [20] and Plank [21]. LeMone's [20] suggested that the formation of the cloud streets in PBL is caused by inflection point of the Ekman spiral. However, the Richardson number in her study is from -0.01 to -0.03, the orientation of bands from the geostrophic wind should be from 5^0 left to 20^0 right instead of deviating about 10^0 - 20^0 left, as shown in Figure 6.

It is reasonable to raise the question whether the Ekman spiral really exists in the atmosphere to produce the cloud

streets. Unfortunately, it seems not very likely. There is no
consistent rotation of the wind vector in cases reported by
Kuettner [9], LeMone and Pennell [22] and others. The directional
variation of the wind in PBL is usually too small compared with
the original Ekman spiral to produce any significant inflectional
instability. It should be noted that the observed wind profile
in Figure 4, which is closed to the modified Ekman spiral was
taken at 0700 (EST), three hours before the picture (Figure 1)
was taken. The Richardson number at 0700 was much larger than
the critical Richardson number, 0.25, such that no roll can develop
at this moment (Brown [15]), although its profile looks like the
modified Ekman spiral. Furthermore, since the observed wind does
not possess a strong inflection point, initially the cloud streets
shown in Figure 1 may just be caused by thermal instability with
shear as predicted by Kuo. The cloud streets observed over the
equatorial region (Malkus and Riehl, [1]; Kuettner, [9] and others)
and those reported by LeMone and Pennel [22] near Puerto Rico are
more likely produced by Kuo's theory, although it has difficulty
to explain the orientation angle and the horizontal wavelength in
some cases.

Observations of the cloud streets from satellites are sufficient for us to determine the spacing interval, but it requires
more data in the temperature and velocity profiles to determine
the orientation angle and the ratio of the horizontal wavelegth
to the depth of the mixed layer. It is also necessary to calculate the energy conversion from the wind shear and from the buoyancy force as LeMone did. A three dimensional numerical model
with a better turbulence parameterization is also required to
study the inflectional instability and thermal instability, and
their interaction under a well developed situation.

3. DEEP CLOUD BANDS

The latent heat is negligible for shallow cloud streets shown
in Figure 1 (Konard, [23]), but it may become important for those
shown in Figure 2 and for the cloud bands around the hurricane
center. Sun's stability analysis of the deep cloud street [19]
in a conditional unstable atmosphere includes the effect of condensation based upon the wave-CISK hypothesis and the effect of
wind shear. The latter is concerned only with conditions favorable for cloud streets to develop; i.e., the wind shear direction
is either constant with height or the wind speed remains constant
but the direction has a 90° shift within a relative shallow layer
in the convective atmosphere. He found that the growth rate and
the propagating speed are strongly dependent on the static stability, moisture supply, the strength of the wind shear and viscosity. Since convectional Rayleigh number and Richardson number are

not good parameters for measuring the strength of the convection
and wind shear in a penetrative convection, especially when it
occurs in a conditional unstable atmosphere, Sun has defined the
flux Richardson number Rf as follows:

$$Rf = \frac{\left| \int_0^H \rho_0 g\alpha \overline{w\,\Theta}\ dz \right|}{\left| \int_0^H [\rho_0 \overline{uw} \frac{dU}{dz} + \rho_0 \overline{vw} \frac{dV}{dz}]\ dz \right|}$$

(3.1)

to measure the relative importance of the buoyancy term compared
with the shear-production term. Large flux Richardson numbers
signify that the buoyancy dominates the perturbation structure,
whereas small flux Richardson numbers signify that the buoyancy
has little effect on the flow. A similar parameter was also used
by Asai [9] and [16]. It is well known that even in Benard convection, weak convection produces a slightly unstable mixed layer,
while strong one can produce a slightly stable mixed layer. Hence,
a conventional Richardson number is not a good parameter in an
unstable atmosphere, either.

In addition to the flux Richardson number, Sun also calculated the total buoyancy force released by the latent heat in the
entire cloud layer and the damping effect coming from the vertical
motion in a stably stratified atmosphere. It is found that if
the buoyancy force released is much larger than the damping effect,
the most unstable mode will be parallel to the wind shear and
cloud streets will be stationary to the mean wind in the cloud
layer. On the other hand, if the buoyancy force is weak and the
conversion of the kinetic energy from the mean flow becomes
stronger, the most unstable mode turns out to be perpendicular to
the wind shear vector and the cloud streets propagate relatively
to the mean flow as internal gravity waves. For a finite amount
of the moisture supply, it is more likely to produce a longitudinal mode if the cloud layer is not deep, but it may become a
transverse mode if the cloud top extends to the upper troposphere
because the total amount of the released latent heat remains the
same but the damping effect becomes much larger for the deeper
clouds due to a stronger stratification of the upper troposphere.
Hence, in a moderately disturbed trough region those medium clouds
with tops at about 5-7 km or less build up in lines parallel to
the wind shear with a wavelength of about 30 km. while, the deep
clouds align perpendicular to the wind shear with a larger wavelength of about 50-100 km. These results may have provided the
explanation of the coexistence of two different cloud streets
observed by Malkus and Riehl in the east side of a moderately
tropical trough.

Syono and Yamasaki [24] in their two-layer hurricane model
found that when the latent heat releases mainly in the lower layer

(at 725 mb), it produces unstable stationary waves, while unstable, propagating waves appear when the heating in the upper layer (at 325 mb) is greater than that in the lower layer. These results are consistent with Sun's results. The development of internal gravity waves may be explained by comparing the diabatic heating effect with the cooling effect due to a vertical motion in a stably stratified fluid. When the total effect of cooling is greater than that of warming in that vertical column, the surface pressure in the ascending region turns out to be higher and re- duces the convergence in the lower layer. The decreasing horizon- tal convergence in the lower layer decreases the amount of the heat released in the column. Thus the temperature in the upper cloud layer cannot continually increase but eventually tends to decrease, and triggers the oscillatory wave. Hence, it can be expected to obtain a stationary, longitudinal mode when the adia- batic heating becomes large due to the increase of moisture supply or condensation efficiency; but a propagating, transverse mode if the diabatic heating is small.

The transverse cloud bands are often observed in the strato- cumulus or altocumulus in a fair sky (Figures 2.28-2.30 of Gedzel- man's; [25]). But the rain bands in a severe weather are more likely parallel to the wind shear. For instance, as in Figure 7, (Browning and Ludlam, [26]) it shows that the convective cells align parallel to and move along with the wind.

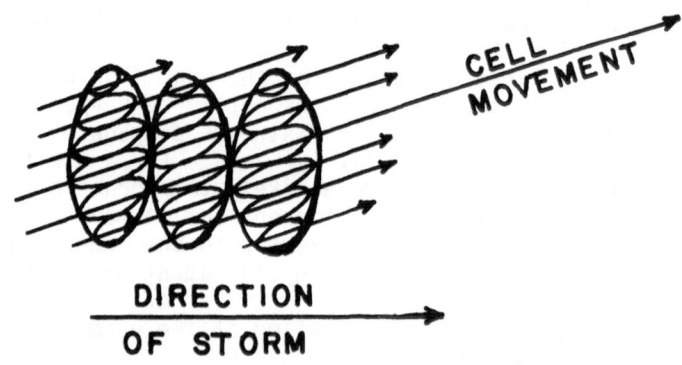

Figure 7. The convection cells in a moving storm (Browning and Ludlam, [26]).

Chen et al. [27] analyzed the radar echoes in the Texas High Plain Experiment area, they found that the weak echoes, originally in a transverse mode, developed and switched into a longitudinal mode with the increase of the latent heat on July 8-9, 1977.

While, in another case, initially a longitudinal mode with well
developed storms, changed into a transverse mode in the dissipating stage on June 23-24, 1977. Those discoveries are also consistent with Sun's results. However, some squall lines observed in
the mid-latitudes are a type of transverse mode. They propagate
relative to the mean flow and may be considered as internal
gravity waves (Raymond, [28]). Unfortunately, the observations
are still not adequate enough to calculate the phase speed,
energy conversions from the mean flow or from diabatic heating,
nor the structure of those deep cloud bands.

The spiral bands around the hurricane has been extensively
investigated by Anthens [29], Kurihara and Tubeya [30] and
Kurihara [31] and others. The numerical results show those
bands propagate radially outward as observed by Senn and Hiser
[32]. The passage of a band is accompanied by a corresponding
drop and rise of the surface pressure; by a wind shift indicating low-level convergence and upward motion; and by a temperature
drop. The phase speed is approximately 40-100 km hr^{-1}. The
source point of the spiral band is in a region surrounding the
center. It moves cyclonically while it produces a new portion of
band. The rainfall within the band is generally small and not
of primary importance for the maintenance and propagation of a
band. The energetic analysis also suggest that no kinetic energy
conversion between the band system and the mean basic flow with
vertical shear. These results suggest that once the band is
formed, it propagates outward as an internal gravity wave without
appreciable further supply of energy. On the other hand, the
cloud band in the eye wall region seems stationary with respective
to the mean flow and produces lots of precipitation. A stationary,
longitudinal mode around eye wall is consistent with Sun's results, but a propagating, longitudinal mode for the spiral bands
cannot be explained with his model, since there is not any horizontal variation in the basic flow and pressure field in Sun's
model. In addition to the bands propagating outward, Kurihara
[31], from solving linearized equations, obtains two more bands
propagating inward, which have never been observed in a real storm
or a numerical simulation. Kurihara pointed out that analysis of
the time variation in rainfall intensity or the tracing of groups
of radar rain bands is required to reveal the mesoscale spiral
structure in a tropical cyclone.

SUMMARY

Theory, numerical simulation and observation of the shallow
and deep cloud streets are now discussed. Those cloud bands may
be triggered by inflectional instability, convection or internal
gravity waves modified by latent heat. Inflectional instability
and convection will produce stationary bands with respect to the

mean flow, but bands associated with internal gravity waves propagate very rapidly. They are in good agreement with either the
experiment or observations, but the horizontal wavelength and
orientation angle of some cloud bands have discrepencies from
theoretical results. They may come from many unrealistic conditions used in theoretical model, for example; the basic state is
homogeneous horizontally, a constant eddy diffusivity, a constant
lapse rate and/or the wave-CISK hypothesis used to calculate the
diabatic heating. The numerical simulation of the shallow cloud
bands have only been studied in two-dimensional models. The three-
dimensional hurricane model is still suffering from a poor resolution, and poor eddy transport of momentum, temperature and moisture. On the other hand, when cloud bands occur, the observational sounding may have been drastically modified by the secondary
flow associated with cloud streets, it may not represent the
basic condition at all. Hence, it needs not only more theoretical studies and numerical simulations, but also more observational data in a carefully designed mesoscale network.

REFERENCES

1. Malkus, J. S., and Riehl, H.: 1964, Cloud Structure and
       Distribution over the Tropical Pacific Ocean. University
       of California Press, Berkey and Los Angeles, 229 pp.

2. Sun, W. Y. and I. Orlanski: 1981, J. Atmos. Sci., 38,
       pp. 1675-1693.

3. Sun, W. Y. and I. Orlanski: 1981, J. Atm. Sci., 38, 1694-1706.

4. Faller, A. J.: 1965, J. Atmos. Sci., 22, pp. 176-184.

5. Lilly, D. K.: 1966, J. Atmos. Sci., 23, pp. 481-494.

6. Brown, R. A.: 1970, J. Atmos. Sci., 27, pp. 742-757.

7. Kuo, H. L.: 1963, Phys. Fluid, 6, pp. 195-211.

8. Asai, T.: 1970 , J. Meteor. Soc. Japan, 48, pp. 18-29.

9. Kuettner, J. P.: 1971, Tellus, 23, pp. 404-425.

10. Faller, A. J.: 1963, J. Fluid Mech., 15, pp. 560-570.

11. Faller, A. J. and R. E. Kaylor: 1966, J. Atmos. Sci., 23,
       pp. 466-480.

12. Gregory, N; J. T. Stuart and W. S. Walker: 1955, Phil. Trans.
       Roy. Soc. London, A 248, pp. 155-199.

13. Brown, W. B.: 1961, Boundary Layer and Flow Control, 2, pergamon Press, New York, N.Y. pp. 913-923.

14. Barcilon, V.: 1965, Tellus, 17, pp. 53-68.

15. Brown, R. A.: 1972, J. Atmos. Sci. 29, pp. 850-859.

16. Asai, T.: 1970, J. Meteor. Soc. Japan, 48, pp. 129-139.

17. Jeffreys, H.: 1938: Proc. Roy. Soc., A 118, pp. 195-208.

18. Gage, K. S. and W. H. Reid: 1968, J. Fluid Mech., 33, pp. 21-32.

19. Sun, W. Y.: 1978, J. Atmos. Sci., 35, pp. 466-483.

20. LeMone, M. A.: 1973, J. Atmos. Sci., 30, pp. 1077-1091.

21. Plank, V. G.: 1966, Tellus, 18, pp. 1-12.

22. LeMone, M. A. and W. T. Pennell: 1976, J. Atmos. Sci., pp. 524-539.

23. Konard, T. G.: 1968, Proc. Intern. Conf. Cloud Physics, Toronto, Canada, pp. 539-543.

24. Syono, S. and M. Yamasaki: 1966, J. Meteor. Soc., Japan, 44, pp. 353-375.

25. Gedzelman, S.: 1980, The science and wonders of the atmosphere. John Wiley & Sons, N.Y. pp. 535.

26. Browning, K. A. and F. H. Ludlam: 1960, Radar analysis of a hailstorm. Tech. Note No. 5, Contract AF61.

27. Chen, P. C., M. E. Humbert and T. B. Smith: 1978, Radar echo organization and development in the mesoscale environment- a case study approach. Meteorology Res., Inc. Atadena, CA.

28. Raymond, D. J.: 1975, J. Atmos. Sci., 32, pp. 1308-1317.

29. Anthens, R. A.: 1972, Mon. Wea. Rev., 100, pp. 461-476.

30. Kurihara, Y. and R. E. Tuleya: 1974, J. Atmos. Sci., 31, pp. 893-919.

31. Kurihara, Y.: 1976, J. Atmos. Sci., 33, pp. 940-958.

32. Senn, H. V. and H. W. Hiser: 1959, J. Meteor., 16, pp. 419-426.

DEEP CONVECTIVE SYSTEMS

OBSERVATIONS AND MODELS

AN INTRODUCTION TO DEEP CONVECTIVE SYSTEMS

Ernest M. Agee

Department of Geosciences
Purdue University, West Lafayette, Indiana 47907 USA

ABSTRACT

The focus of this introduction is the thunderstorm and its ability to produce severe local weather. Particular attention is given to the climatology and morphology of both the thunderstorm and the tornado. Some basic concepts of thunderstorm formation and structure are summarized, with emphasis on the observational and modeled characteristics of the severe thunderstorm cell. A brief treatment is also given to the tornado cyclone, tornado and suction vortex phenomena.

1. INTRODUCTION

The concept of deep convective systems was introduced in the beginning section on shallow convection, based primarily on the vertical extent of the cloud system through the tropospheric depth. As previously stated, the thunderstorm is the best example of a deep system (extending the full depth of the troposphere), and accordingly it has received the most attention in the second part of this volume. Particular attention is given to the thunderstorm that is producing severe local weather, especially tornadoes.

The approach and passage of a thunderstorm on a warm and humid summer afternoon represents one of nature's more spectacular events, especially when observed in close detail. The formation and development of congestive cumulus clouds into a thunderstorm anvil, the arrival of cooler gusty winds from the outflow at the bottom of the cloud, and the subsequent rainfall are nearly always welcomed events. The thunderstorm can bring needed relief on a sultry day, as well as often provide much needed rainfall

195

E. M. Agee and T. Asai (eds.), Cloud Dynamics, 195–232.
Copyright © 1982 by D. Reidel Publishing Company.

for farmland, crops and reservoirs that support water supplies
to both urban and rural areas. According to Circular N of the
National Weather Service (USA), a rainshower becomes classified
as a *thunderstorm* when audible thunder is heard at the observing
station. Beginning with this simple observational criterion, a
considerable amount of information about the characteristics and
behavior of thunderstorms has accumulated over the years. A
thunderstorm is labeled *severe* when one or more of the following
criteria are met: (1) surface wind speeds $\geq$ 50 knots, (2) hail
of diameter $\geq$ 3/4 inch, and (3) a funnel cloud or tornado of any
size or intensity.

Thunderstorms can usually be classified as either *air mass
thunderstorms* or as *frontal thunderstorms*, and it is the latter
class that usually has the best chance of becoming severe. An
exception to this general statement might be the thunderstorm
associated with a tropical storm or hurricane, which can also produce tornadoes. The three types of air mass thunderstorms are
convective, orographic and nocturnal. The *convective* type is
produced in rather moist, potentially unstable air masses as the
result of convective buoyancy induced by daytime surface heating.
These storms typically occur in the late afternoon during the
summertime, are relatively small and exhibit slow movement. *Orographic* thunderstorms are produced by intense daytime solar heating (especially during the summer months) and/or by the lifting
of convectively unstable air over mountainous terrain. Such
storms can become rather large and frequently are accompanied by
hail, but can often remain rather stationary producing intense
rainfall amounts and a threat of flash floods. The third type
of air mass thunderstorm is the *nocturnal* variety, which is
triggered by the effects of the diurnal oscillation of the ageo-
strophic jet and generally occurs between the hours of 2200 and
0600 LST. These storms are seen to occur most likely in the
central plain states of the USA during the summer months. These
storms may sometimes be scattered, but frequently organize as
squall lines (and often are unpredicted).

The class of *frontal* thunderstorms are those that result from
the forced lifting of moist air at the discontinuity surface between air masses. These can occur with all types of frontal systems, but are most often associated with the passage of a cold
front. However, warm fronts can be just as active as cold fronts
in helping trigger the thunderstorm event. These storms often
become severe during the spring and early summer months, and the
intense squall lines that form along strong cold fronts are a
major source of tornado-producing storms. Another frontal related
type of thunderstorm is the *pre-frontal* squall lines (or instability line) which is triggered by the advancing cold front. Gravity
waves may emanate from the frontal-push and travel ahead of the
surface cold front, passing through moist potentially unstable

air where the thunderstorm line can form.   Even though these
squall lines occur in the warm air mass sector of an extratropical
cyclone, they are still regarded as frontal-induced thunderstorms
because of their basic causative mechanism. As in the case of
the frontal squall line, the pre-frontal instability line is also
quite capable of producing severe thunderstorms and tornadoes. A
good example of such an event is the 3 April 1974 tornado outbreak (see Figure 1) where three pre-frontal squall lines developed
ahead of a rapidly moving cold front, all of which contained tornado-producing thunderstorms. This case is somewhat unusual,
however, in that some of the squall line activity was likely
attributed to the juxtaposition of the upper level jetstream and
the warm moist low level flow from the south, resulting in formation far ahead of the cold front.

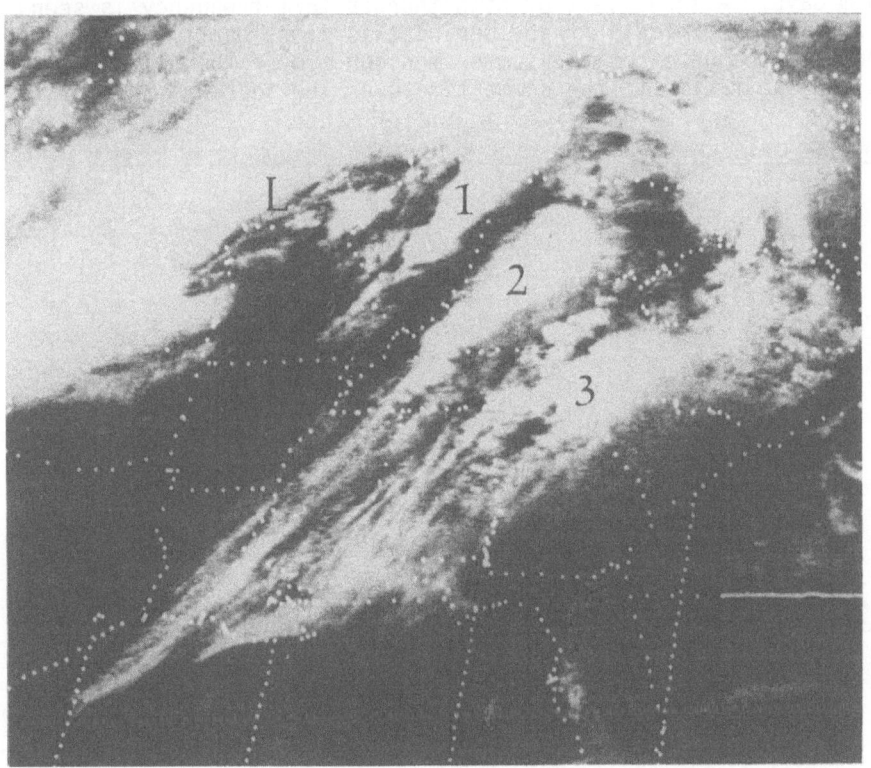

Figure 1. ATS III visible satellite imagery at 1953 GMT on 3
April 1974 showing three prominent tornado-producing squall lines
over Illinois, Indiana-Kentucky, and Tennessee-Georgia. These
instability lines developed in the warm sector of the deep low
pressure region (center at L of 982 mb).

Figure 2 shows the global distribution of *thunderstorm days*
for the month of January. As to be expected, most of the thunderstorms are occurring in the Southern Hemisphere, over continental
regions, however data are more sparse over oceanic regions. Some
regions in South America and the southern portion of Africa have
thunderstorms almost daily. Whereas a large part of the Northern
Hemisphere, being in the middle of the winter season may not have
a single thunderstorm day on the average. Figure 3 shows a similar global distribution of thunderstorm days, except it is for
the month of July. A reversal of the pattern from the previous
figure is evident, with only a few thunderstorm days in the
Southern Hemisphere, and a significant increase in thunderstorm
days over the continental regions of the Northern Hemisphere.
The onset of summer has brought increased activity, but still
much of North America and Europe-Asia have less than 10 thunderstorm days. A large belt of high thunderstorm frequency is seen
to persist over middle Africa during this time, partly in association with the onset of the summer monsoon. Over the United States,
one can see maxima in the summertime over the south-eastern states

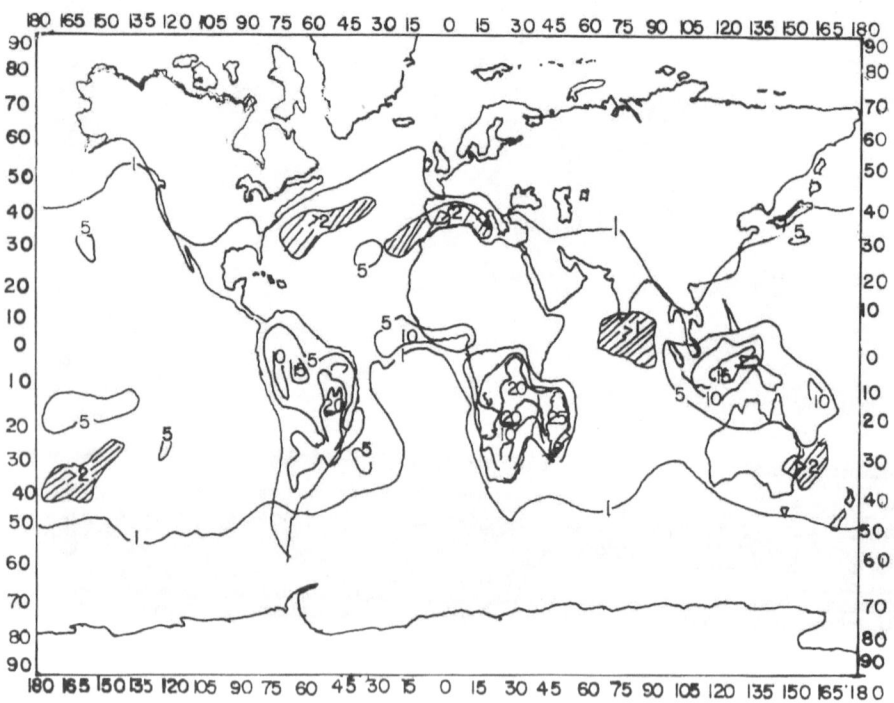

Figure 2. Average number of thunderstorm days, depicted globally,
for the month of January (taken from the Handbook of Geophysics,
United States Air Force, 1960).

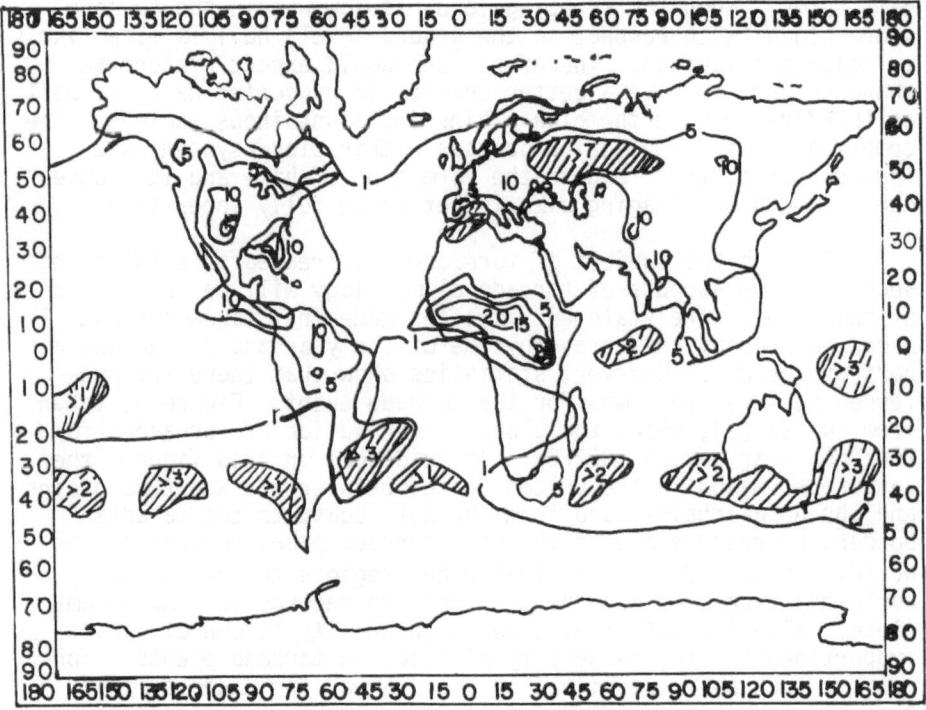

Figure 3. Average number of thunderstorm days, depicted globally, for the month of July (taken from the Handbook of Geophysics, United States Air Force, 1960).

and the Rocky Mountain region. This is attributed to the frequent occurrence of convective air mass thunderstorms and orographically-induced storms, respectively.

Gokhale [1] and Appleman [2] have compiled statistics to show the average monthly distribution of *thunderstorm days* and *hailstorm days* for the USA, including the ratio of hailstorm frequency to thunderstorm frequency in percent. These data show the preponderance of thunderstorms between the months of April and September, with the greatest frequency nationally during the month of July. The peak season for hailstorms is the month of May, but the largest percentage of thunderstorms that produce hail occurs during the winter months. The highest incidence of hail in the USA is in the Central and High Plains states, and the Rocky Mountain region. Ironically the thunderstorm maximum nationally, near Tampa, Florida, has an average of less than one day per year with hail. The city of Greeley in northeast Colorado, on the other hand, has an annual average of over seven days with

hailstorms. When the freezing level is low within the thunderstorm cloud (with respect to the ground level) hail is more likely
to reach the surface. Therefore, one would expect wintertime
thunderstorms to have a better chance for producing hail, as well
as thunderstorms in the high plains and mountainous regions. The
complicated microphysical processes within cloud systems and
thunderstorms, that lead to the formation of hail and its subsequent growth to damaging size is yet to be fully understood.

Although the subject of tornadoes is treated in a following
section, some aspects of tornado climatology will be introduced
at this time. The statement is often made that tornadoes can
occur just about anywhere, anytime of the year and during any
part of the day. However, statistics show that there are preferred regions and times for the tornado event. Figure 4, taken
from Fujita [3], shows the global distribution of tornadoes for
the four year period 1963-66. As evidenced in this figure, the
cancerous region in the United States between the Rocky Mountains
and the Appalachains, and from the Gulf Coast to the Canadian
border, represents one of the most tornado prone regions of the
world. It should be noted that other regions throughout the
world might be more prominent if tornado records were more complete. Also the effect of population density is one of direct
proportionality to the density of recorded tornado events. For
reasons such as these it can be speculated, for example, that
Australia may have many more tornado events than represented by
the results shown in Figure 4, especially since favorable synoptic
conditions should by likely during the spring months.

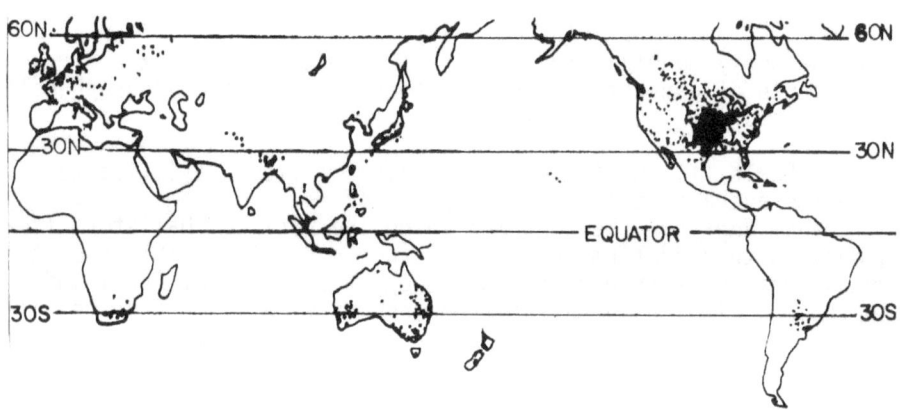

Figure 4. Global distribution of tornado events for the period
1963-66, as determined by Fujita [3].

Data tabulated by the Department of Commerce, NOAA, (USA) for the period 1953-79 show that the U.S. has an average of 722 tornadoes and 111 deaths per year, with approximately two tornado events per 100 x 100 square mile region. The monthly distribution of these events is shown in Figure 5. The first five states with the greatest annual number of tornadoes per equal area

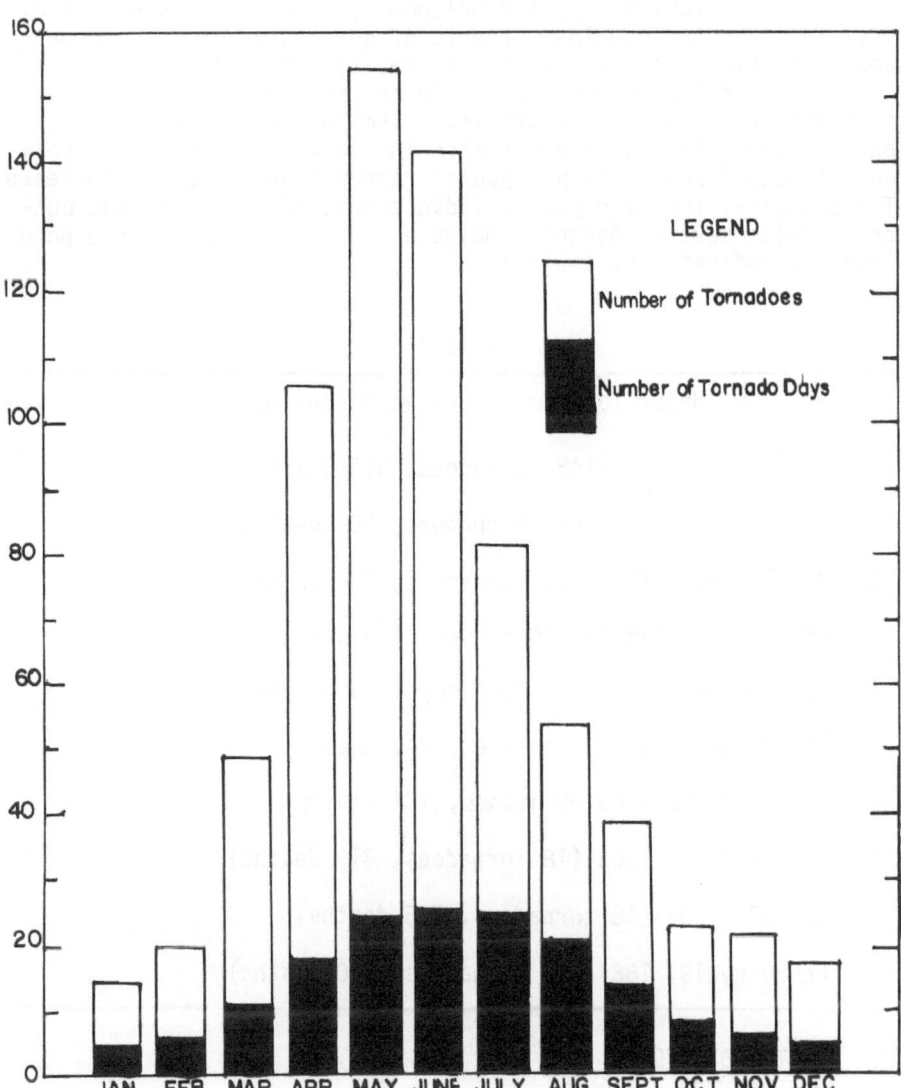

Figure 5. Average number of tornadoes and tornado days per month for the United States, based on 19,493 tornadoes that occurred from 1953-1979.

(10,000 square miles) are Oklahoma (7.69), Florida (6.93), Indiana
(6.15), Virgin Islands (5.57), and Kansas (5.83). Many of the
tornadoes recorded for Florida and the Virgin Islands, however,
are actually waterspouts (weak tornado-like vortices that form
over water in shallow cumulus cloud systems) that move on shore
and quickly dissipate. The states listed with the most tornado
deaths per year are Texas (14), Mississippi (12), Michigan (8),
Indiana (7), Alabama (7), and Oklahoma (7), which give a general
indication of the combined effects of state size, tornado density
and intensity, and the population density. A small decrease in
the number of tornado deaths during the period has been noted,
in spite of increasing population. The number of tornadoes reported during this period has also increased, but this is likely
due to better reporting procedures rather than an actual increase.
The potential for rare yet periodic events of major tornado outbreaks with several deaths remains a definite threat to the popu-
lace, as indicated by Table I.

Table 1

Major Tornado Outbreaks in the USA

1. April 3-4, 1974 (148 tornadoes, 315 deaths)

2. April 11-12, 1965 (51 tornadoes, 265 deaths)

3. March 21-22, 1952 (28 tornadoes, 204 deaths)

4. April 5-6, 1936 (17 tornadoes, 446 deaths)

5. March 21-23, 1932 (33 tornadoes, 334 deaths)

6. March 18, 1925 (7 tornadoes, 740 deaths)

7. March 13, 1913 (8 tornadoes, 181 deaths)

8. April 24-25, 1908 (18 tornadoes, 310 deaths)

9. May 27, 1896 (18 tornadoes, 306 deaths)

10. February 19, 1884 (60 tornadoes, 420 deaths)

2. THE THUNDERSTORM

During the World War II era it had become increasingly
apparent that the thunderstorm represented a severe hazard to
aviation, both to military and commercial aircraft, and that a
serious effort should be undertaken to learn more about

thunderstorms. By 2 August 1945 plans culminated within the USA to establish a Thunderstorm Project, for conducting field investigations of summertime thunderstorms in Ohio and Florida. Some highlights of scientific findings from these field programs are now briefly discussed, but a more detailed account of the Thunderstorm Project can be found in the extensive compilation by Byers and Braham [4]. The season and geographic location of the field investigations in the Thunderstorm Project indicate that this effort concentrated largely on the study of convective air mass thunderstorms, a common occurrence during the summer months in the mid-afternoon over a continental region. This would be particularly true for the majority of the Ohio storms, but to a somewhat lesser extent over central Florida where sea breeze circulations can help force the convection over land during the daytime.

Generally, the Thunderstorm Project has given us some very useful information that describes the various stages of air mass thunderstorm development and its associated features, that stands today as representative of most thunderstorms. These storms are not expected to take on severe characteristics, and may appear in their early development as a beautiful and innocent convective ensemble of cloudiness like that shown in Figure 6. Radar and other observations showed that the thunderstorm did not represent a single convection cell, but actually consisted of from three to five units of concentrated circulation existing at various times and of various strengths throughout the lifetime of the thunderstorm. One of the major findings of the Thunderstorm Project was the determination of the three characteristic stages of the thunderstorm cell, as summarized below.

The *cumulus stage* is characterized by updrafts throughout the cell, and its onset is marked by the first appearance of a radar (10 cm) echo. This echo notes the presence of numerous sizeable water droplets, which is obtained when the cloud top extends above the freezing level. The cumulus stage visibly appears as a rapid vertical development of swelling cumulii with sharp boundaries, as shown for example in Figure 6. Precipitation is not associated with the cumulus stage. The horizontal extent may be as small as 1 to 2 km at the beginning but can extend out to sizes of 10 km or more along the major axis of the developing storm. The general shape of a horizontal cross-section through the thunderstorm is elliptical, with the major axis or elongation of the storm aligned with the direction of the environmental wind at that level. The vertical extent of the cumulus stage can vary from 2-3,000 feet above the freezing level ($\sim$ 15,000 feet above the ground) to as much as 10-20,000 feet above the freezing level.

The structure of the cumulus stage is dominated by updrafts from the lowest to the highest levels. The maximum updrafts in

Figure 6.   A cumulus congestus ensemble in the process of growing
into a cumulonimbus.   Glaciation appears to be emanate.

this stage occur in the upper portion of the cloud, and can obtain
magnitudes up to 15 ms-1.   The temperature within the cell is
everywhere greater than the environment, with the magnitude of
the positive temperature anomalies reaching a maximum at the end
of the cumulus stage.   The pressure at the surface beneath the
cloud is lower than the environmental pressure, with convergence
of the horizontal wind occurring at all levels.   This kinematic
property results in the entrainment of environmental air and
mixing, as well as an exchange of momentum between the updraft
within the cloud and the adjacent relatively still air.

The quantity of visible water and the size of water particles
increases throughout the cumulus stage.   Since the hydrometeors
are in an updraft, their fall relative to the earth is slight or
even negative.   The greatest concentration of hydrometeors, liquid
and/or solid, occur at the freezing level and above.   In updrafts,
liquid water may exist several thousand feet above the freezing

level, with a gradual transition to solid hydrometeors with increasing height (determined in part by the speed of the updraft).
The duration of the cumulus stage is approximately 10 to 15 minutes, the time from the first appearance of a radar echo until
rain reaches the surface.

The arrival of rain at the ground level marks the beginning
of the *mature stage*. As water condenses during the cumulus stage
the drops and ice crystals become more numerous and increase in
size. As the mass of the particles increases, their upward motion may no longer be sustained and they start to fall relative
to the earth and the surrounding updraft. The drag on the air
by the precipitation particles induces a *downdraft*, aided in part
by evaporative cooling, which starts in the middle and lower
levels of the cell. Rain at the surface signifies a downdraft
aloft, which spreads in both vertical and horizontal extent with
time. At the same time the updraft continues, which increases
in speed with altitude through the lower 25,000 feet of the storm.
The strongest updraft in the thunderstorm occurs early in this
stage and may achieve values of 20 to 30 ms^{-1}. The strongest
downdraft may only be one-half of this value, and occurs at a
much lower level in the cloud. The vertical motion of the downdraft is transformed into horizontal motion at the earth's surface, resulting in the characteristic gusty current flowing outward from the area of rainfall. This outflow is represented by
strong horizontal divergence in the surface wind field, and the
pressure field is anomalously high. The advancing edge of the
horizontal outflow is marked by a sharp discontinuity (a micro-
cold front or gust front), where the wind changes direction, is
gusty and increases in speed, the pressure rises rapidly and the
temperature drops sharply. The coexistence of an updraft and
downdraft in the thunderstorm cell, results in relatively warm,
low pressure air converging into the front of the storm and ascending through the cloud base into the updraft region. In close
proximity is the downdraft air descending through the cloud base,
resulting in relatively cold, high pressure air spreading over the
surface throughout the region of precipitation. The passage of a
mature thunderstorm cell would result in a sinusoidal change of
the synoptic scale surface pressure tendency, characterized by a
fall ($\sim$ 1 mb), then a sharp rise with the passage of the gust
front ($\sim$ 2 mb) and then a general decline ($\sim$ 1 mb) back to the
synoptic scale tendency.

The temperature within the cloud during the mature stage is
anomalously warm and cold in the updraft and downdraft regions,
respectively. The largest positive anomaly is near the middle
level of the storm, with the greatest negative departure in the
boundary layer (actually beneath cloud base). Turbulence reaches
its maximum intensity in this stage and is strongest in the regions of maximum updraft and downdraft. The vertical extent of

the cloud can approach the 40,000 foot level; with rain at the
lower levels, mixed rain and sonw at the middle and upper levels,
and primarily snow at the highest level. Small hail can occur
during this stage, but it is not found in every storm. Liquid
water in some cases may be carried well above the freezing level
in the region of unusually strong and localized updrafts.

Heaviest surface rain is coincident with the area of maximum
divergence in the surface winds, but preceeds the maximum divergence since the rain is falling with respect to the downdraft.
As the rainfall continues, the downdraft extends over the entire
cell at the lower levels, which signifies the end of the mature
stage. The duration of this stage varies from 15 to 30 minutes,
and represents the time that the storm is most active and intense
in essentially all respects.

The *dissipative stage* begins when the downdraft and outflow
cover the entire lower region of the storm. This is to be expected since the updraft has been eliminated, which is the "life-
blood" of any thunderstorm. As the dissipative stage continues
the downdraft overspreads the entire area of the cell at successively higher and higher altitudes. With this process the amount
of liquid water being released is less and less, since no moisture
laden updrafts are entering the storm's base. Thus the mass of
water available to sustain the downdraft is reduced and the vertical motion eventually ceases, as well as the precipitation. With
the abatement of rainfall and downdrafts, the velocity divergence
at the surface also dissipates. At the end of the dissipative
stage no distortion of the surface wind field is attributable to
the storm. The air brought down to the surface and spread out is
now cooler than the environment, but eventually it is warmed by
mixing due to the prevailing synoptic scale flow. Turbulence
during the dissipating stage goes from moderate to light and
eventually ceases. For these kind of reasons the duration of the
dissipative stage is rather indefinite. Intermittent light rain
can continue for an hour or more, falling from layers of stratus
formed by the degeneration of the thunderstorm. From the beginning of the dissipative stage to the cessation of significant
downward vertical motion, the average time is approximately 30
minutes.

The success of the Thunderstorm Project seemed to breed more
success in the understanding of thunderstorm systems, and the
prediction of their behavior and features. An historic event
took place at Tinker Air Force Base in Oklahoma on 25 March 1948,
in that Captain Robert Miller and Major E. W. Fawbush, together
with their team of weather analysts, made the first successful
prediction of a tornado-producing thunderstorm. The detailed
account of events that led to this prediction are given by Newton
et al. [5], which centers around the tornado event at Tinker Air

Base five days earlier (March 20th) and the subsequent events
leading up to the successful forecast. To forecast another tornado at the same location five days later certainly was demonstra-
tive of the weather team's confidence in their analysis techniques
and interpretations. Their effort was aided largely be previous
work by Showalter and Fulks [6], but emphasized the importance
of a high-speed mid-tropospheric flow as a necessary condition
for the development of severe and tornadic thunderstorms. Aircraft reports of strong localized jets on days of severe local
storms were helpful in calling attention to this feature in the
synoptic-scale wind field. In addition to this pioneering work
(see Fawbush and Miller [7], Fawbush [8] and Miller [9]) successful post-war efforts were also being made in Germany. One of the
notable contributions was the thunderstorm model proposed by
Wichmann [10]. This model was based upon a wealth of information
provided by German sailplane pilots that dared to soar inside of
thunderstorm systems. Their observations provided useful information on the vertical velocity field (w) and the location of
very strong and narrow or localized "chimneys" of updraft, sometimes hardly the width of a glider's turning circle ($\sim$ 200 m).

Such features were observed to be characteristic of thunderstorms
that produced large hailstones. Typically, w was 2 to 6 ms^{-1} at
cloud base and often exceeded 30 ms^{-1} at heights of 8 km within
the updraft "chimneys". Wichmann's model recognized the presence
of strong vertical shear in the horizontal wind field and the resulting asymmetry in the thunderstorm system and tilt in the updraft chimney. Such an asymmetry can allow for the deposition
of hail and other hydrometeors outside the region of strong
local updraft. Later work by Dessens [11] also showed that a
very strong wind or jet stream (> 25 ms^{-1}) in the upper troposphere always accompanied severe hailstorms, and he proposed a
model expressing the idea that the updraft chimney does not "draw"
well unless a strong horizontal draft blows over it at high levels.

The above comments on observed features of thunderstorms
serve to illustrate the fine line of distinction between a discussion of severe and non-severe thunderstorm events, pursued in
more detail below. However, one additional concept as to how a
thunderstorm interacts dynamically with its environment will be
introduced at this point, even though the implication of the results apply largely to the severe and long-lasting thunderstorm
event. There is common recognition among meteorologists today
that the cumulonimbus acts as a semi-barrier to the environmental
wind flow. Small cumulus clouds are seen to tilt with the wind
shear (and even break apart); however the cumulonimbus tends to
stand more erect in defiance of the wind shear. Aircraft observations of the environmental winds in close proximity to thunderstorms have shown how the windfield can actually split and move
around the storm, in a manner somewhat similar to a river flowing

around the pillar of a bridge. Given that the storm splits the
environmental flow, one can then consider the possible effects
of induced dynamic pressures and the corresponding changes in the
vertical accelerations. Newton and Newton [12] were the first
to examine the dynamical implication of this phenomenon on the
vertical motion field in the vicinity of the storm. This and the
subsequent work by Newton [13], [14] are now briefly discussed.

The characteristic relative inflow and outflow for a typical
(USA) midwest thunderstorm is conducive for new convective growth
on the right flank of the storm. The principal updraft core of
the cumulonimbus (viewed as cylindrical in shape) has been idealized to serve as the main blocking element to the environmental
flow. In fact this updraft can actually tilt against the horizontal flow as one moves up in the vertical direction. Low level
flow from the south and upper level flow from the west, are conducive to the establishment of vertical gradients of dynamic
pressure, especially for conditions of subcritical Reynolds flow.
Newton argues that if the dynamic pressure decreases rapidly
enough with height, air could be lifted even though it could be
cooler than its environment and thus negatively buoyant. It is
important to note that whatever the shape of the blocking element,
positive dynamic pressure exists near the forward stagnation
point and negative dynamic pressures exist on both flanks. Thus,
the directional changes of wind with height would tend to establish vertical gradients of dynamic pressure. By using the vertical component equation of motion and appropriately partitioning
the pressure field, Newton showed that both buoyancy and dynamic
pressure effects could contribute to vertical accelerations. In
fact it is conceivable that (under favorable conditions) positive
dynamic pressure accelerations could even override the contribution by negative buoyancy. This may be particularly true at times
along the right flank of existing thunderstorms, where the relative low level winds induce positive dynamic pressure that is
beneath an upper level negative dynamic pressure. An important
kinematic feature of the wind field that would allow for this to
occur, is that the wind typically veers or turns clockwise with
height in the pre-storm environmental flow.

In summary, three important features of the environmental
flow and thunderstorm structure have been identified, which will
be considered further in the section to follow on severe thunderstorms. Firstly, is the strong local updraft (SLU) in the storm
and its persistence as a key feature to the production of severe
local weather. Secondly, is the occurrence of a strong mid and
upper troposphere jet stream in the synoptic   scale flow, which
provides strong magnitude shear and thus ample horizontal vorticity for converting into vertical vorticity. Thirdly, is the
veering of the winds with height, especially between the 850 mb
and the 300 mb levels, and the effect that this can have upon the

dynamic pressure field and the steering of individual storm cells as well as the growth of new convective regions on the right flank.

3. THE SEVERE THUNDERSTORM

As a result of the Thunderstorm Project and the subsequent work by several other investigators, the characteristics of multi-cellular air mass storms and the effects of their presence in a sheared environmental flow at middle latitudes were largely known. An aspect of these findings, as pointed out by Browning and Ludlam [15], addresses the influence of cell propagation on movement of the thunderstorm. Specifically in multi-cellular storms each subsequent cell development occurs to the right of the previous cells giving an effective right-hand propagation of the thunderstorm echo-mass. Earlier work had drawn specific attention to such behavior in certain rainstorms, which propagated to the right of the winds in the lower and middle troposphere. Subsequently, however, evidence began to accumulate to show that large, intense and persistent *individual storm cells* could also propagate to the right of the environmental winds and virtually at all levels of the troposphere. This result (discussed in more detail later) meant that strong single-celled thunderstorms could truly deviate to the right of the environmental flow as an entity, rather than through an "artificial" propagation of the thunderstorm echo mass due to the sequencing of cell dissipation and generation. The timely occurrence and field monitoring of a severe thunderstorm at Wokingham, England, on 9 July 1959 represented another milestone in the study of thunderstorm events, and this was largely responsible for the Browning and Ludlam [16] severe thunderstorm model briefly discussed below. The Wokingham storm had a translational speed of 35 knots and a direction of advance nearly 15° to the right of the predominant direction of the tropospheric winds. Hail fell nearly continuously along the storm's 130 mile long path and often exceeded one inch in diameter.

3.1 The Wokingham Storm.

On the morning of 9 July 1959, a cold front extending from the central North Sea across southeast England and northwest France was moving slowly to the east. This frontal zone was characterized by a strong horizontal temperature gradient and was accompanied by a horizontal wind field possessing a strong vertical wind shear in association with an upper tropospheric jet. The wind speeds associated with this jet went from 15 knots at 6,000 feet to 35 knots at 13,000 feet, with a maximum of nearly 70 knots near the tropopause at 37,000 feet. At about 0800 BST the Wokingham storm had left the coast of France to cross the English channel, and subsequently traveled within the cold front

zone across southeast England. Most fortuitously this storm was
to come under the observation of five meteorological radars based
near Dunstable, England. Six other thunderstorms came under
radar survey during the course of the day, but none were as persistent and as intense as the Wokingham storm.

The detailed radar echo structure observed in this storm
provided considerable new information on the characteristics of
severe thunderstorms, which remain today as important and recog-
nizable features in almost all mid-latitude severe thunderstorms.
Figure 7 shows this characteristic structure, as determined in
the Wokingham storm, with its *echo free vault, wall, forward overhang, anvil,* and *penetrative convective dome.*

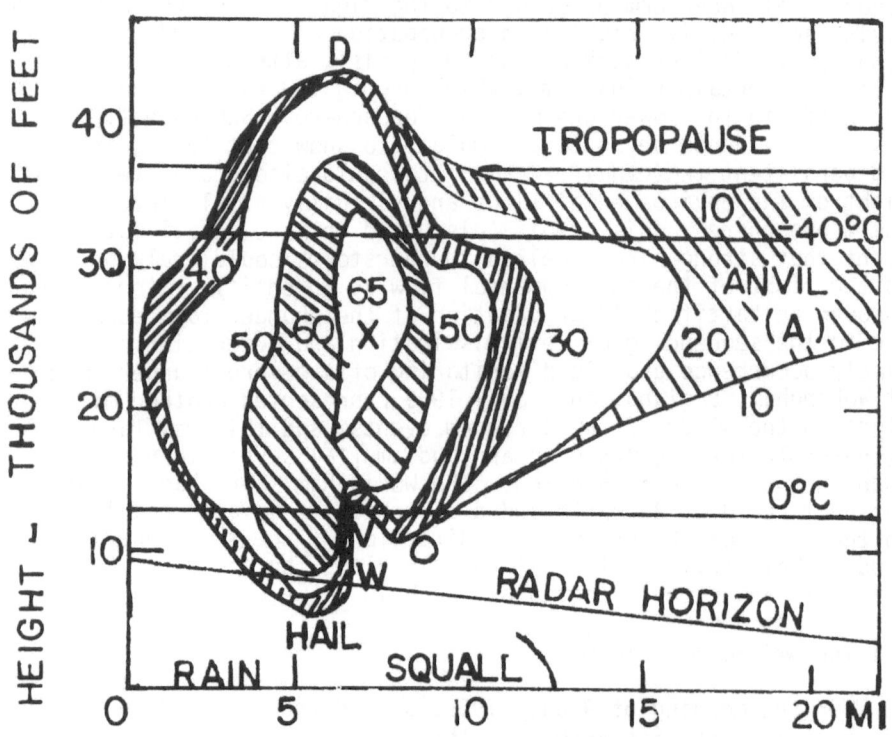

Figure 7. Radar echo structure in the Wokingham storm at its
most intense phase, shown on the range-height cross-section from
the 4.7 cm radar. Isopleths drawn express reflectivity as 10 log
$Z_e(mm^6m^{-3})$. Features noted are the anvil (A), the penetrative convective dome (D), the forward overhang (O), the echo free vault
(V), the wall (W), and the highest echo intensity (X). (After
Browning and Ludlam, [15]. Also, see Ludlam [16].)

The *echo free vault* (V), as shown in Figure 7, identifies the zone in which a strong local updraft enters the front of the storm and prevents the fall of hydrometeors into the region (resulting in low echo intensity cloud material). This feature has been detected in numerous subsequent severe storm events, and identifies the presence of a strong local updraft (SLU). The SLU is recognized as a suspect feature in thunderstorms in identifying their likelihood for producing severe local weather. More recent work by Chisholm [17] had identified a characteristic "weak echo range" in hailstorms over Alberta, Canada, but this feature can be interpreted as an indication of the same phenomenon as the echo free vault; namely, the SLU. The detection of the vault in the Wokingham storm was a major original finding in the field investigation of thunderstorm events.

The *wall* (W) is depicted in the central and right-hand parts of the echo mass, and is associated with the leading precipitation near the ground (as determined from ground observations). This precipitation consists largely of hail, with a sharply defined upright front face (the wall) that is perpendicular to the direction of the storm's motion. As shown in Figure 7, the wall is directly beneath the highest echo top (D) and about one mile behind the maximum echo intensity (X), which is centered at about 25,000 feet with a dbz value of 65. The wall aslo marks a region of strong gradients of echo intensity.

The *forward overhang* (O), induced in part by the presence of a SLU and vault, is shown to extend down to a level of 12,000 feet and is located about 1-3 miles ahead of the wall with a parallel orientation. Between the forward overhang and the wall, the base of the echo rises again to 15,000 feet, but in some cases this base has been reported to be as high as 43,000 feet (e.g. see Browning [18]).

The *anvil* (A) represents a diffuse echo region extending forward of the storm and should correspond to the visible anvil cloud, which can be frequently seen more than 50 miles ahead of the thunderstorm event at ground level. The anvil represents a downstream transport of cloud material and hydrometeors that is carried along by the upper tropospheric jet. Also, the top of the anvil rides along at the base of the stable stratosphere ($\sim$ 37,000 feet in the case of the Wokingham storm).

The *penetrative dome* (D) depicted in Figure 7 is an indication of the region of most intense convection, which has resulted from a strong local concentration of upward momentum capable of penetrating the strong negatively buoyant stratosphere. Essentially, one can view the SLU carrying hydrometeors above the level of zero buoyancy and suspending them in that region until growth to a critical mass is achieved. As already stated, the highest echo top D ($\sim$ 44,000 feet) is directly above the wall of precipitation, which is mostly hail. Also, in close proximity to this is the echo free vault, slightly ahead of the wall and essentially beneath the highest echo intensity (X). All of the features (A, D, O, V, W, X) observed in the Wokingham storm are improtant and unique features in large single-cell and long-lived thunderstorms that are capable of producing severe weather. These and other findings in this study had an immediate impact on efforts to model the severe thunderstorm.

3.2 The Browning-Ludlam Model.

The three-dimensional model of the airflow within the Wokingham storm was presented by Browning and Ludlam [19]. This model was particularly unique and informative at that time because it displayed in a physically logical manner those features believed to be characteristic of a severe single-cell thunderstorm at middle latitudes. Many features of this model have been confirmed on many occasions since the observations were made in the Wokingham storm. Notable in their model are the trajectories of particles which become large hailstones. These particles are those that fall forward into the SLU, and are subsequently carried to great heights in the anvil dome. Such large hail is deposited on the right flank of the swath of precipitation laid down by the storm as it travels along. Although no tornado accompanied the Wokingham storm, the model proposes the expected position. This location is near the interface of the updraft and downdraft on the storm, and essentially rides along the righthand boundary (or south side) of the hail swath. More specifically the tornado position proposed in this model is in the echo free vault region, but in close proximity to the downdraft and wall structure discussed above.

The Browning-Ludlam model also shows the tornado riding along the micro cold front, resulting from the outflow at the surface associated with the downdraft. This micro-frontal boundary represents essentially a vortex sheet that is capable of becoming unstable to perturbations that can wrap-up into a vortex or tornado. Browning and Ludlam expressed the view (also see Ludlam, [16]) that the tornado can form in response to a mesolow (discussed below), analagous to the genesis of the synoptic scale cyclone from a large-scale baroclinic disturbance. Their model also shows the severe thunderstorm as a semi-barrier to the environmental flow, where flow divides and moves around the main updraft core. Air is also entrained into the storm, and from the back side as well, which tends to enhance the downdraft process (through evaporative cooling that can take place). Browning and and Ludlam felt that their model would serve to stimulate the theoretical study of the severe storm, which certainly has been the case.

3.3 The Tornado Cyclone.

Discussions of the Wokingham storm and the Browning-Ludlam model were based on a storm that became severe because of its hail production. As noted in the previous definition, however, any thunderstorm that produces a tornado or funnel cloud of any proportion is also labeled severe. Interestingly, storms that do become tornadic generally produce hail stones greater than 3/4 inch in diameter. It is now appropriate to give some attention to additional features in thunderstorms that relate to the likelihood that the tornado event is eminent.

Brooks [20] was the first to introduce the term *tornado cyclone* to refer to the mesoscale low pressure system that develops within a thunderstorm (generally in the right-rear quadrant), which can be interpreted as a precursive state to the tornado event. If a tornado cyclone is sufficiently intense and persistent, it can concentrate the vorticity in a core region resulting in the tornado event. In this context all tornadic thunderstorms have tornado cyclones, but not all thunderstorms with tornado cyclones actually produce tornadoes. It bears noting again, however, that thunderstorms can produce other vortices that are often labeled as tornadoes, which are not attributable to a tornado cyclone. Vortices that form locally on a strong gust front would be a good example. Brook's observation of a tornadic thunderstorm near St. Louis, Missouri (USA) on 19 March 1948 proved to be a good example of a tornado cyclone event because of the circumstantial evidence gleaned from the analysis of the wind and pressure fields associated with the storm. At the St. Louis airport, 2.5 miles southeast of the tornado path, a Friez microbarograph registered a sudden drop of 3 mb followed by a sudden rise of 5 mb within a 15-minute period. During this period (time-centered at about 6:30 a.m. CST) the highest wind gusts at the St. Louis airport were 22 mph from the southeast, while Brooks some 2 miles southeast of the airport station observed gusts up to 50 mph. Similarly, west winds were observed on the north side of the tornado track. Brooks thus deduced a mesoscale low pressure of cyclonically rotating winds, nearly 10 miles in diameter, embedded in the thunderstorm and directly associated with this tornado event. As mentioned above, this mesolow was named the *tornado cyclone*.

It was not until 9 April 1953 that the radar signature of a tornado cyclone was detected. This milestone in severe storm detection was accomplished by the atmospheric science group at the Illinois State Water Survey, (USA), and was reported by Stout and Huff [21]. An electrical engineer, Donald A. Staggs, operating the APS-15 (3 cm) radar on this date called to the attention of the meteorologists the presence of an unusually shaped *hook-echo* structure that he unassumingly called a tornado. Later, reports

gathered proved that this feature was indeed associated with a
tornado event ($50 million in property damage and two deaths).
The sequence of six photos in Figure 8 shows the evolution of the
first documented radar hook-echo, the signature of a tornado
cyclone, as it occurred from 1755 CST to 1907 CST in east-central
Illinois.

Surface weather maps on that day indicated that a cold front
was located about 70 miles to the west of the area of tornado
formation. This front extended from a deepening low pressure
center ($\sim$ 995 mb), about 100 miles to the northwest of the tornado
touchdown. A warm front extended eastward from this low, across
northern Illinois and northern Indiana. The thunderstorm shown
in Figure 8 was located in a broken squall line ahead of the front,
and the tornado reportedly was moving to the east northeast at 45
mph. Reports from the Weather Bureau as well as eyewitness
accounts, indicated that rain and hail occurred to the immediate
north of the tornado track, but no precipitation was occurring
with the tornado. As shown in Figure 8, the apparent center of
the tornado cyclone circulation came to within 10 miles of the
radar site, a most fortuitous event.

From the early 1950's to present it has been determined by
innumerable observational studies (e.g. Agee, et al. [22]) that
there is a strong and persistent relationship between the rain and
hail swaths, and the location of the tornado track within the tornado cyclone. Nearly all thunderstorms with a well-defined tornado
cyclone, that also become tornadic, produce rain, hail and tornadoes in the proximity shown in Figure 9. Here the parent storm
is depicted moving toward the northeast. This example was characteristic of 18 tornado events in Indiana (USA) on 3 April 1974,
but has proven to be generally true in over thirty cases of tornadic thunderstorms studied by the author.

To the naked eye, the visual appearance of the core region
of the tornado cyclone would be somewhat like that shown in Figure
10. This core is typically about 2 to 4 km in diameter and is
characterized by a lower extension of the cloud base, known as the
pedestal cloud (P), within which the tornado funnel cloud (T) can
emerge, as shown. As indicated in this photo, and consistent
with the above discussion, it is common for tornadoes to occur in
the absence of rainfall, although hail frequently occurs. Even
if hail does not occur simultaneously with the tornado event, it
is nearly always present to the immediate north of the track (as
schematized in Figure 9). Of course other directions of travel
are possible, but the proximity of the hail and tornado is generally preserved as shown. The basis for this relationship becomes
more evident, when the additional work by Browning is considered.

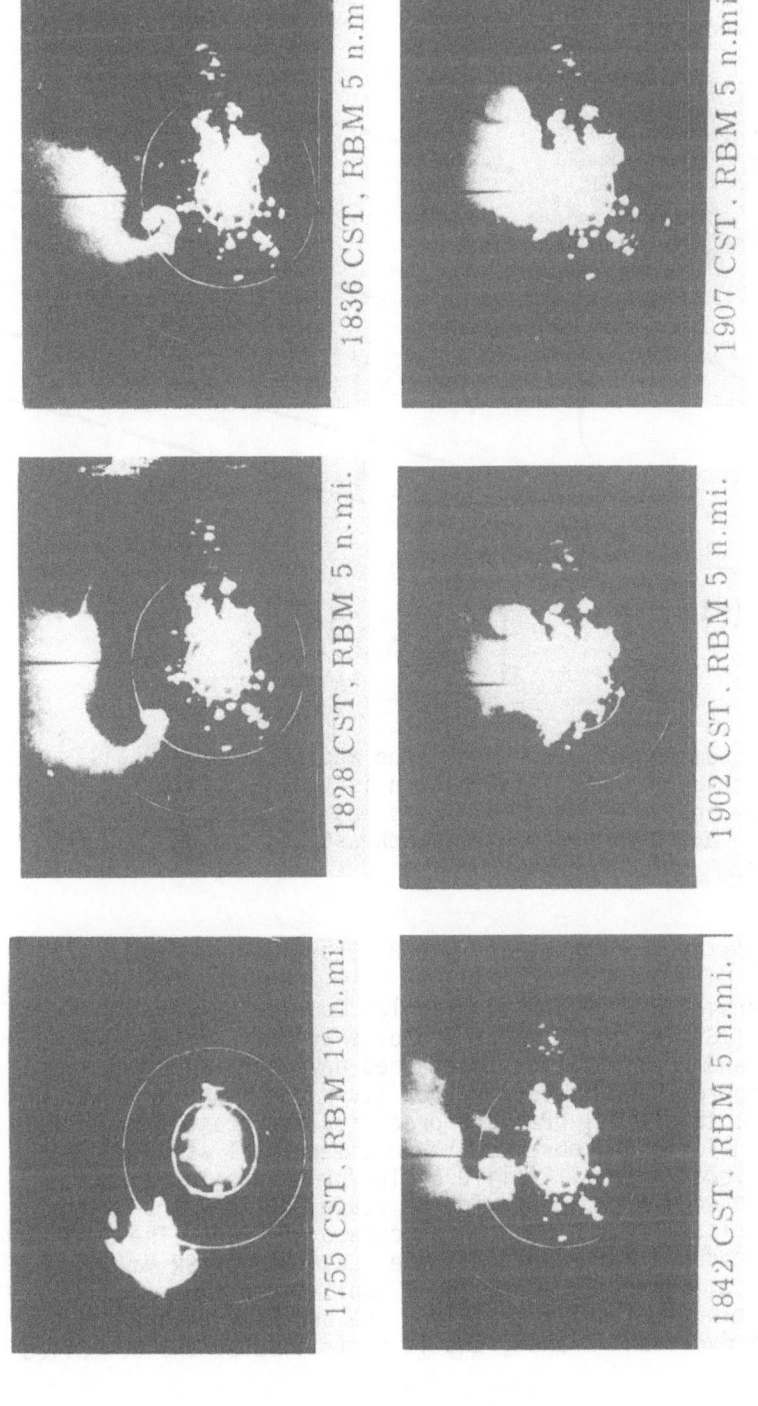

Figure 8. APS-15 (3 cm) radar echoes of a severe thunderstorm in east-central Illinois on 9 April 1953. These PPI photos with 10 and 5-mile range markers show the *first* documented detection of a *hook-echo* by radar, signifying the organized structure of the tornado cyclone (located in the right-rear quadrant of the thunderstorm). See article by Stout and Huff [21].

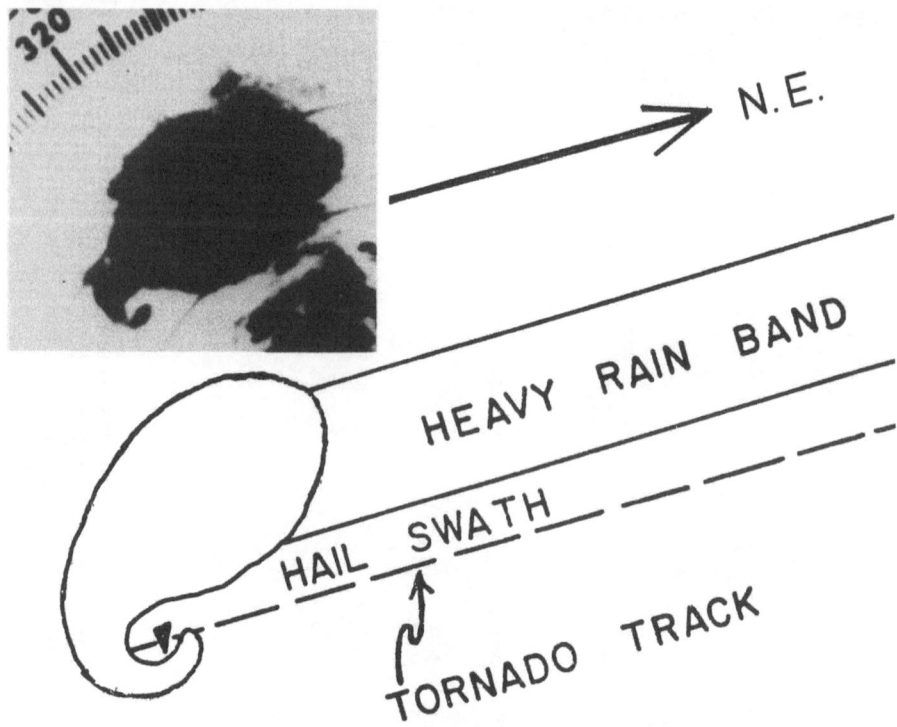

Figure 9. Schematic of radar hook echo and accompaning hail
swath and tornado track. This pattern is typical of tornadic
thunderstorms in the midwest (USA), as was the case in this
example shown on 10 cm radar over southeast Indiana at 2032 GMT,
3 April 1974.

3.4 Browning's Severe Right-Moving (SR) Storm Model.

An extension of the Browning-Ludlam model [15, 19] was made
by Browning [18] and [23], which proposes an open system where
the updraft and precipitation-maintained downdraft are continously
fed by air approaching the storm from its right flank. As discussed below, this model proposes precipitation trajectories that
are consistent with the observed three-dimensional structure of
many severe local storms. A life cycle for severe local storms
is also proposed, which adds a fourth stage to the Byers-Braham
three stage classification scheme discussed earlier. The additional stage introduced by Browning is the quasi-steady SR *Mature
stage*, where S indicates severe and R implies right-moving. This
stage is distinctively different from the ordinary mature stage
presented by Byers and Braham, and is indicative of the so-called

Figure 10. Pedestal cloud (P) within the core region of a tornado cyclone over central Indiana (USA at 2130 GMT, 20 March 1976. The pedestal usually extends below the cloud base and is precursive to the formation of the tornado funnel(T), as shown.

supercell structure with its echo free vault and hook echo features.
The SR storms propagate to the right and usually at speeds slower
than the mean wind. These storms are also believed to possess
strong cyclonic rotation, which adds to their ability to produce
severe local weather.

     As previously noted, the occurrence of severe local storms
is generally associated with winds which not only increase
strongly with height but also veer with height (see Newton, [24]).
As presented by Browning [25], the environmental airflow relative
to an SR storm for such wind shear conditions is essentially that
given in Figure 11. This shows the wind hodograph relative to
the ground and to the storm, where L, M and H represent tropospheric winds at low, middle and high levels, respectively.
Ludlam argues that the storm's motion to the right of these winds,
particularly at the lower and middle levels, is responsible for
the organization and structure, characteristic of the SR-type of
storm. The updraft and downdraft in an SR storm are maintained
as shown in Figure 11. The warm and potentially unstable air at
low levels approaches the storm from the right forward quadrant.
After ascending through the updraft, this air spreads out at the
high levels (near the base of the stratosphere) within the anvil
cloud and is carried downstream by the strong environmental winds.
The potentially coldest air that becomes a downdraft is generally
located at the middle levels (see Fawbush and Miller, [26]) and
it is usually argued that the most efficient way to make it negatively buoyant is through evaporative cooling. Even though
environmental air at the middle level is not readily entrained
into the interior of the storm (due in part to the semi-barrier
effect already noted) the SR type of storm does lend itself to an
appreciable component toward the right rear quadrant of the storm
(see Figure 11). This relatively dry air undergoes vigorous cooling and descends through the region of small precipitation descending on the downshear side of the updraft. Continuity principles
require that the downdraft air upon reaching the surface, exits
toward the left rear flank of the storm. Another schematic further
illustrating this circulation was prepared by Browning, and is
given in Figure 12. This figure also shows three different precipitation trajectories (see #1, #2, #3 also in Figure 13), the
approximate extent of the precipitation at the surface (see
hatched area), and the positions of the gust front and the tornado
(if present). The model depicted in Figure 12 is very similar to
that of Browning and Ludlam [15], except that the updraft and
downdraft in the SR storm are derived from air approaching with
strong components on the right flank. As a result of the winds
(relative to the storm) veering with height, air leaves the updraft roughly at right angles to the direction from which it
approached at low levels. Radar information suggests that this
turning with height is accomplished through a 270 degree turn in
a cyclonic sense (although this feature could be accomplished

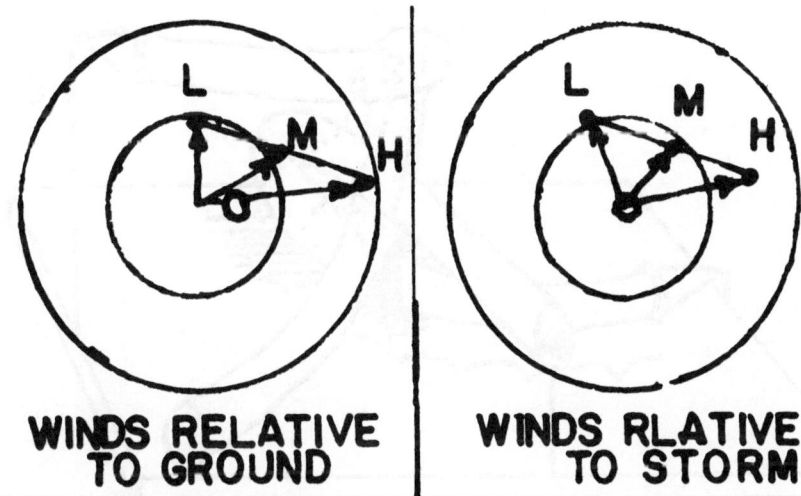

Figure 11. Browning's [25] illustration of how airflow within
an SR storm is governed by the environmental wind at low (L),
middle (M) and high (H) levels and its own velocity (denoted by
an open circle).

through a 90 degree anticyclonic turn).

    Figure 13 is intended to show the relationship between the
hook echo (a radar feature seen on the plan-position-indicator or
PPI) and the echo free vault (a radar feature most prominently
seen on the range-height-indicator or RHI). As noted by Browning,
when the SR storm is viewed at close range by conventional radar,
the following features can be noted: (1) an extensive forward
overhang, which slopes downward to the ground toward the left

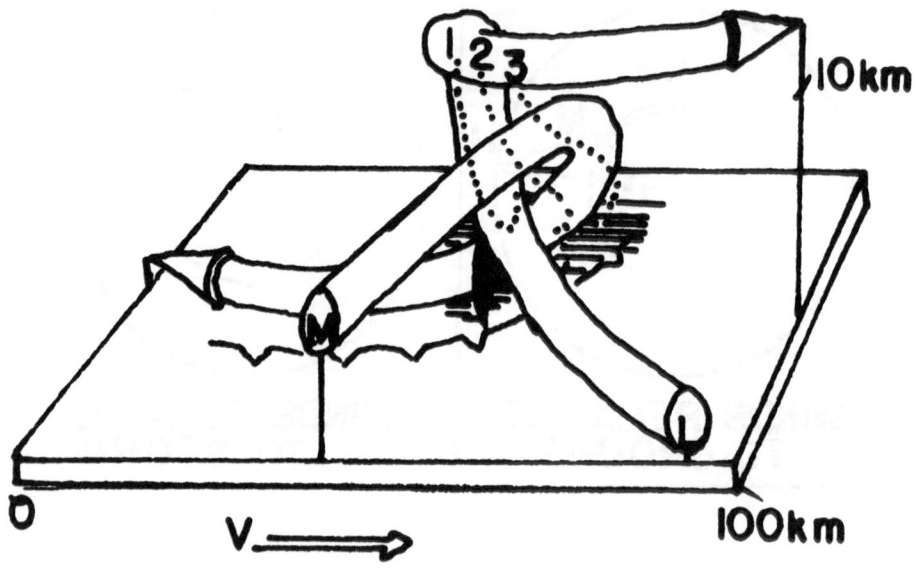

Figure 12. Three-dimensional model of airflow within an SR storm, after Browning [25]. Three precipitation trajectories are shown, as well as the position of the gust front and tornado (when present). The hatched area shows the extent of precipitation at the surface.

flank of the storm; and often with cyclonically curved streamers extending beneath it, (2) a region of low reflectivity, called the vault, which penetrates into the interior of the storm positioned beneath its highest echo top, and (3) an intense hook-shaped echo that partly surrounds the vault at low levels. As shown in Figure 13, the sense of rotation about the vault and hook echo features is cyclonic as noted by the small arrows around the periphery of the echo. The most likely position for the tornado is along the leading edge of the hook, as indicated. Hail falls from the echo surrounding the vault, but usually not from the hook end of the echo. Water droplets, however, can sometimes wrap around the center of a mature tornado cyclone manifesting a spiraling thin rain curtain surrounding the tornado located within the central core region of the tornado cyclone.

Figure 13 also shows the updraft air arriving at the top of the anvil and diverging outward in all directions, but symetrically about the line from A to B. The streamlines turn cyclonically on the south side of the anvil, with air subsequently moving downstream in the extension of the anvil. Updraft air depicted

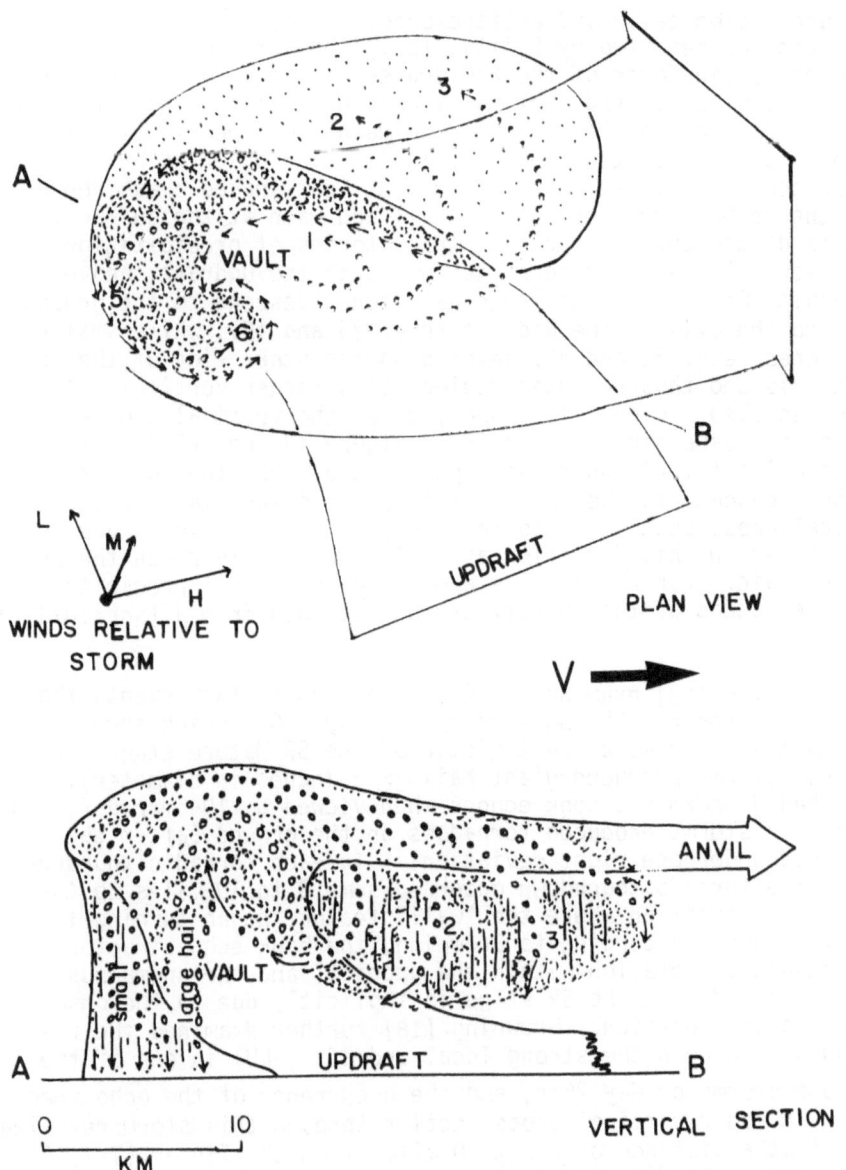

Figure 13. Schemata of horizontal cross-section (upper) and vertical cross-section (lower) as given by Browning [25] for the SR
storm model. In both cross-sections the updraft is represented
by solid curves and the precipitation trajectories by dotted
curves. The vertical section along AB is oriented in the direction of the mean wind shear.

can turn cyclonically and anticyclonically, however, as already
noted (and as reported by Fujita, [27]) the anticyclonic turn can
be ruled out in favor of the 270 degree cyclonic turn. The cyclonic turn is also favored from the viewpoint that the updraft is in-
clined upward at low levels *into* the mean tropospheric wind shear,
which is also consistent with the inclination observed by radar
in the echo free vault. The heavy and light shaded areas represent the respective positions of hail and rain at the surface.
The six dotted curves represent trajectories of precipitation
particles released from near the summit of the updraft, above
the vault, Generally, the large hailstones descend to the ground
close to the axis of the updraft (e.g. #5 and #6), while smaller
hailstones (e.g. #2 and #3) descend on the other side of the updraft axis and thus re-enter regions of stronger vertical velocity
to be recycled. Figure 13 (lower) gives the vertical cross-
section corresponding to line AB in Figure 13 (upper). Particle
trajectories #1, #2 and #3 as shown, can be recycled, but many
of these descend to the ground as they cross the plane of the
vertical cross-section (into the page). These particles are
size sorted in this direction as well, and usually reach the surface as rain. Further details about the sorting of precipitation
particles ahead of the updraft can be obtained from Hitschfield
[28].

Browning [23] examined five *severe* thunderstorm events that
occurred in central Oklahoma on the 26 May 1963, which seem to
display the features characteristic of the SR Mature stage. Each
of these storms produced giant hail ($\geq$ 2 inches in diameter),
three had discernible hook echoes when viewed by 10-cm radar, and
two of the storms produced tornadoes on the ground (after the
storm had undergone right deviation). In each tornadic thunderstorm, the tornado touchdown occurred some 8 to 10 miles to the
east of the position where the "wall" feature began turning to
form the hook. According to Browning, the hook echo structure
is a result of rotation within the updraft, and the anomalous
storm motion during the SR stage is implicitly due to the onset
of this storm rotation. Browning [18] further examined the relationship between the strong local updraft (SLU) in one of the
Oklahoma storms on May 26th, and the occurrence of the echo free
vault. A radar vertical cross-section through this storm revealed
a vault at a distance of about 20 miles from the Tinker (TKR) Air
Base radar, extending to an almost unbelievable height of *43,000
feet*. This particular case study by Browning is of further interest because it shows the apparent effect of the strong rotating
updraft. It so happened that another storm cell to the south
of this particular storm was influenced by the close proximity
of the hook echo and the echo free vault, such that a curtain of
rain was drawn northward into the tornado cyclone.

3.5 Tornadoes and Multiple Vortices

As already noted, the thunderstorm is viewed as the parent of the tornado, and a supercell storm is capable of producing several different tornadoes during a prolonged severe stage. In extreme cases, as on 3 April 1974 in the USA, a single thunderstorm produced eight separate tornadoes (commonly referred to as a *tornado family*). However, it is more typical that a tornadic thunderstorm produces only one relatively weak short-lived tornado. The scenario for tornado formation is summarized in general terms as follows: 1) favorable synoptic scale conditions lead to the formation of thunderstorms (usually as squall lines, but isolated events are possible), 2) favorable conditions also exist at the sub-synoptic scale, allowing the thunderstorm to become severe (typically a supercell storm), 3) the tornado cyclone develops as a precursive stage (although exceptions have already been noted of tornado-like vortex events, requiring no tornado cyclone), and 4) a sufficiently strong and persistent tornado cyclone maintained by the parent thunderstorm can concentrate vorticity in a localized region, resulting in the tornado formation at the central core. The vorticity is concentrated through both tilting and stretching processes, accomplished by the development of a strong local updraft in an environment that possesses ample amounts of both horizontal and vertical vorticity.

The character of the tornado vortex that forms can range over a spectrum of configurations, namely, 1) a single *laminar* vortex, 2) a single vortex undergoing *transition* from the laminar to the turbulent stage (commonly referred to as the breakdown configuration), 3) a fully-developed *turbulent* vortex, and 4) the evolution of multiple *subsidiary vortices* (or suction vortices) within the tornado, ranging from two up to possibly as many as six vortices. Figures 14, 15 and 16 show, respectively, examples of laminar, transition and multiple vortex configurations. Even though the visible laminar condensation funnel shown in Figure 14 is not fully attached to the surface, the strength of the columnar rotating wind field is evident by the presence of surface debris. Figure 15 is somewhat unique in that it shows the turbulent structure in the upper funnel as well as the helical spiral structure beneath it, a characteristic feature of the transition stage. Figure 16 shows the evolution of three subsidiary vortices that evolved from a previously large single turbulent funnel cloud. For a more thorough treatment of the vortex breakdown phenomenon and the evolution of multiple vortices see the papers by Church et al. [29], Snow [30] and Church, et al. [31].

Observational evidence collected by the author suggests that the more destructive tornado events are of the multiple vortex variety, as shown in Figure 17. The well-known Fujita intensity scale used to describe the magnitude of tornado wind speeds in

Figure 14. The formative stage of a tornado event near Marrero,
Louisiana (USA), on 21 September 1978. The laminar structure of
the condensation funnel is characteristic of a weaker tornado
event, however, debris can be noted at the surface even though
the funnel does not extend completely to the ground. This photograph is somewhat unique in that the tornado was observed to have
clockwise rotation, an uncommon event for the Northern Hemisphere.

terms of structural damage is given as follows: F0 (40 to 72 mph),
F1 (73 to 112 mph), F2 (113 to 157 mph), F3 (158 to 206 mph), F4
(207 to 260 mph), F5 (261 to 318 mph), and F6 (319 mph to Mach 1),
The tornado outbreak on 3 April 1974 in the midwestern states of
the USA resulted in 148 tornadoes, which were identified by consecutive numbers from 1 to 148. Twenty three of these events have

Figure 15. Tornado near Salina, Kansas (USA), on 25 September
1973. The funnel consists of a turbulent upper portion and a
helical spiral structure at the lower level, which is characteristic of the vortex as it undergoes transition from the laminar
to the turbulent stage.

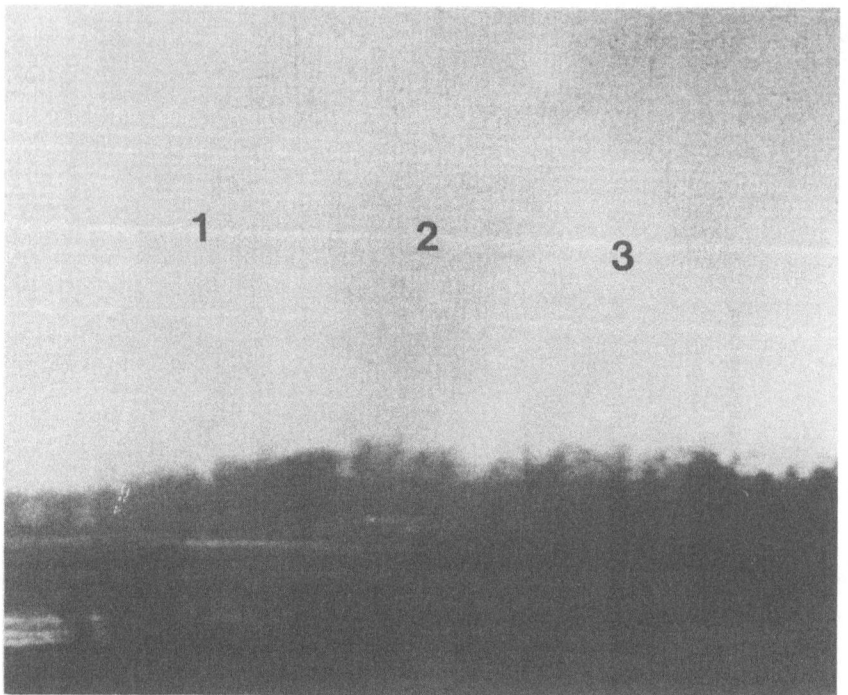

Figure 16. A multiple vortex configuration in a tornado near
Muncie, Indiana (USA), on 3 April 1974.

been categorized in Figure 17 according to their F-scale intensity.
The dichotomy in this sample, as well as subsequent events that
have been studied, strongly suggest greater severity in the multiple (suction) vortex tornado. Six tornadoes (#'s 14, 34, 45, 11,
9 and 31) of F1-F2-F3 intensity had no evidence of suction vortices,
while 17 cases of F3-F4-F5 intensity manifested multiple vortex
features. A thorough field investigation of a suction vortex
tornado event on 20 March 1976 and its relationship to F-scale
damage has been done by Agee et al. [32], which further substan-
tiates the relationship proposed here.

       Equally interesting as the multiple suction vortex phenomenon,
yet on a larger scale, is the concept of the *parallel* mode tornado
family discussed by Agee, et al. [22]. Here, it has been proposed
that the tornado cyclone can form two or more centers of action
called mini-tornado cyclones that can produce (simultaneously) two
or more different tornadoes. This is illustrated in Figure 18,
where the individual tornadoes follow the projected curtate cycloidal paths. In such cases, eyewitness reports often can see
both tornadoes that are produced by the same severe thunderstorm.

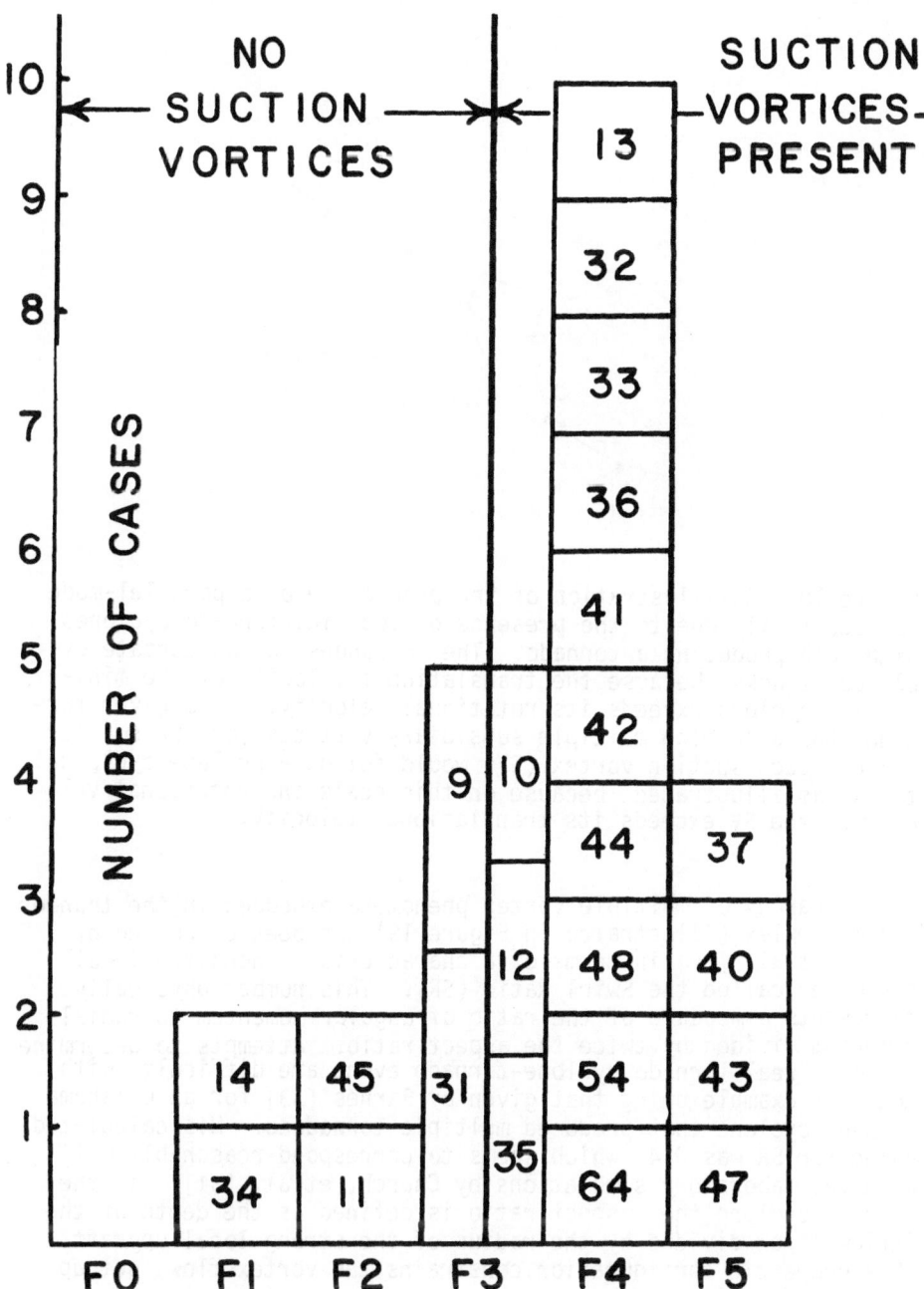

Figure 17.   USA tornadoes on 3 April 1974 (identified by number), plotted as a function of F-scale intensity.   Stronger tornadoes are characterized by the multiple vortex structure.

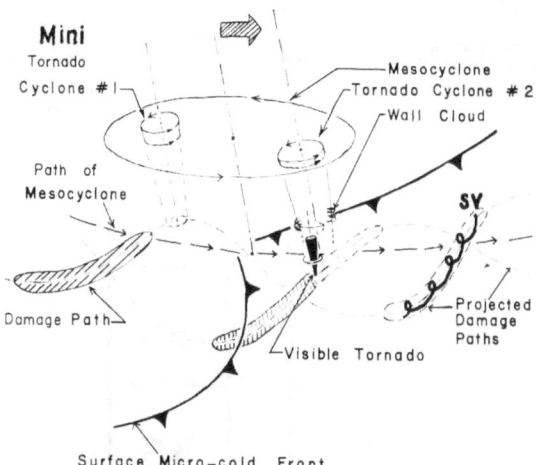

Figure 18. An illustration of the production of a parallel-mode
tornado family due to the presence of two mini-tornado cyclones,
with each producing a tornado. The tornadoes follow curtate cycloidal tracks, because the translational velocity of the mini-
tornado cyclone exceeds its rotational velocity. If a given tornado should develop multiple subsidiary vortices (on the smaller
scale), each suction vortex (SV) would follow a prolate cycloidal
track, as illustrated, because on this scale the rotational velocity of the SV exceeds its translational velocity.

The hierarchy of multiple vortex phenomena embedded in the thunderstorm complex (illustrated in Figure 19) has been discussed by
Church et al. [29] in terms of a characteristic nondimensional
parameter called the Swirl Ratio (SR). This number physically
represents a measure of the ratio of angular momentum to radial
momentum divided by twice the aspect ratio. Attempts to determine
SR for a real tornado cyclone-tornado event are difficult, with
the best example being that given by Barnes [33] for an Oklahoma
tornado cyclone that produced multiple tornadoes. His calculated
value for SR was 0.4, which seems to correspond reasonably well
with the laboratory simulations by Church, et al. [31]. In the
tornado cyclone the aspect ratio is defined as the depth of the
inflow layer divided by the radius of the strong local updraft.
This geometric configuration constrains the vortex flow, set up
by the inflow induced by the thunderstorm updraft in the presence
of background rotation. Preliminary indications are, at least in
nature, that vortices become more intense as SR increases.

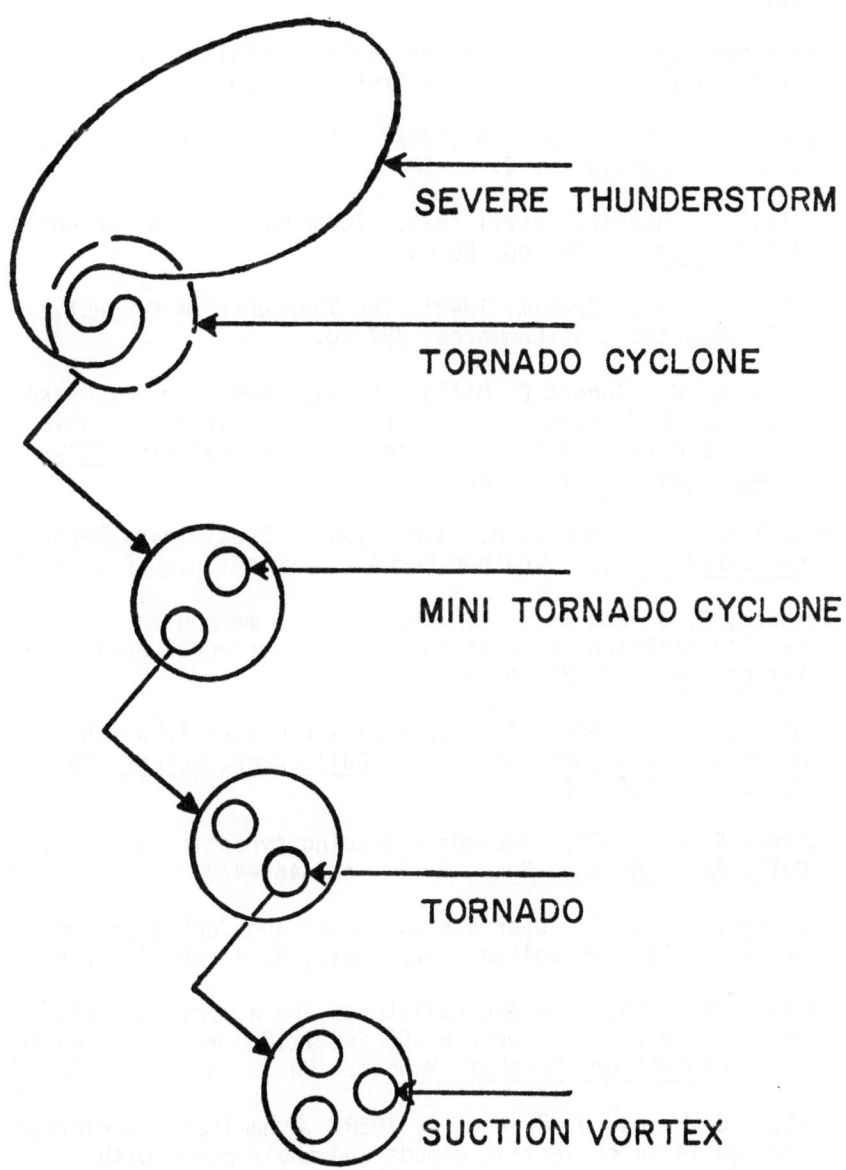

SEVERE THUNDERSTORM

TORNADO CYCLONE

MINI TORNADO CYCLONE

TORNADO

SUCTION VORTEX

Figure 19. The hierachy of multiple vortex phenomena in the
thunderstorm-tornado complex. Both theoretical and experimental
studies of vortex breakdown and the formation of a multiple vortex
structure is largely controlled by the Swirl Ratio (see study by
Church, et al. [31]).

REFERENCES

1. Hailstorms and Hailstone Growth, 1975: Gohale, Narayan R.;
   State University of New York Press, 465 pp.

2. Appleman, H. S., 1959: An investigation into the formation
   of hail. Nubila, V. 2, p. 28.

3. Fujita, T. Theodore, April 1973: Tornadoes around the world.
   Weatherwise, V. 26, pp. 56-83.

4. Byers, H. and D. Braham, 1949: The Thunderstorm Project,
   U.S. Department of Commerce, 287 pp.

5. Newton, C. W., Robert C. Miller, E. Ray Fosse, D. Ray Booker,
   and Peter McManamon, 1978: Severe thunderstorms: Their
   nature and their effects on society. Interdisciplinary
   Science Reviews, V. 3, pp. 71-85.

6. Showalter, A. K. and J. R. Fulks, 1943: Preliminary Report
   on Tornadoes, U.S. Weather Bureau, Washington, D.C., 162 pp.

7. Fawbush, E. J. and R. C. Miller, 1953: A method for forecasting hailstone size at the earth's surface. Bull. Amer.
   Meteor. Soc., V. 34, p. 235.

8. Fawbush, E. J., 1954: The types of air masses in which
   North American tornadoes form. Bull. Amer. Meteor. Soc.,
   V. 35, pp. 154-165.

9. Miller, R. C., 1959: Tornado-producing synoptic patterns.
   Bull. Amer. Meteor. Soc., V. 40, pp. 465-472.

10. Wichmann, H., 1951: Uber das Vorkommen and Verhalben des
    Hagels in Gewitterwolken. Ann. Met., V. 4, pp. 218-225.

11. Dessens, H., 1960: Severe hailstorms are associated with
    very strong winds between 6,000 and 12,000 meters. Physics
    of Precipitation, Geophys. Monogr., No. 5, p. 333.

12. Newton, C. W. and H. R. Newton, 1959: Dynamical interactions
    between large convective clouds and environment with vert-
    ical shear. Jr. Meteor., V. 16, pp. 483-496.

13. Newton, C. W., 1960: Hydrodynamic interactions with ambient
    wind field as a factor in cumulus development. Cumulus
    Dynamics, edited by C. E. Anderson, pp. 135-143.

14. Newton, C. W., 1962: Dynamics of severe convective storms.
    Nat. Sev. Storms Project No. 9, U.S. Weather Bureau.

15. Browning, K. A. and F. H. Ludlam, 1960: Radar Analysis of
     a hailstorm. Technical Note No. 5, Dept. Met., Imperial
     College London.

16. Ludlam, F. H., 1963: Severe local storms: A review. Met.
     Monogr., Amer. Meteor. Soc., Boston, 5, p. 1.

17. Chisholm, A. J., 1968: Observations by 10-cm radar of an
     Alberta hailstorm in a sheared environment. Proc. 13th
     Radar Met. Conf., Montreal, p. 82.

18. Browning, Keith A., 1965: Some inferences about the updraft
     within a severe local storm. J. Atmos. Sci., 22, 669-677.

19. Browning K. A. and F. H. Ludlam, 1962: Airflow in convective
     storms. Qrtly. Jr. Royal Meteor. Soc., 88, pp. 117-135.

20. Brooks, E. M., 1949: The tornado cyclone, Weatherwise, 2,
     32-33.

21. Stout, G. E. and F. A. Huff, 1953: Radar records Illinois
     tornado genesis. Bull. Amer. Meteor. Soc., 34, 281-284.

22. Agee, E. M., J. T. Snow and P. R. Clare, 1976: Multiple
     vortex features in the tornado cyclone and the occurence
     of tornado families. Mon. Wea. Rev., 104, 552-563.

23. Browning, K. A. 1965: The evolution of tornadic storms.
     J. Atmos. Sci., 22, 664-668.

24. Newton, C. W., 1963: Dynamics of severe convective storms.
     Severe Local Storms, Meteor. Monograph, 5, No. 27, 33-58.

25. Browning, K. A., 1964: Airflow and precipitation trajectories
     within severe local storms which travel to the right of the
     winds. Jour. Atmos. Sci., 21, pp. 634-639.

26. Fawbush E. J., and R. C. Miller, 1954: The types of air
     masses in which North American tornadoes form. Bull Amer.
     Meteor. Soc., 35, 154-165.

27. Fujita, T., 1958: Mesoanalysis of the Illinois tornadoes
     of April 9, 1953. J. Meteor., 15, 288-296.

28. Hitschfeld, W., 1960: The motion and erosion of convective
     storms in severe vertical wind shear. J. Meteor., 17,
     270-282.

29. Church, C. R., J. T. Snow and E. M. Agee, 1977: Tornado
     vortex simulation at Purdue University. Bull. Amer. Meteor.
     Soc., 58, 900-908.

30. Snow, J. T., 1978: On inertial instability as related to
     the multiple vortex phenomena. J. Atmos. Sci., 35,
     1660-1677.

31. Church, C. R., J. T. Snow, G. L. Baker, and E. M. Agee, 1979:
     Characteristics of tornado-like vortices as a function of
     swirl ratio: A laboratory investigation. J. Atmos. Sci.,
     36, 1755-1776.

32. Agee, E. M., J. T. Snow, F. S. Nickerson, P. R. Clare, C. R.
     Church, and L. A. Schaal, 1977: An observational study
     of the West Lafayette Indiana tornado of 20 March 1976.
     Mon. Wea. Rev., 105, 893-907.

33. Barnes, Stanley L., 1978: Oklahoma thunderstorms on 29-30
     April 1970, Part II: Radar observed merger of twin hook
     echoes. Mon. wea. Rev., 106, 685-696.

THE VARIABLE NATURE OF THUNDERSTORM UPDRAFTS AND PRECIPITATION

Louis J. Battan

Institute of Atmospheric Physics
The University of Arizona, Tucson 85721, U.S.A.

ABSTRACT

Observations of rain and hail falling from thunderstorms show a high degree of variability over time periods of the order of minutes. Measurements of the water loading and vertical motion profiles by means of techniques having appropriately small spacial resolutions, of the order of 100 m or less, show a high degree of variability over space scales of the order of about a kilometer or less. It is argued that the temporal variations of precipitation are related, to a significant degree, to the spacial variations in draft velocities. It is concluded that observations of the structure of thunderstorms and theoretical models of thunderstorms should seek to examine the effects on precipitation and on the storm itself of small-scale variations of vertical air motions and water content.

1. INTRODUCTION

Observations of rain and hail falling to the ground from thunderstorms show a high degree of variability in time and in space. At any one location, large variations of rainfall intensity and of the rate and character of hail occur over periods of minutes. At any instant, there often are large differences in rain and hail intensities over distances of a few hundred meters. These variations are easily observed visually during periods of thunderstorm precipitation and have been reported by various authors. The highly variable nature of showery precipitation has not been explained adequately.

E. M. Agee and T. Asai (eds.), Cloud Dynamics, 233–242.
Copyright © 1982 by D. Reidel Publishing Company.

Over the last decade, high-resolution radar observations by means of zenith-pointing radars have revealed that cumulonimbus clouds exhibit great variability over distances of a few hundred meters. These data confirm earlier observations by means of instrumented airplanes [*e.g.*, Ackerman (1)].

It is proposed in this article that large variations of radar reflectivity, and hence liquid water, and of the vertical air motions are characteristics of all thunderstorms and account, to a significant extent, for the variable nature of thunderstorm precipitation.

2. OBSERVATIONS OF PRECIPITATION

The variability of thunderstorm rainfall is so easily observed visually and recorded on rapidly responding gauges that it is taken for granted.

Comprehensive studies of the spacial variations of hail have been made by the staff of the Illinois State Water Survey. By means of hailpad data and surveys of damage to vegetation, detailed analyses have been made of hailfall patterns within a

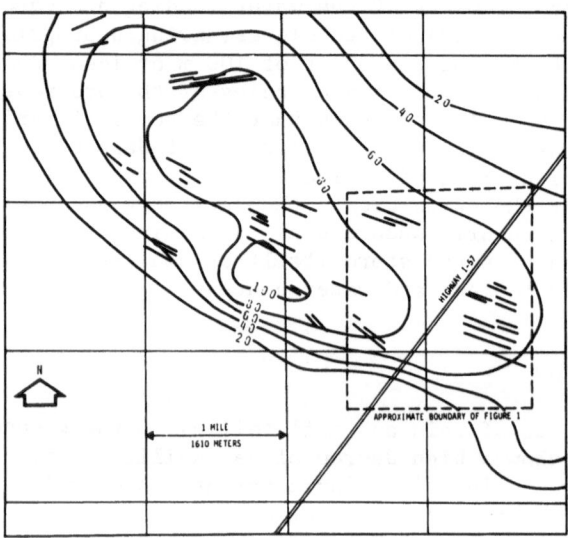

Fig. 1. Distribution of damage (total loss of yield in percent for corn) from the storm of 29 June 1976. Tracings of the striped patterns are overlaid on the damage pattern. Greatest damage is 8 km north of Champaign, Ill. From Towery and Morgan (3).

hailswath, specified by earlier investigators as areas perhaps 30 km by 150 km and sometimes represented as being nearly continuous in hail damage. But this certainly is not usually the case.

Changnon (2) proposed the term *hailstreak* to represent an area wherein the hail is nearly continuous. The typical dimensions of a hailstreak are relatively small, with 80 percent covering an area of less than 40 km^2. Most hailstreaks have widths of about 2 km and lengths of about 10 km.

Towery and Morgan (3) observed that within hailstreaks there are even smaller identifiable features that they call *hailstripes* (Fig. 1). Their typical dimensions are 20 m by 300 m. The authors speculate that the "hail distribution at the ground is a random one with maximums and minimums due to redistribution... processes...aloft" associated with turbulent wind fields.

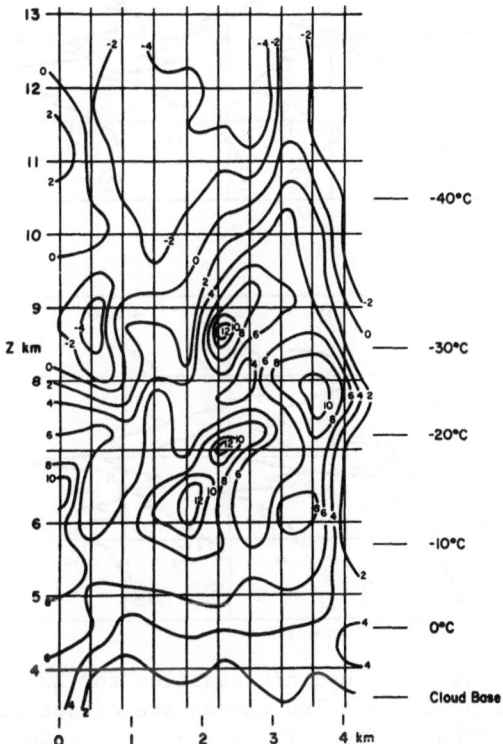

Fig. 2. Contours of updraft speed for 22 July 1972. Ordinate is height above mean sea level, and abscissa is approximate horizontal spacing. Contours are at 2 m sec^{-1} intervals. From Bushnell (4).

3. OBSERVATIONS OF SPACIAL VARIATIONS IN THUNDERSTORMS

Instrumented airplanes flying through convective clouds commonly observe temperature, liquid water content and vertical air motions or a related variable. Ackerman (1) found, from spectrum analyses, peaks in the power spectrum at wavelengths of about 300 m and 600 m. More recent measurements by investigators flying through cumulonimbus clouds also show large variations of updraft and water loading over distances of hundreds of meters.

Bushnell (4) made a series of dropsonde measurements through a thunderstorm and published the results in Figs. 2 and 3. Large

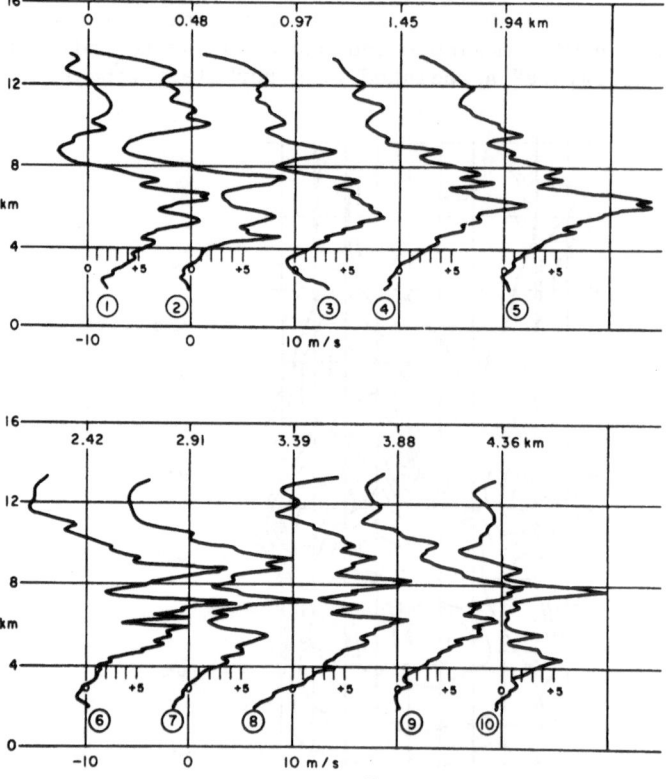

Fig. 3. Individual updraft profiles for 22 July 1972. Updraft is positive. The curves are offset 10 m sec^{-1} from each other. The partial scales show the location of zero for each sounding. The circled numbers give the order in which the dropsondes were launched. The numbers at the top give the horizontal distance of each launch from the first one. Heights are above mean sea level. From Bushnell (4).

changes of air motion were observed over distances of hundreds
of meters.

Over the last two decades, zenith-pointing doppler radars
have been making detailed measurements of the structure of
thunderstorms. Typically the vertical resolutions have been
about 150 m and the time between vertical profiles about 1 min.
Figure 4 from Battan (5) shows the patterns of updrafts and radar
reflectivity factor in a long-lasting severe hailstorm in Arizona.
Figure 5 shows similar patterns in a Colorado thunderstorm that
produced very little small hail. Vertical profiles of various
properties, at about 1542 MDT and 1546 MDT, through the storm
portrayed in Fig. 5, are displayed in Fig. 6.

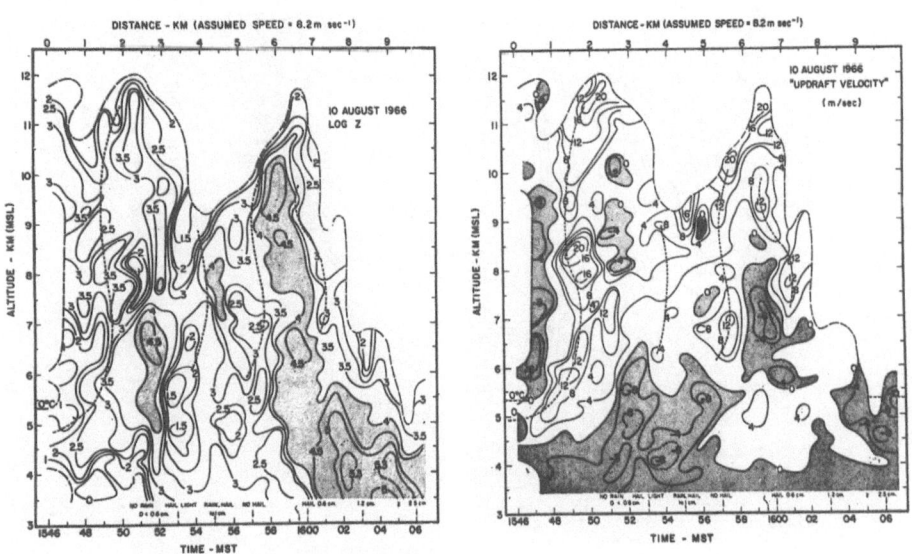

Fig. 4. Time-height distributions of radar reflectivity and
updraft velocity in a severe hailstorm in Arizona. At 1606 MST,
hailstones were about 2.5 cm in diameter. Shaded areas in the
right-hand drawing represent *downdrafts*. From Battan (5).

The data presented in Figs. 3 to 6 clearly illustrate the
highly variable character of vertical motions and water loadings
over distances of hundreds of meters and periods of minutes.

Many observations of thunderstorm properties have been made
by means of vertically or azimuthally scanning radars. Over
recent years, multi-doppler radar systems have yielded magnifi-
cent measurements of the structure of thunderstorms. But it is

important to recognize the degree to which these techniques
smooth the data. Figures 7, 8 and 9 are examples of the vertical
structure through active thunderstorms, constructed on the basis
of data obtained by means of radar techniques whose effective
resolutions are perhaps an order of magnitude or more greater
than the one that yielded Figs. 4-6.

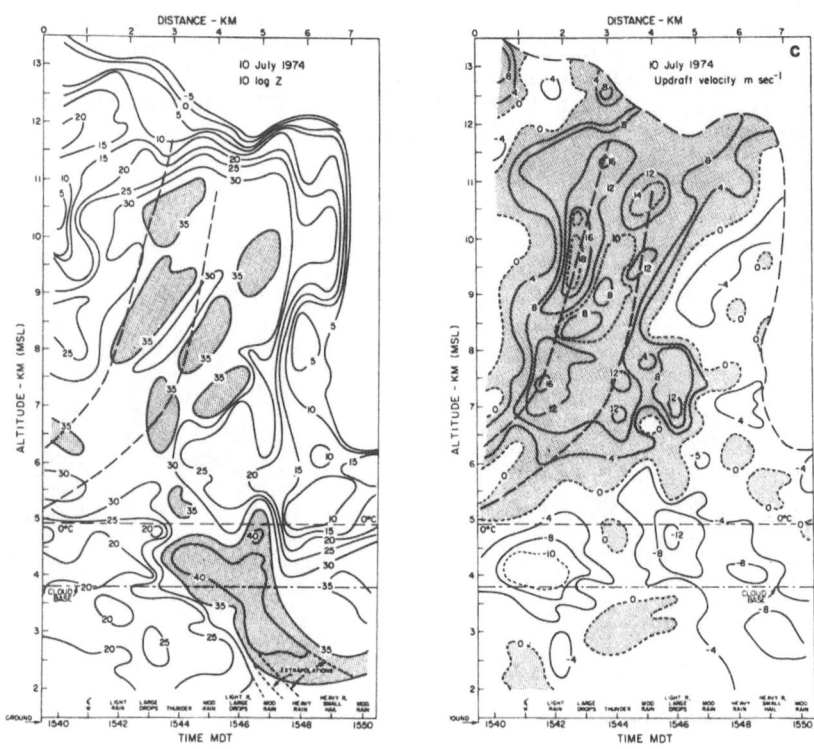

Fig. 5. Time-height distributions of radar reflectivity and updraft velocity in a storm in Colorado on
10 July 1974. Note that the shaded areas in the right-
hand drawing are regions of *updrafts*. From Battan (6).

It is clear that, in the process of producing diagrams such
as Figs. 7-9, a great deal of detail has been averaged out of
the data. Whether or not this is important depends, obviously,
on the role played by eddies, having dimensions of hundreds of
meters, in the processes under consideration.

4. SOME SPECULATIONS

It seems evident to the author that the observed temporal and spacial variations of rain and hail from thunderstorms can be explained, to a significant extent, in terms of the spacial and temporal variations of the vertical air motions. Most microphysical models used to study the evaluation of precipitation assume patterns of vertical motion that are smooth, such as those in Figs. 7 and 8, and that exhibit essentially a single updraft maximum aloft. It would be informative to examine the evolution of precipitation particles in a cloud having vertical velocity profiles resembling those shown in Figs. 4-6.

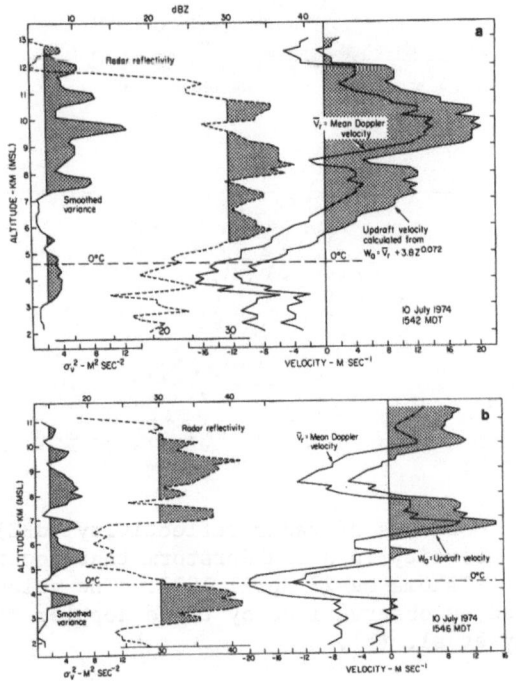

Fig. 6. Quasi-vertical profiles of various quantities at 1542 and 1546 MST in the storm shown in Fig. 5. From Battan (6).

Most dynamical models of thunderstorms also exhibit patterns of vertical motion much smoother than those in Figs. 4 and 5. The highly turbulent nature of updrafts composed of a series of eddies and the indication of a series of closely spaced updraft cores would produce a different pattern of buoyancy and pattern

of energy, moisture and momentum transfers than would be the
case in a thunderstorm whose structures resemble those shown in
Figs. 7-9.

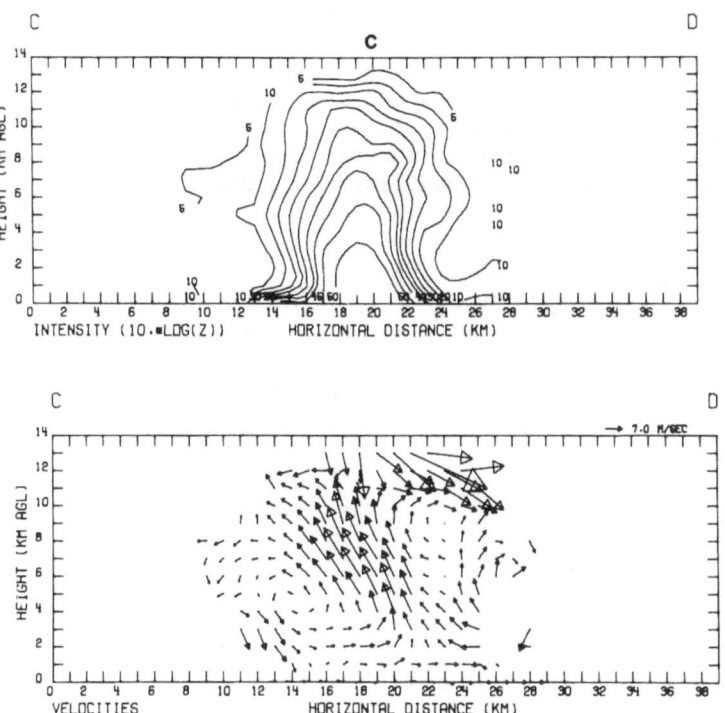

Fig. 7. Contours of radar reflectivity (dBZ) and
updraft velocity in a thunderstorm that occurred in
central Oklahoma on 29 April 1977. The velocities
are based on observations by three doppler radars.
From Ray et al. (7).

5. CONCLUSION

The observed patterns of vertical air motion in a thunderstorm obviously depend on the temporal and spacial resolutions
of the observational methods. Scanning techniques of the type
used in multi-doppler systems smooth out fluctuations having
dimensions of hundreds of meters and periods of minutes. High
resolution measurements, in a limited number of thunderstorms,
by means of vertically directed radars show that over such resolution increments there are large variations of updrafts and
water loading.

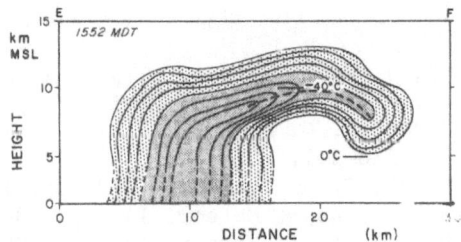

Fig. 8. (a) Pattern of radar reflectivity in a storm on 21 June 1972 near Fleming, Colorado, observed by a radar located about 95 km west of the storm. The radar beam was 1° wide. Reflectivity contours are at intervals of 5 dBZ with outer contour representing 30 dBZ. From Browning and Foote (8).

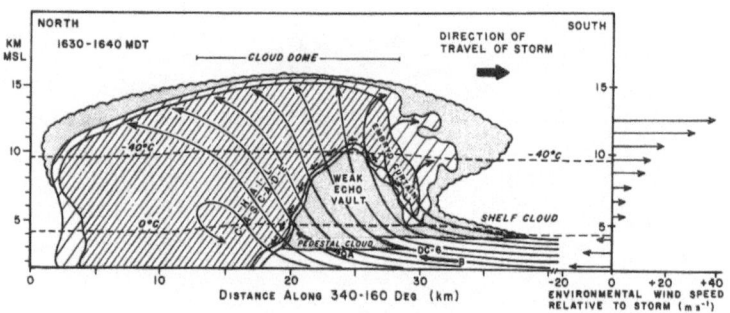

Fig. 9. Vertical section through the Fleming storm (see Fig. 8) showing the visual cloud boundaries, the pattern of radar echo (hatched) and the location of hail. The thin lines are streamlines of airflow relative to the storm. From Browning and Foote (8).

It has been said by some experts that supercell thunderstorms are, at times, in a quasi-steady state. Until such time as high resolution observations are made, the author leans towards the expectation that such storms also are composed of highly variable updrafts.

ACKNOWLEDGMENTS

This research was supported by the Atmospheric Sciences Section of the National Science Foundation under Grant No. ATM-78-10082.

REFERENCES

1. Ackerman, B.: 1967, J. Appl. Meteor. 6, pp. 61–71.
2. Changnon, S. A., Jr.: 1970, J. Atmos. Sci. 27, pp. 109–125.
3. Towery, N. G., and Morgan, G. M., Jr.: 1977, Bull. Amer. Meteor. Soc. 58, pp. 588–591.
4. Bushnell, R. H.: 1973, J. Appl. Meteor. 12, pp. 1371–1374.
5. Battan, L. J.: 1975, J. Appl. Meteor. 14, pp. 98–108.
6. Battan, L. J.: 1980, J. Appl. Meteor. 19, pp. 580–592.
7. Ray, P., Wagner, K. K., Johnson, K. W., Stephens, J. J., Bumgarner, W. C., and Mueller, E. A.: 1978, J. Appl. Meteor. 17, pp. 1201–1212.
8. Browning, K. A., and Foote, G. B.: 1976, Quart. J. Roy. Meteor. Soc. 102, pp. 499–534.

CHARACTERISTICS OF VERTICAL WIND PROFILES NEAR THE MOST SEVERE THUNDERSTORMS OF SOUTH-WESTERN FRANCE.

J. DESSENS and S. GODARD

Observatoire du Puy de Dôme et Laboratoire Associé
de Météorologie Physique
Université de Clermont-Ferrand II, FRANCE.

ABSTRACT

    South-Western France is a region frequently damaged by severe travelling hailstorms, and sometimes also by stationary rainstorms leading to flash floods. The main difference between the mesoscale meteorological conditions associated with these two types of convective storms appears to be in the vertical wind profile in the environment of the storm.

    Wind profiles are given for the most severe cases of hailstorms or rainstorms which have occured in the last thirty years, and it is shown that the vertical gradient of the vectorial mean wind between the surface and 3 to 12 km is a distinctive parameter for both kinds of storms. For hailstorms, this mean wind regularly increases with the altitude at a typical rate of $1.6 \text{ m s}^{-1} \text{ km}^{-1}$ without shear in direction ; for rainstorms, the mean wind velocity increases very slowly with the altitude above 3 km at a rate of $0.16 \text{ m s}^{-1} \text{ km}^{-1}$ which is not so regular as in the case of hailstorms because there are often wind shears in direction.

    These results have applications in the forecasting of severe thunderstorms for an experiment of hail prevention by ground seeding, and they may also have implications in cloud modelling and physical understanding.

INTRODUCTION

    Every year, from may to october, the northern region of the Pyrénées is regularly affected by one or the other of the two following types of thunderstorms :

E. M. Agee and T. Asai (eds.), Cloud Dynamics, 243–257.

Table 1   Major damaging hailstorms

N°	Day	Time	Hailpath length and width (km)	Storm velocity (m s⁻¹)	Hail diam. (mm)	Other hail-paths	Sounding Bx : Bordeaux Tse: Toulouse	m	b	r^2
1	7 May 1952		150	9		1	Bx, 12.00	0.92	11.30	.98
2	18 July 1955	18.00	80		50	1	Bx, 12.00	1.44	12.20	.97
3	9 Sept 1956	20.00	150, 5				Bx, 12.00	2.08	4.74	.95
4*	18 Aug 1958	16.00	90		60	1	Bx, 12.00	1.25	7.92	.99
5	1 June 1964	14.00	120		25	2	Tse, 12.00	2.12	3.03	.98
6	5 July 1964	15.00	45, 6				Tse, 18.00	1.27	3.51	.99
7	26 July 1964	16.00	75, 2		60		Bx, 18.00	0.91	2.45	.99
8	26 Sept 1965	16.00	100		60		Tse, 12.00	1.03	4.44	.99
9	27 Sept 1965	15.00	80		15		Tse, 12.00	1.44	6.35	.93
10	6 July 1967	18.00	90		80	2	Tse, 18.00	1.45	0.31	.95
11*	10 July 1968	18.00	120		60		Tse, 18.00	2.28	2.09	.93
12	13 Aug 1969	21.00	70	14	60	2	Bx, 24.00	1.59	7.77	.98
13	14 June 1970	20.00	65, 8		30	2	Tse, 18.00	1.44	2.60	.98
14*	16 May 1971	17.00	85, 9	11	50		Bx, 12.00	1.62	11.55	.98
15*	19 Aug 1971	17.00	140, 8	17	30	1	Tse, 18.00	2.06	6.93	.96
16	10 Aug 1972	21.00	27	14	50		Bx, 18.00	1.85	7.29	.99

17	2 May 1973	13.30	55, 6		60		Bx, 12.00	1.20	8.09	.98
18*	27 June 1973	17.30	140, 10	17	70	2	Bx, 12.00	0.59	14.20	.80
19	6 Sept 1975	14.00	25, 5		30		Tse, 06.00	2.24	1.13	.99
20	11 June 1980	18.00	90		20		Bx, 12.00	1.72	11.38	.94
21*	14 June 1980	18.00	100, 5	14	40	1	Tse, 06.00	2.83	4.03	.98
22	8 May 1981	15.30	40, 8		30		Bx, 12.00	0.96	10.19	.96
Average			92.6 6.5	13.7	46			1.56	6.52	.96

Notes :   * Extreme damage
Cases 7, 10, 19 : also wind damage
Case 11 : also rain damage
Sounding n° 3 has been done on 10 sept 1956

m, b and r^2 are the coefficients of the linear regression equation giving the mean wind velocity as a function of the altitude.

Table 2 Damaging rainstorms.

Ref	Day	Time	Area damaged, rainfall rate	Sounding	m	b	r²
A	5 June 1960	16.00	Sainte-Livrade, 111 mm/135 min.	Bx, 12.00	0.22	8.25	.80
B	8 Oct 1962	20.00	Thuir	Tse, 18.00	1.26	0.57	.95
C	1 June 1963	15.00	Haute-Garonne	Tse, 12.00	0.20	0.24	.57
D	9 June 1966	15.00	Sarrant, 109 mm/120 min.	Tse, 12.00	0.38	4.66	.90
E	14 June 1966		Monclar d'Armagnac	Tse, 12.00	0.20	2.38	.65
F	8 June 1970	17.00	Miélan, 137 mm/24 hr	Tse, 18.00	-0.20	6.89	.84
G	18 June 1972	17.00	Lannemezan	Tse, 18.00	-0.39	8.72	.52
H*	23 Aug 1973	22.00	Bonnemazon, 86 mm, 2 persons killed	Tse, 06.00	0.03	0.75	.05
I	21 Aug 1974	18.00	Segonzac, 130 mm. Pyrénées-Orientales	Bx, 12.00	0.09	4.43	.36
J	19 May 1977		Several flash floods along the Pyrénées	Bx, 12.00	0.04	13.90	.04
K*	8 July 1977		Labéjan, 225 mm, 18 persons killed	Bx, 18.00	-0.31	6.85	.98
L	15 July 1977		Hautes-Pyrénées	Tse, 06.00	-0.57	5.90	.95
M	10 June 1978		Chelle-Debat. Campistrous.	Tse, 06.00	0.32	0.96	.66
N	18 Aug 1980	19.00	Cassagne, 90 mm. Hautes-Pyrénées	Tse, 06.00	0.27	3.36	.56
O	5 July 1981	18.00	Ardiège, 85 mm/105 min.	Tse, 06.00	0.98	0.93	.98
Average					0.17	4.59	.65

Notes : * Extreme damage
Cases D, G, O : also hail damage
m, b and r² : see Table 1.

- Hailstorms (HS) with hailstones of more than 15 mm diameter, leading to intensive damage on crops ; these storms travel from south-west to north-east at a typical velocity of 14 m s^{-1} during about 4 hours for the most severe of them. The hail problem in South-Western France has initiated there in 1952 a project of hail prevention by ground seeding, silver iodide being released before the formation of the storms (1).

- Rainstorms (RS) produced by local thunderstorms moving very slowly and giving excessive rain at the same place for several hours. The consequences are "true flash floods" which are not to be mistaken with "hybrid floods" produced by larger cloud systems (2).

In order to avoid any artificial increase of precipitations, the seeding for hail prevention must not be operated when there are flood probabilities ; for that reason it is necessary to forecast distinctively both types of thunderstorms.

As shown by Dessens (3), a parameter which enables to differenciate severe HS situations from ordinary ones is the maximum wind velocity in the tropopause, "severe hailstorms being associated with very strong winds between 6,000 and 12,000 m". This observation has been confirmed in South-Western France by several case studies of damaging hailstorms from 1960 to 1981, and it is now statistically possible to define a typical wind profile of this kind of storms. We shall try to do the same about the "opposite" RS type of storms to compare HS and RS profiles.

Apart from the fact that the wind is probably the best distinctive parameter, it is interesting to support the forecasting with wind observations because wind soundings are more numerous in time than temperature-humidity ones, at least in France.

THE CASE STUDIES

For the listing of the case studies we have mainly used the annual report made by the "Association Nationale d'Etude et de Lutte contre les Fléaux Atmosphériques" (4).

22 cases of major hailstorms (or haildays) have been selected for the severity of the damage (Table 1). When several hailstorms have occured on a same day, Table 1 gives for the most damaging of them the time, the hailpath dimensions, the travelling velocity, and the hailstones diameter ; the number of other long hailpaths in the same day is also given.

15 cases of flash flood rainstorms have also been selected for the severity of the damage (Table 2) ; the floods have been

caused by localized thunderstorms, and not by excessive rains like
those generated on large areas by extratropical hurricanes. According to the classification by Crysler et al. (5), all these cases
belong to the class intitled "heavy damage" except for cases H*,K*
and B, C which respectively belong to the classes intitled "extreme"
and "light damage". For the most severe case of July 8, 1977, the
total damage has been evaluated to about one milliard of francs.

The wind data used for this study are those given by the
soundings operated four times a day at Bordeaux (since 1950) and
Toulouse (since 1962). In general the sounding selected (Tables 1
and 2) is the nearest to the storm in space (Figure 1) and time,
but there are some exceptions when the data are not complete, or

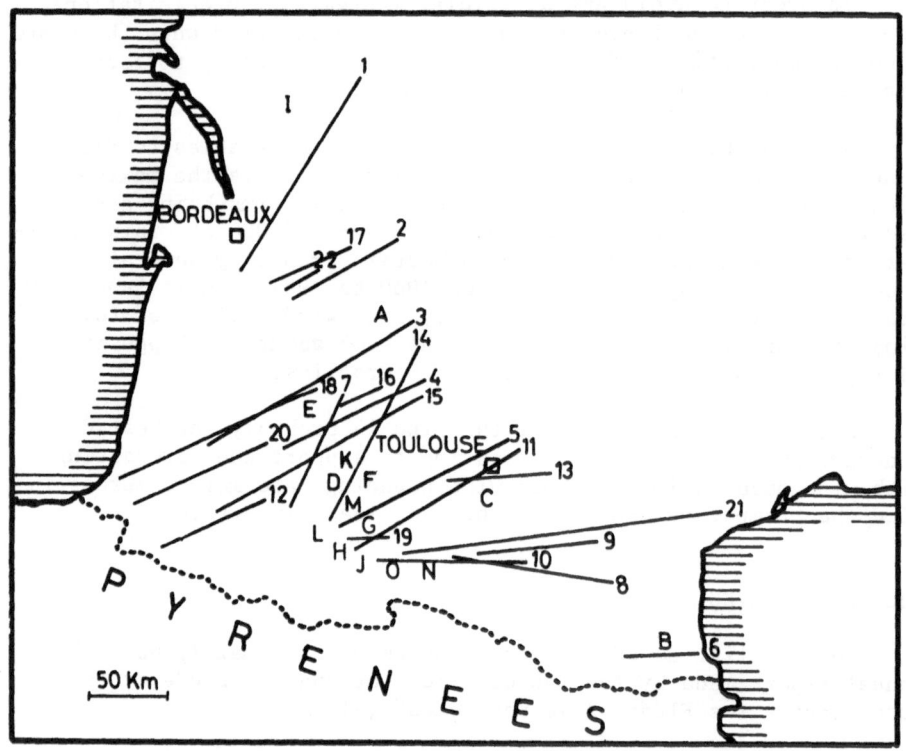

Figure 1 Geographical location of the 22 hailstorms (1 to 22) and
    the 15 rainstorms (A to O) studied.

when the meteorological analysis assigns to take a sounding which
is more representative of the atmosphere near the storm. One HS
case has been suppressed (7 sept 1972, Pyrénées Orientales) because
the nearest sounding (Toulouse, 18.00) was evidently not representative of the atmosphere around the hailstorm (directions of the
wind and of the hailstorm path were opposite); all the other cases

have been kept.

THE MAXIMAL TROPOSPHERIC WIND

The first attempts to use the wind data to forecast hail in South-Western France have concerned the maximum wind in the troposphere : Dessens (3) has shown that the frequency distributions of the maximum wind velocity for days with heavy destructive hailstorms and for control days with non damaging thunderstorms are different, and Molénat (6) has formulated a hail forecasting procedure which takes into account the direction of the maximum wind : this wind never blows from SSE to NW (via E and N) when hailstorms occur.

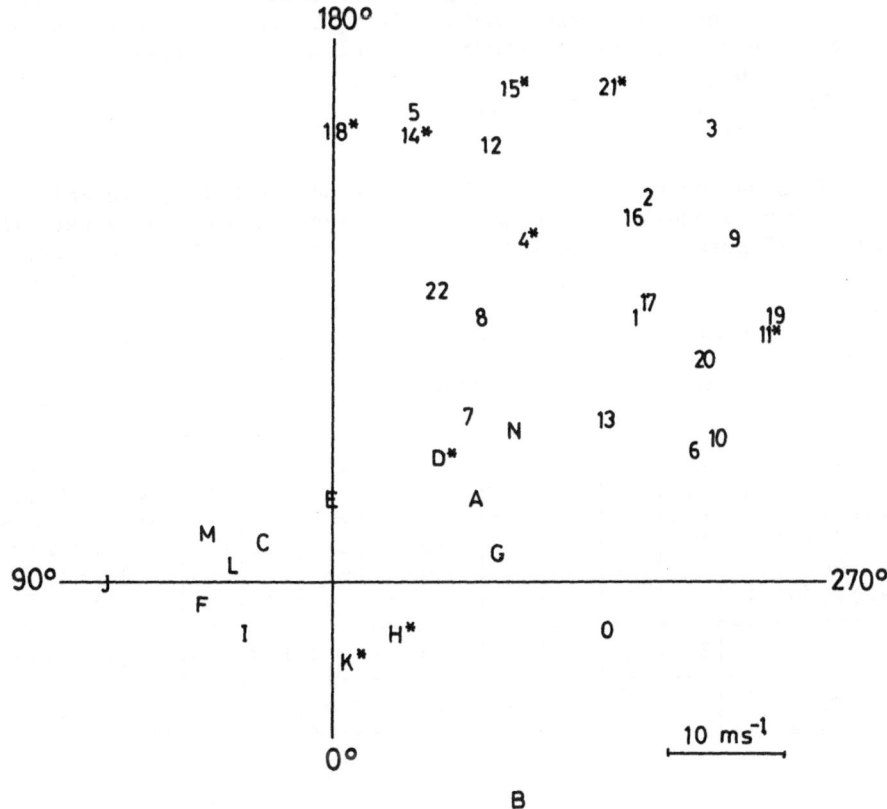

Figure 2 Extremities of the vectors representing the maximum of the wind in the troposphere near hailstorms (1 to 22) and rainstorms (A to O).

The extremity of the vector representing the maximum wind in the troposphere has been plotted on Figure 2 for the HS cases (numbers) and for the RS ones (letters). Numbers and letters are well separated, particularly for the most damaging cases. The fact that there was hail in case D is worth noting so that it is not surprising to have D in the neigbourhood of the HS area.

In the following sections, we will more exhaustively examine the wind data for the two types of thunderstorms, studying successively the shears in directions and velocities.

THE HODOGRAPH OF THE CUMULATED WIND VECTOR

Instead of the classical hodograph giving the vertical distribution of the horizontal wind, we will consider here the hodograph of the cumulated wind vector. If $\vec{v}_i$ is the wind vector at i km, this vector for the altitude h is defined as follows :

$$\vec{V}_h = \sum_{i=1}^{h} \vec{v}_i \qquad i \in \{1,2,\ldots,h\}$$

This vector has been calculated for h=1 to h=12 (h in km) for the 37 wind soundings. An example of presentation of these data is given on Figure 3. The surface wind has not been taken into account

```
   TOULOUSE          DATE: 23-08-73          HEURE: 06H00

   CALCULS DU VECTEUR VENT EFFECTUES ENTRE : 01 ET 12 Km

*****************************************************************
*        |SOL| 1 | 2 | 3 | 4 | 5 | 6 | 7 | 8 | 9 | 10| 11| 12*
*======= |===|===|===|===|===|===|===|===|===|===|===|===|===*
* VENT   |200|310|120|110|120|120|150|290|290|230|230|220|210*
*  AU    |---|---|---|---|---|---|---|---|---|---|---|---|---*
*NIVEAU  | 2 | 7 | 4 | 3 | 4 | 3 | 1 | 3 | 4 | 5 | 4 | 5 | 5 *
*=========================================================*
* VENT   |   |310|323| 32| 97|106|111|112|119|218|224|222|219*
*        |---|---|---|---|---|---|---|---|---|---|---|---|---*
*CUMULE  |   | 7 | 3 | 2 | 4 | 7 | 8 | 5 | 1 | 5 | 9 | 14| 19*
*****************************************************************
```

Figure 3 Example of presentation of the wind sounding data (above) and of the mean wind calculated (below). Rainstorm case H[*].

in the determination of the mean wind because when it is measured at Bordeaux, near the Atlantic Ocean, or Toulouse, in the Garonne

valley, it is generally not well representative of the surface
wind in the proximity of the storm.

The $\vec{V}_h$ hodograph represents the horizontal projection of an
air parcel ascending with constant velocity. Differently from the
$\vec{v}_i$ hodograph, this one gives a smoothed representation of the wind
aloft, which is more convenient for the study of several cases together.

Figure 4 and Figure 5 give the $\vec{V}_h$ hodographs of the winds
associated with the two types of thunderstorms ; for a best reading
of these figures, only the extremities of the vectors $\vec{V}_h$ at 3,6,9
and 12 km have been plotted. The two families of hodographs are
different : first, there is only one quadrant used for the HS winds,
while the four quadrants are nearly equally used for the RS winds ;
then, shears in direction are more frequent and more important for
the RS winds than for the HS ones ; and finally winds above 3 km
are greater and increase with the altitude. This last point will
be studied in the next section.

THE VERTICAL GRADIENT OF THE MEAN WIND VELOCITY

We define the mean velocity of the wind between the surface
and the altitude h by :

$$v_h = (1/h) \; || \; \vec{V}_h ||$$

Figure 6 and Figure 7 give the values of v_h as a function of
h for the two types of thunderstorms. As for Figure 4 and Figure 5,
only the values for h = 3,6,9 and 12 km have been plotted. The
slopes of the lines between two successive altitudes represent the
inverses of the vertical gradient of the mean velocity.

The values of v_h from the surface to 3 km are not very different in the HS and RS cases, but they rapidly increase above
this altitude for the HS soundings while they are nearly constant
—or sometimes decrease— for the RS ones. Above 3 km it is possible
to represent the variations of v_h with h by the linear regression
equation :

$$v_h = b + m(h - 3)$$

b (in m s^{-1}) approximately corresponds to v_3 and m (in m s^{-1} km^{-1})
is the vertical gradient of the mean wind velocity between the altitudes 3 and h (in km).

The values of b, m and r^2 (correlation coefficient) are given
respectively in Tables 1 and 2 for the HS and RS soundings.

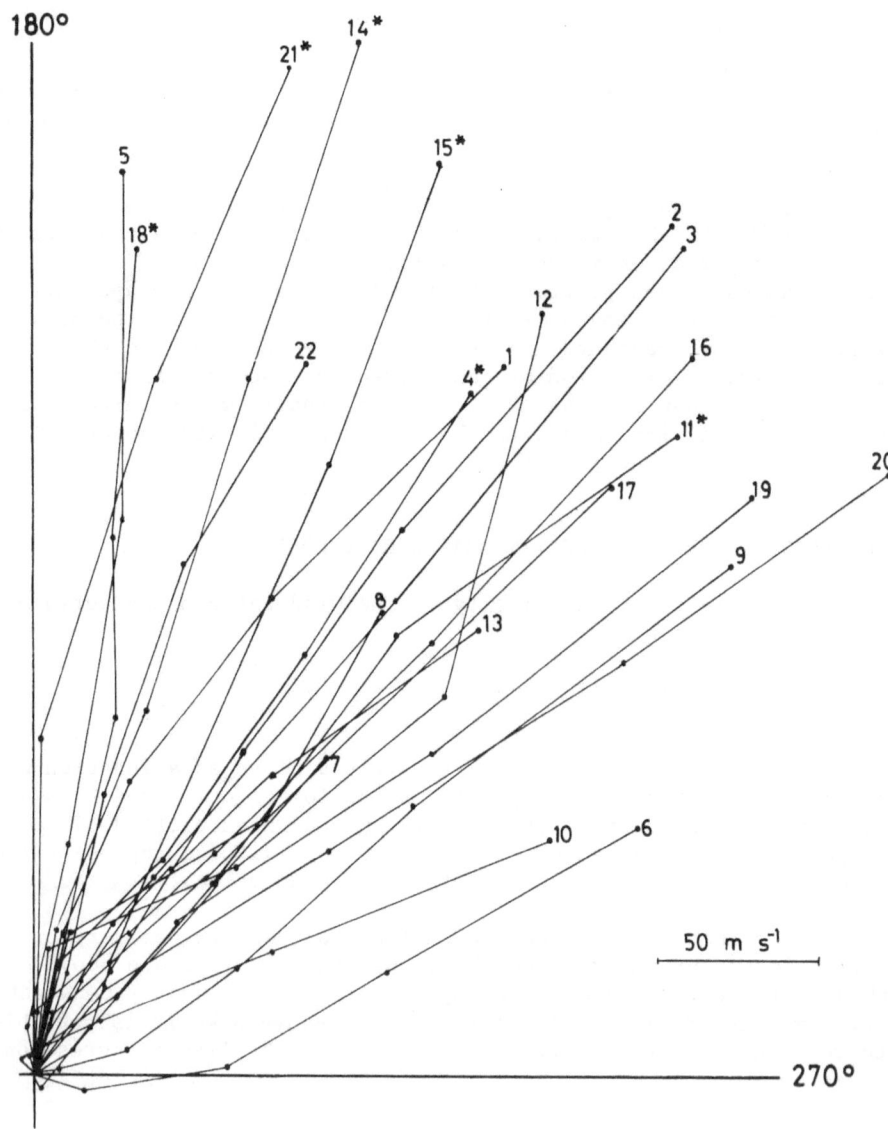

Figure 4 Hodographs of the cumulated wind vector $\vec{V}_h = \sum\limits_{i=1}^{h} \vec{v}_i$ for the hailstorms soundings.
The extremities of this vector are pointed for h = 3,6,9 and 12 km.

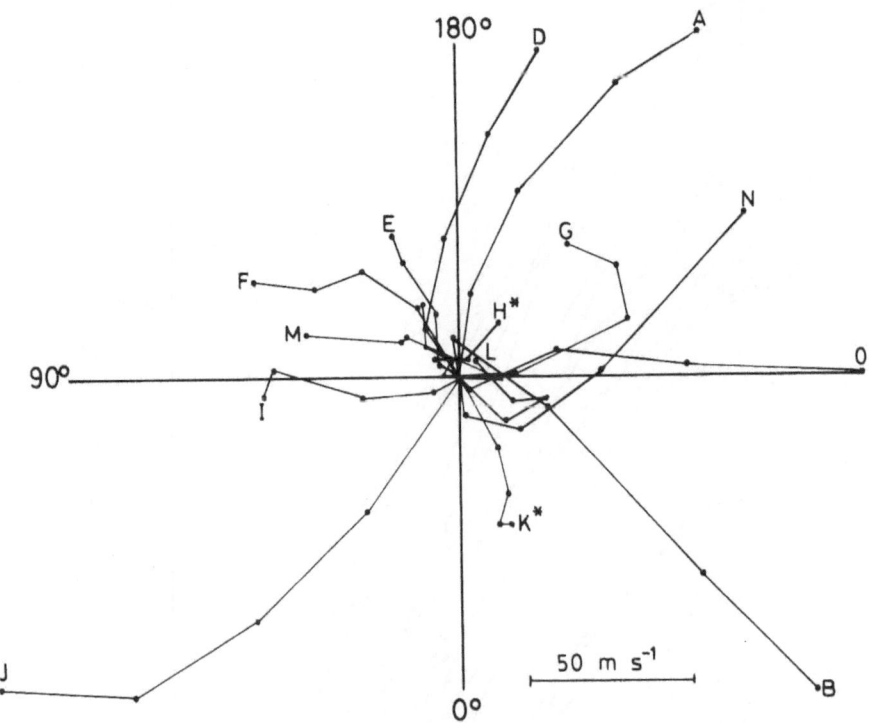

Figure 5 Same as Figure 4 for the rainstorms soundings.

Standard values of b, m and r^2 for both types of soundings may be obtained by averaging the values in the tables, but for b and m it seems more correct to calculate first the averages of v_h for each family of soundings for h = 1, h = 2, ..., h = 12 (Figure 8), then to write the regression equation of these averages. The results for h > 3 km are the following :

$$(v_h)HS = 7.8 + 1.6(h - 3)$$

$$(v_h)RS = 4.7 + 0.16(h - 3)$$

These equations must be completed with the values of r^2 found precedently :

$$(r^2)\ HS = .96$$

$$(r^2)\ RS = .65$$

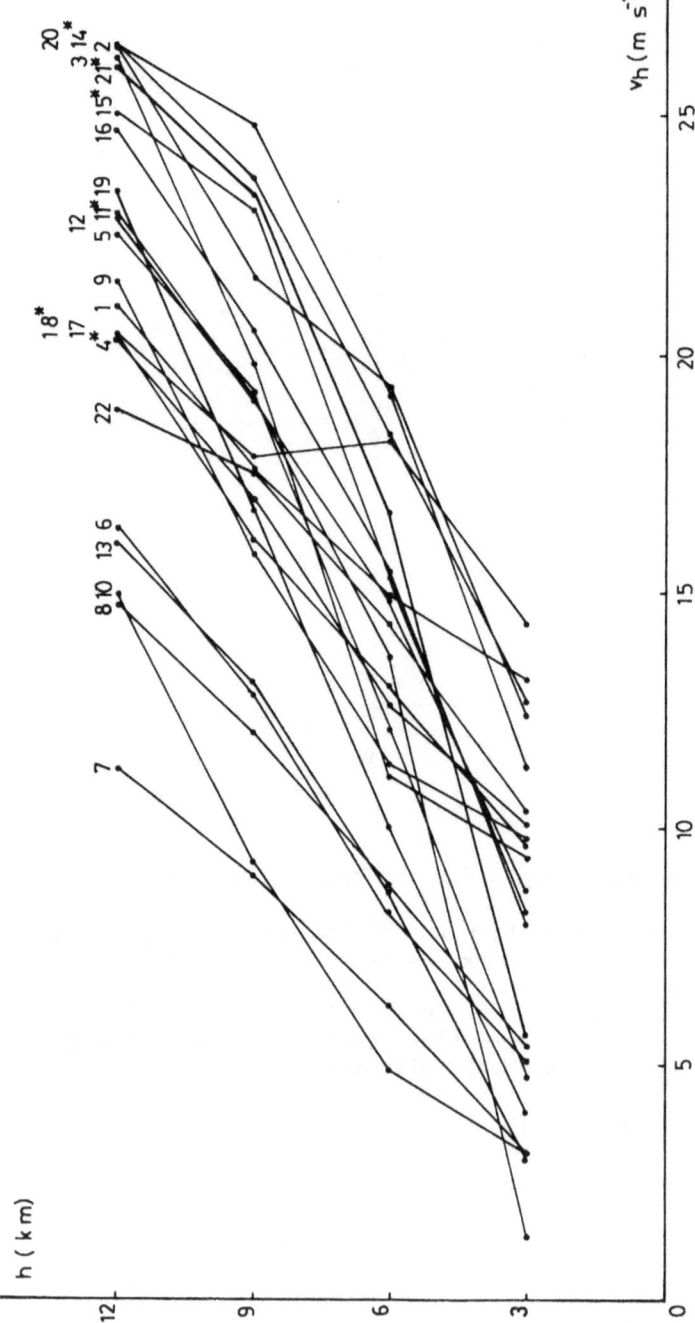

Figure 6   Profiles of the mean values of the wind velocity for the hailstorms
soundings.

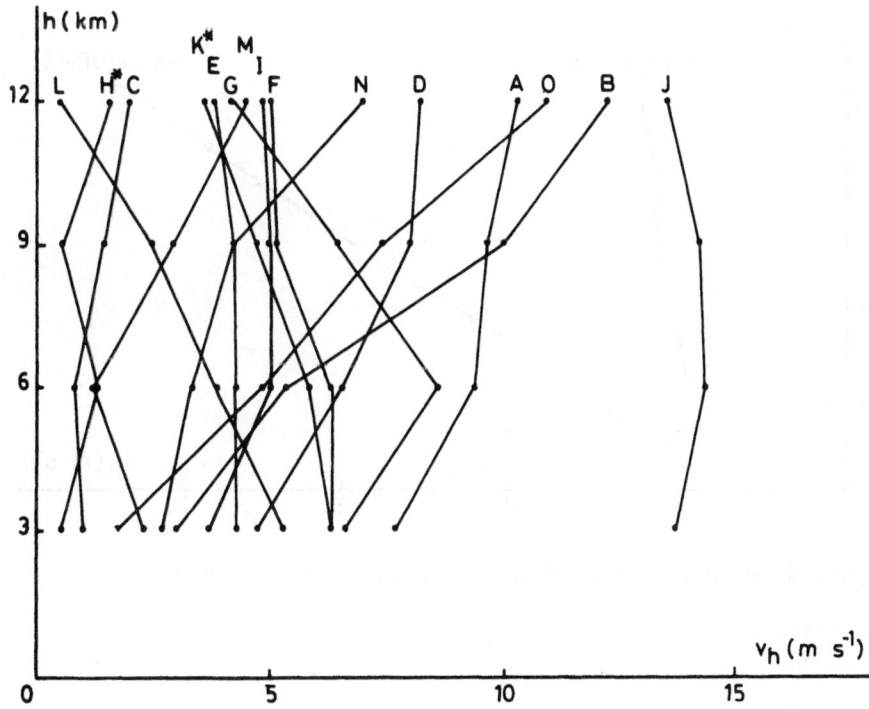

Figure 7   Same as Figure 6 for the rainstorms soundings.

The values of b do not allow to classify a sounding in one of the two categories, even if these values are generally higher in the case of hailstorms ; two main exceptions are case J (rainstorm with high value of b, but this value is associated with a low value of m), and case 10 (hailstorm with low value of b, but this value is associated with a high value of m).

The values of m differ of one order of magnitude from one type of storm to the other ; once again there are exceptions :
– For the HS type, case 18* has a low value of m ; nevertheless this case belongs to the class intitled "extreme damage". But we observe that the value of b is the largest of all the values of Tables 1 and 2 ; an unusual wind profile explains these deviations and also the low value of r : the maximum wind velocity was 38 m s⁻¹ at 6 km, when it was only 10 and 13 m s⁻¹ respectively at 5 and 8 km.
– For the RS type, cases B and O have high values of m ; but case B was only in the "light damage" class, and case O had both rain

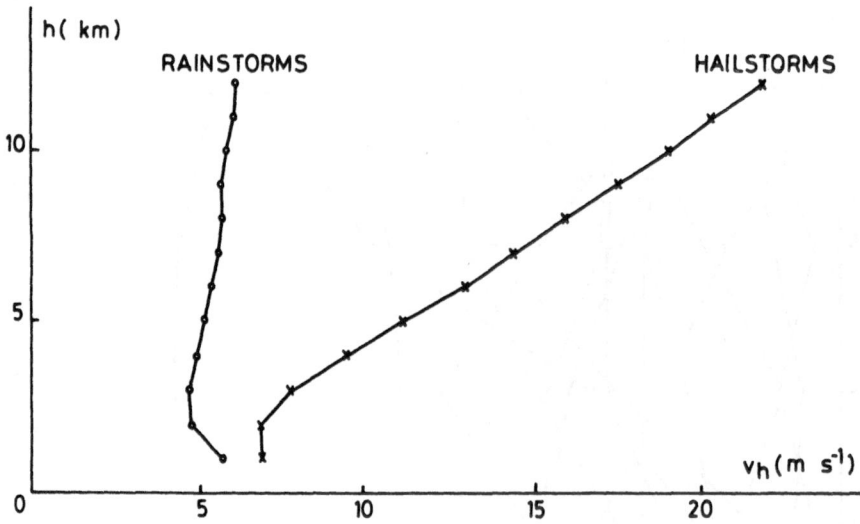

Figure 8   Mean profiles of the mean values of the wind.

and hail.

CONCLUSION

In case of severe thunderstorm probability in South-Western France, we may forecast with a high percentage of success the type of expected disaster -hailfall or flash flood-, and then operate or not the hail prevention by ground seeding, with the following analysis of the wind sounding :
1. $\vec{v}_i$ being the wind vector at the altitude i (in km), we first calculate the mean wind velocity v_h between the surface and the altitude h (in km) for h = 3, h = 4, ..., h = 12 :

$$v_h = (1/h) \left\| \sum_{i=1}^{h} \vec{v}_i \right\| \quad , \quad i \in \{1, 2, ..., h\}$$

2. With these 10 values of v_h, we then write the regression equation of v_h as a function of h :

$$v_h = b + m (h - 3), \text{ correlation coefficient } r^2.$$

3. We finally compare the values of b, m and r^2 with the values found as typical of "hailstorm sounding" or "rainstorm sounding" :
       hailstorm : b = 7.8 m s^{-1}, m = 1.6 m s^{-1} km^{-1}, r^2 = .96
       rainstorm : b = 4.7 m s^{-1}, m = 0.16 m s^{-1}km^{-1}, r^2 = .65

The hailstorm sounding with an increase of the wind above 3 km is characteristic of a frontal synoptic weather situation, and in fact we observe in South-Western France severe hailstorms being associated with cold fronts.

In order to explain physically the effect of the wind on the development of a thunderstorm into either a hailstorm or a rainstorm, we can suggest the following process :
- When the wind at about 3 km is low, and when it does not increase above this altitude, the new cells of the storm penetrate the older ones and are naturally seeded with ice crystals. The efficiency in rain production is high, and as the storm moves slowly, the amount of rain in a given area is important.
- When the wind increases with the altitude above 3 km, the new cells of the storm grow in clear air, and they may produce hailstones in the first stage of their life ; the continuous production of new cells and the rapid moving of the storm lead to long hail-paths.

ACKNOWLEDGMENTS

Thanks are due to the Météorologie Nationale for giving the sounding data, and to François Romeuf and Solange Vidal for preparing this manuscript.

REFERENCES

1. Dessens, J. : 1979, "Ground seeding hail prevention experiment in France". J. Weather Modif., 11, pp. 4-17.
2. Schwartz, G. and D.R. Dingle : 1980, "The not quite flash flood-The hybrid". Second Conf. on flash floods, Amer. Meteor. Soc. pp. 254-258.
3. Dessens, H. : 1960, "Severe hailstorms are associated with very strong winds between 6000 and 12000 meters". Amer. Geo. Union Monograph N° 5, pp. 333-338.
4. A N E L F A : 1953 to 1982, Annual Reports n° 1 to 30. Available on request to A N E L F A, 52 rue A. Duméril, 31400 Toulouse, France.
5. Crysler, K.A., L.R. Hoxit anc R.A. Maddox : 1980, "A climatology of the flash flood hazard in a four state region of Appalachia". Second Conf. on flash floods, Amer. Meteor. Soc., pp. 62-69.
6. Molénat, J. : 1975, "Technique de la prévision des risques de grêle dans le Sud-ouest de la France". A N E L F A, n° 23, pp. 19-20.

THE DEVELOPMENT OF THE CUMULONIMBUS CLOUDS WHICH MOVE ALONG A VALLEY

Mladjen Ćurić

Institute of Meteorology, University of Belgrade
Yugoslavia

ABSTRACT

On the base of the numerous radars data and of the observations made at the ground we separated individual cumulonimbus
with high hail production. A model of the development described
in this paper primary is based on the radars data of the damaging
cumulonimbus which occured in a Western Morava valley. Hail produced at the ground has the hailstreaks form separated about lo
km. It is shown that hailstreaks are related with periodic change
of the updraft on the inflow side of the cumulonimbus, but this
periodic change is related with cooled air outflow from the base
of the cumulonimbus in a valley direction. This mechanism lead to
the cumulonimbus regeneration.

1. INTRODUCTION

The dynamics as well as mycrophysics of the processes associated with cumulonimbus clouds are expressed in models which
have been schematic drawn by numerous writers. Every study of the
dynamical processes inside cumulonimbus and in its environment
must contain information about the behavior of the air velocity,
temperature, pressure etc. Its temporal evolution are need too.
Such informations are growed so rapidly that now we are in
position to present very elegant schematic descriptions of the
cumulonimbus at various stages of its evolution. This progress
amaze us.

The cumulonimbus models described by Howard (1803), Davis
(1894) or Suckstroff (1939) amaze us due their good concept in

E. M. Agee and T. Asai (eds.), Cloud Dynamics, 259–272.
Copyright © 1982 by D. Reidel Publishing Company.

spite bounded numbers of information. Theirs observation were
made practically only by eye. So, Davis emphasises an important
characteristic of the cumulonimbus (or severe storm as it is
named later), the great asymmetry of the anvil cloud which usually
extends far forward of the storm. His diagram also contain the
mamma which are often seen below the thicker parts of the anvil
cloud. As it can be seen this model was not significantly improved
during the following six decades.

The production of radar, and of robust aircraft capable of
safely traversing storm (such is extensively modified for meteorological research MRF HERCULES XV 208), brought new opportunities
for the systematic exploration of cumulonimbus. This technique
were first intensively used in the well known Thunderstorm Project
(Byers and Braham, 1949). The most important finding were location
of warmer updraft and colder downdraft.

A more detaile measurement of the significant parameters is
resulted in well known three-dimensional model of the Wokingham
storm (Browning and Ludlam, 1962). Cumulonimbus clouds which are
developing and reaching increasing heights can sometimes produce
daughter clouds (Dennis, 1970; Chisholm, 1973; Browning, 1977).
These young daughter clouds plaies important role in regeneration
of the mature, usuale hail, cloud. Its origin is not discussed
yet on the satisfactory way.

As it is shown by Nelson and Braham (1975) a seemigly steady-
-state updraft in the lower portion on the inflow side of the
storm broke into an unsteady structure in the upper levels. They
are interpreted it as indicating that precipitation loading
gradually decelerate the upper reaches of the updraft.

This present paper intend to explain the structure and
development of cumulonimbus, or severe thunderstorms, which move
along the valley in the mountains regions. This consideration
receive considerable attention since in Yugoslavia, as well as in
some other regions in the world, such storms shown a specific
behaviour. Unpredictable development of the storms in these
regions usually produce hail which cause serious damages to agricultural cultivations in spite well organised of the hail pro-
tection. By combining radar measurements with conventional, and
speciable reports of hail, in Western Morava valley the basic
cumulonimbus dynamics is described.

2. WESTERN MORAVA VALLEY AND THE SURROUNDING OROGRAPHY

Western Morava is a river which streaming roughly from WNW
to ESE direction. A mean hight of this valley is 300 m (m.s.l.)
approximately. The topography in the immediate environment of the

Western Morava valley is shown in Fig. 1. General characteristics
of the north portion topography are mountains with heights about
1 km (m.s.1) whose mountain sides are not too steep.

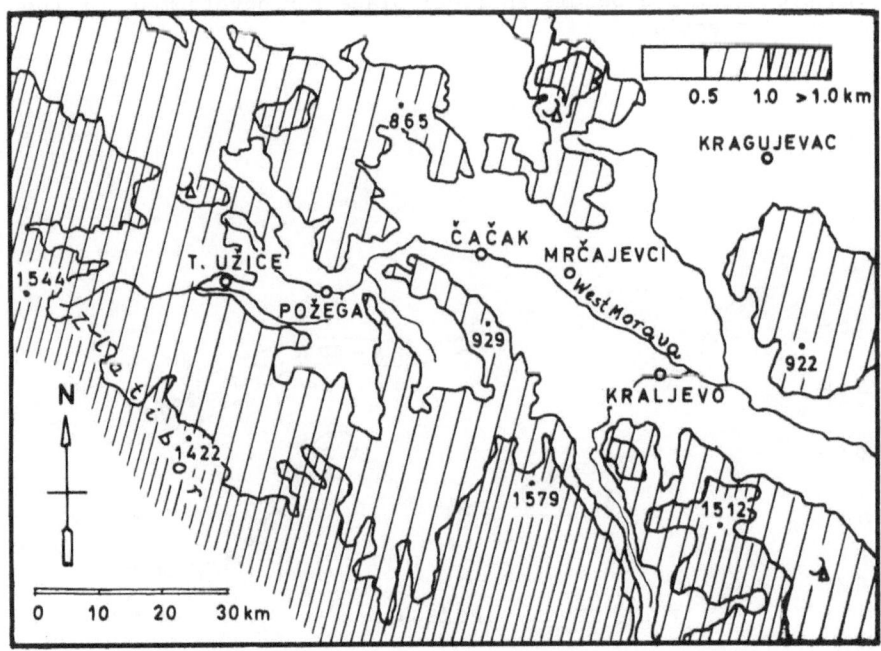

Figure 1. Contour map of Western Morava valley showing the
           location of the some places mentioned in the text and
           of the radar positions.

Over southern portion of a valley the orography is much
more expressed. Peak heights are over 1.5 km. In the region
between Požega and Čačak the valley gets narrower. This narrow
part is known as the Ovčar-Kablar's garge. It is so useful to
bear in mind this gorge since its wide is only a few hundred of
metres. Under such circumstances a cumulonimbus is forced to
cross a mountain.

To illustrate this cross over mountain we present Figure 2.
It contains the profile of topography in direction WNW-ESE,
(general direction of the Western Morava valley), along with
typical direction of motion of the warm air in the planetary
boundary layer. Some other characteristics of the warm air,
temperature pressure and relative humidity are shown too.

On the western portion of the valley is a plateau Zlatibor.
Its mean height is approximately 1 km. Actually, this western part
of the Serbia area usually is a source of the convection. As it

is known, the most common form of convection in midlatitude have
its roots at the surface at all. It is more than evident in this
plateau. Except in some small area it is no afforested, so that
such stony surface become an ideal place for development of the
contact convection.

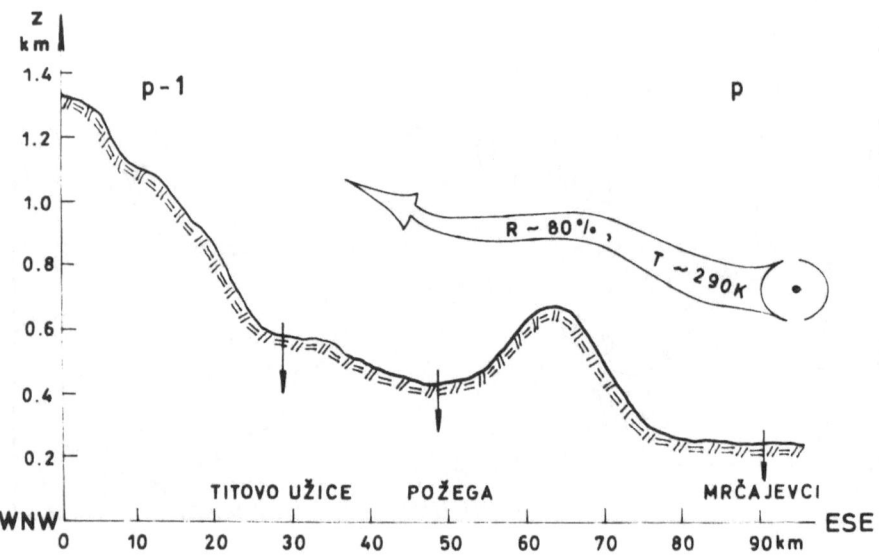

Figure 2. Cross section along the Western Morava valley. The
          arrow represent flow of the warm air in the planetary
          boundary layer. Relative humidity and temperature are
          labelled inside the arrow.

3. RADAR AND OTHER DATA

The primary sources of information used in this paper are
the radars, conventional meteorological network and ground based
launchers stations network (with observation persons on its).

Some of the meteorological services of the Yugoslavia have
established a weather radar network. The Serbia have been almost
completely covered with the weather radar network since 1968. At
the present time 14 radars are in the operation. The use of the
radars is for general weather surveillance and analyses (3 cm
Wave length radars), and the primary, for operational hail pro-
tection project (10 cm wave length radars). Radar network at
Serbia area is very dence, and they are operating during the day
and night time in the summer part of the year. During the storm,

the radars was operated rapidly, so that it was able to follow storm evolution.

As well as in the other region the systematic radar obser-vations of damaging cumulonimbus were conducted in the Western Morava valley region. On the base of the numerous radars data from this period (RHI and PPI pictures, of radar cloud top, heights of the accumulation zone, reflectivity from the accumu-lation zone) and of the observations made at the ground we are separated individual cumulonimbus with high hail production. A typical such cumulonimbus was on 30 July 1980, and it will be described.

The ground based launchers stations network is used for observations of hail, windshift, pressure jump etc. Distance be-tween launchers station is about 10 km (Ćurić, 1980).

4. MESOSCALE CHARACTERISTICS OF THE FACTORS RESPONSIBLE FOR THE DEVELOPMENT OF THE CUMULONIMBUS CLOUDS

Almost all the individual cumulonimbus in Western Morava valley region initiate its development at the same synoptic situations. Fig. 3 shows surface synoptic situation over Yugo-slavia on 30 July 1980 at 13oo GMT. During that day is fair weather. In the early morning there are no clouds. Convective clouds start to increase gradually about looo LMT mainly in the mountain portions (area covered with shading). Surface temperature is about 30°C at noon. Surface situation is without expressed advection. Only slow motion of air occurs toward the Dinaric Alps and the eastern mountain range. This motion toward the mountains is caused by a morning warming of a mountain sides (Čadež, 1964) what result in the pressure gradient force directed as it is represent in Fig. 2.

Abundant moisture at low level maintains during of all day what we see in Table 1.

Table 1. Relative humidity (%) from Beograd radiosonde ascents at 06oo and 12oo GMT on 30 July 1980.

GMT	heights (km)	0.2	0.5	1.0	2.0	3.0	4.0	5.0	6.0
06oo		87	81	6o	59	42	62	62	25
12oo		67	74	81	73	37	35	69	1o

Such relative humidity change with height is caused by local

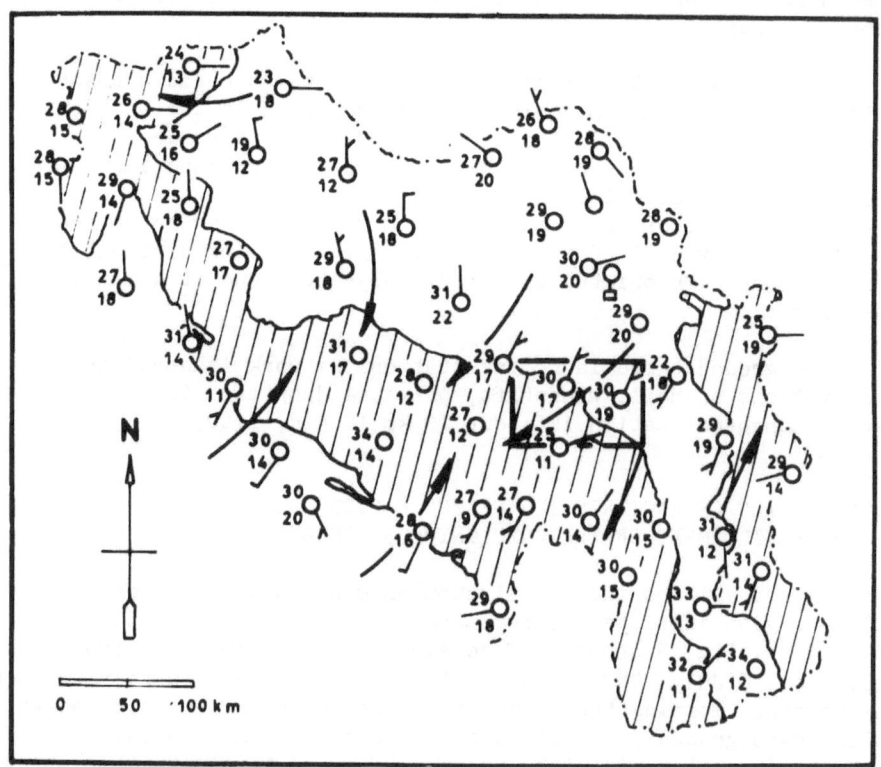

Figure 3. Surface synoptic condition over the Yugoslavia at 12oo
          GMT on 30 July 1980. The quadrilateral is the studied
          area. In the NE direction from it is indicated location
          of the rawinsonde. Arrows represent approximately
          streaming of air at the surface.

cumulus forming at different height.

      Abundant moisture at low level in Western Morava valley is
increased in the morning due to the inversion which occurs at
this day. Inversion occured at the Beograd at the 06oo GMT but it
is desapeared at 12oo GMT, Fig. 4.

      The radiosonde data show that atmosphere on this day is not
very unstable.

      The hodographs (Fig. 5a and 5b) are weakly sheard during of
the day. The weak low level easterly flow shiftes to the nearly
westerly flow (nearly parallel with the Western Morava valley).
Very weak shear is at heighter level (2-5 km). To this time this
weakly sheared hodograph is associated with short lived storm or

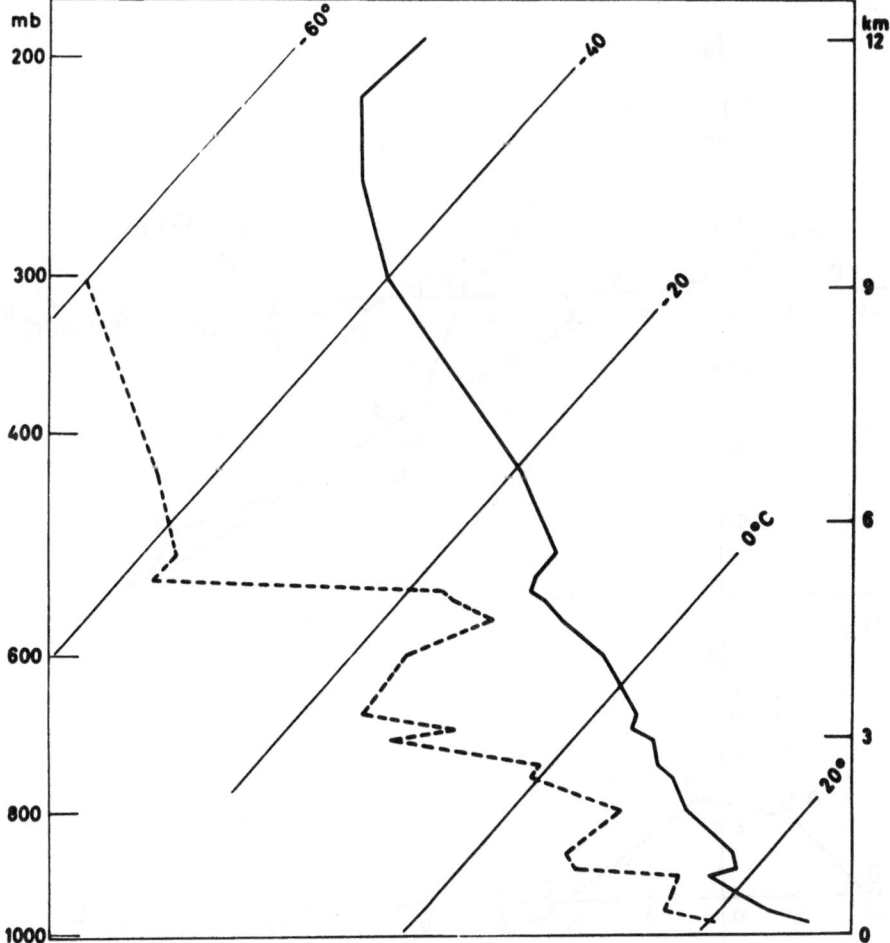

Figure 4. Vertical profiles of temperature (solid line) and dew point (dashed line) from the Beograd sounding at 12oo GMT on the 30 July 1980.

with poorly organized multicell storms which fail to produce large or extensive hail (Browning, 1977). In this paper it is shown that it is no case in the presense of the orography.

The amount of directional shear in our case is very strong. Since the motion of air at the surface in Western Morava region is from ENE, the directional shear is greater than 180°. Shear of the wind velocity in this case is very small.

Summarised said, in the case of the cumulonimbus clouds (storms) which move along a valley, all the usually considered factors responsible for the storm development are not satisfied

a.

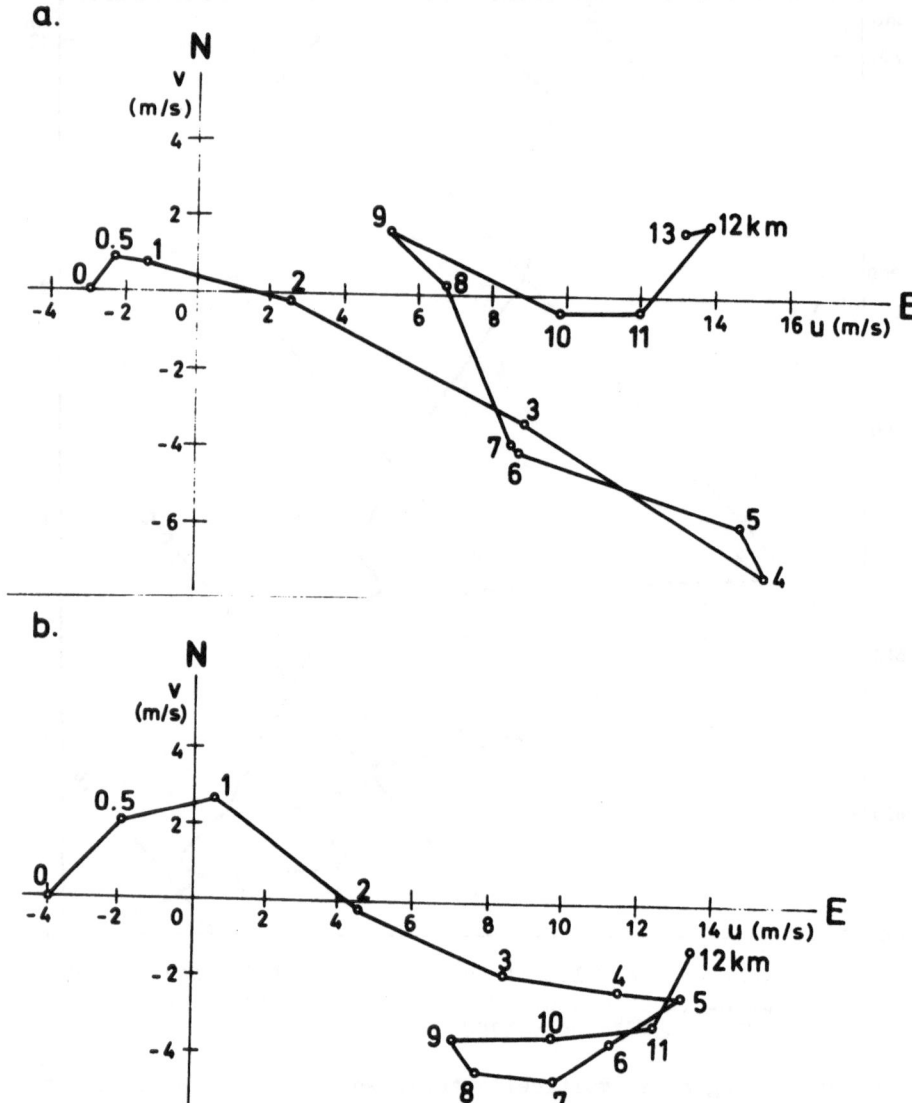

Figure 5. Hodograph of environmental airflow at Beograd on 30
        July 1980 at a) 06oo GMT and b) 12oo GMT. Relative
        heights are in kilometers. Line is the locii of the
        endpoints of the vector winds.

in all. Factor responsible for development of the storm is a
forsed lifting of a warm air on the nose of the outflowing cold
air from the base of the storm.

5. GENERAL AND INTERNAL CUMULONIMBUS CHARACTERISTICS

As a usually the cloud first appeared on the plateau Zlatibor. After that it moved toward east-southeast and reached in the Western Morava valley. The environmental shear was weak with wind varing from southeast at 3 m/s at the height 0.5 km (relative height) to west northwest at only 13.5 m/s at 12 km (Fig. 5b). Wind speed at 2 - 5 km was 5 - 12 m/s without directional shear. Much of the apparent motion of the cloud was due to propagation (about 55 km/h).

In the valley when cross over the mountain between Požega and Čačak (Fig. 2) the cloud become much more vigorous. A strong storm wind and hail caused serious damages to agricultural cultivations. Hail produced at the ground has the hailstreaks form separated about 10 km. In Fig. 6 is shown track of the cloud and the hail-streaks in the phase of its maximal development from 1150-1300 GMT. From 1110 to 1300 it produced heavy rain and hail as it is shown. After 1300 GMT the cumulonimbus continued to dissipate.

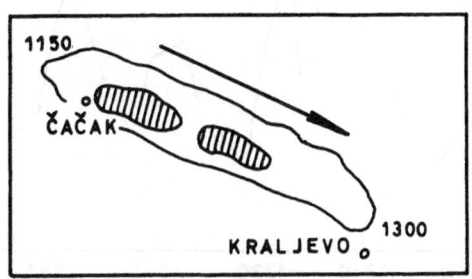

Figure 6. Track of the cloud and the hailstreaks (shaded area) in the phase of the maximal development from 1150-1300 GMT on the 30 July 1980.

The observed internal characteristics by the 10 cm weather radar (radar reflectivity from the accumulation zone, Ze, height of the accumulation zone, Ha, and the radar cloud top, Ht, as a function of time are shown in Fig. 7. Because the cloud crossed over mountain its radar top become, smaller, decreased from a 12 km to a 10.5 km within 8 min interval. At the same time reflectivity are not changed and height of the accumulation zone is decreased. After that the Ha increased, first rapidly and afterwards more slowly, reaching a maximum value of 13 km. Ze first slowly increased and afterwards sharply decreased, until Ha have similar trend as a Ht.

It is very important to emphasize an important fact, Ht again sharply increased at times 1220 and 1245 GMT (at 1320 and

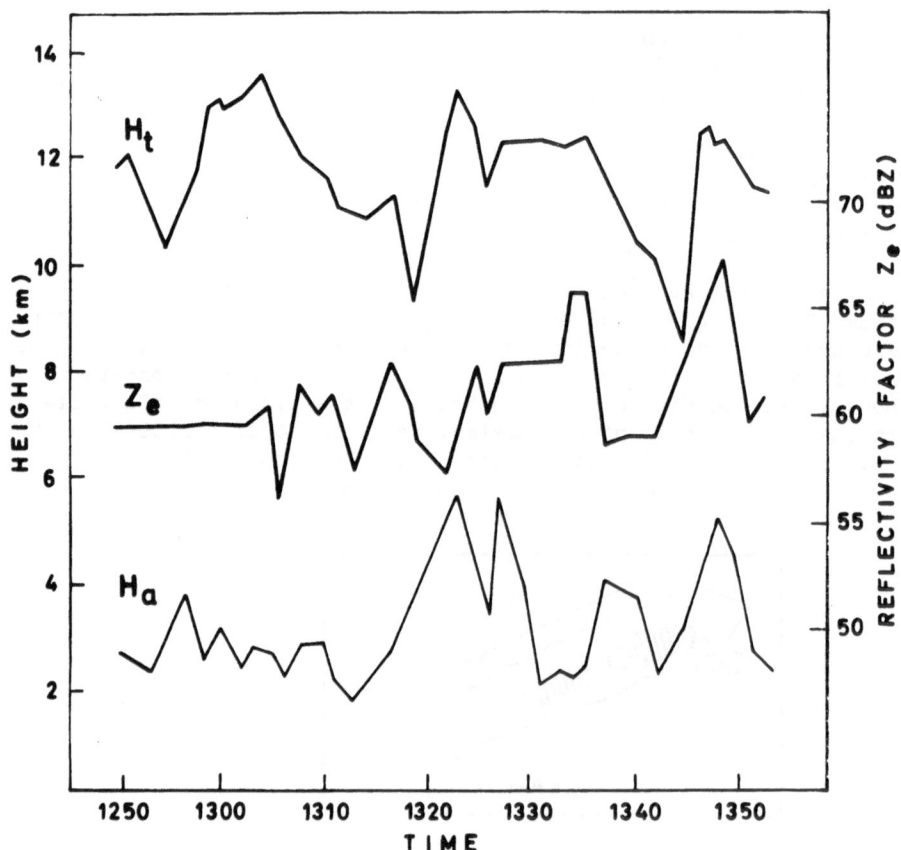

Figure 7. Time variation of the height of radar cloud top (Ht),
          radar reflectivity from accumulation zone (Ze) and
          height of the accumulation zone (Ha) of the cumulo-.
          nimbus which moved along Western Morava valley on the
          30 July 1980.

1345 local time). Ze and Ha have same behavior as it is previously
described. The period between of these successive sharp increases
is about 25 min.

      The peaks of the radar cloud top coincides with the surface
hail.

      Vertical cross-section of the cumulonimbus along of the
Western Morava valley is shown in Fig. 8. Figures a, b, c, d,
corresponds to the positions of the cloud when it has trend of

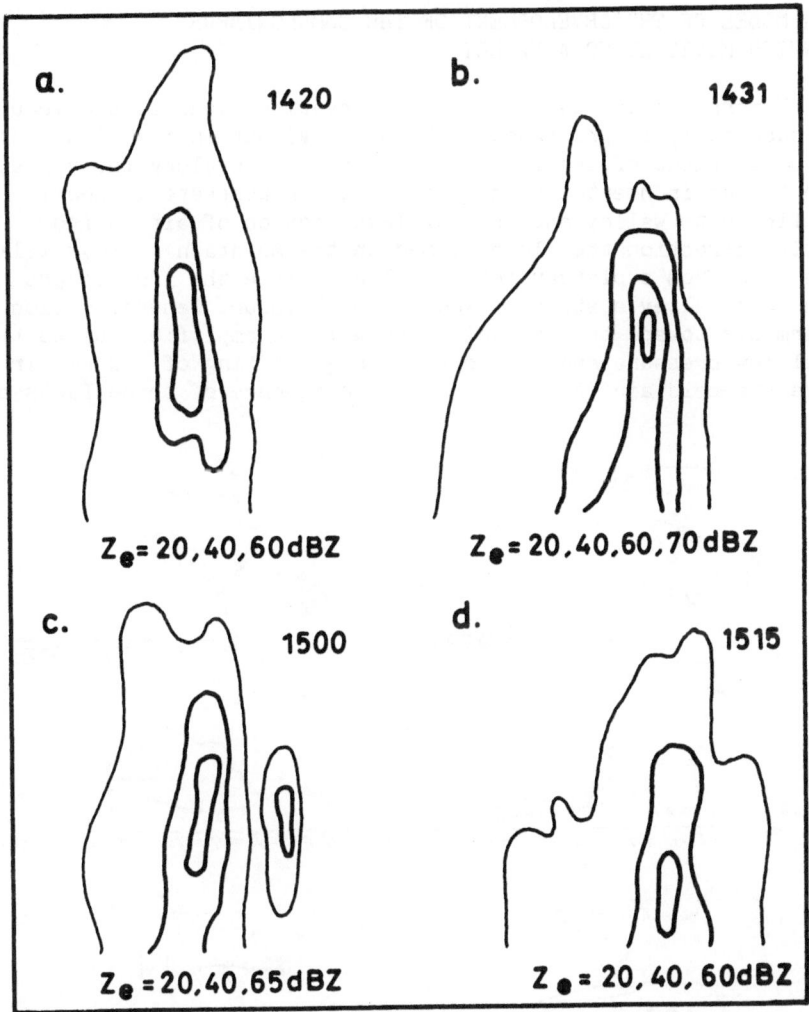

Figure 8. Vertical cross-section of the cumulonimbus along
          direction of the Western Morava valley. Figures a, b,
          c, d, corresponds to the positions of the cloud shown
          in Fig. 9 a, c, e, f, respectively.

the parameters same as they are at the times 114o, 12o5, 1225 and
1245 GMT in the Fig. 7.

   We consider it necessary to pointed out that the new formed
cell (Fig. 8c) is merge later with the man cumulonimbus (Fig. 8d).

6. A MODEL OF THE DEVELOPMENT OF THE CUMULONIMBUS WHICH MOVES ALONG A VALLEY

Numerous radar data and the observations made at the ground (an case study is presented in this place) put on the figure of the development of the cumulonimbus which move along the valley. When the environmental wind in the mean troposphere is nearly parallel with walley and the low-level motion of air is from oposite direction the cloud formed on the mountain plateau will came down from a plateau into a valley. Since the land slopes down to a valley a strong downdraft will occur. General motion of a warm air toward the mountain side will be amplified due to the local low pressure and due to the forced lifting of a warm air above the cold air (Fig. 9b). As a consequence of these facts a

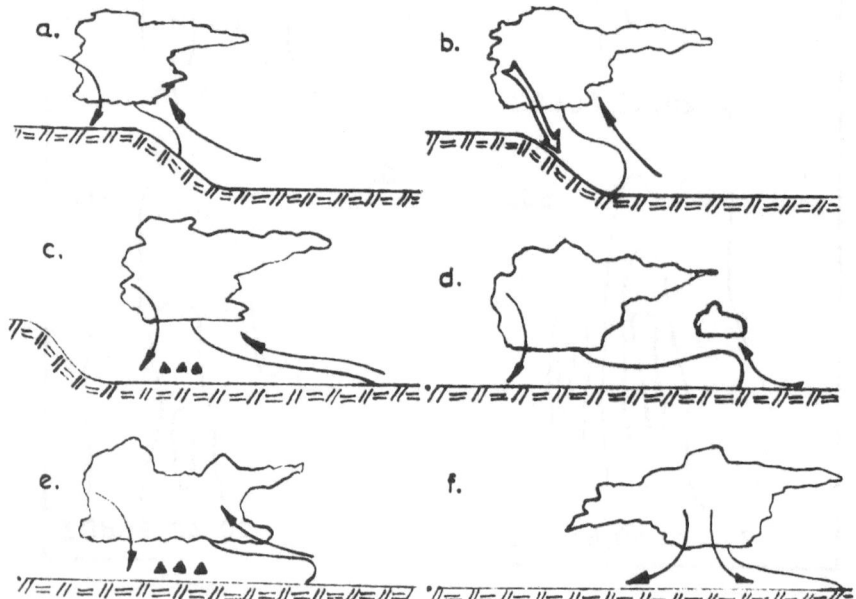

Figure 9. Some phase of the cumulonimbus development with direction of motion along valley. Successive positions are: a) the cloud is on the plateau; b) strong cold air downdraft down mountain side produces strong updraft of the warm air; c) updraft decreases due to the spreading out of the cold air, hail occurs; d) formation of the daughter cloud on the nose; e) daughter and main cloud merge (hail occur); f) cloud continues to dissipate or to regenerate.

strong updraft in the front portion of the clouds will occur.

After that hail reaches to surface. The cold air spreading out in a valley direction. After certain time the nose of a cold air become heighter (Ćurić, 1980). New daughter cloud will form on the nose (Goff, 1976, Browning, 1977), Fig. 9d. The main cumulonimbus and daughter cloud merge since the main cloud is a faster than the daughter (or feeder cloud as it is known earlier). After that the cumulonimbus can to continue its dissipation or regeneration. Sometimes formation of the new daughter clouds lead to the repeating of the phase from Fig. 9c forward. It is observed on 22. June 1977 that by this processe of the cumulonimbus regeneration its life was longer than six hours.

REFERENCES

1. Browning, K.A., 1977: The structure and mechanisms of hailstorms. Meteor. Monogr., No. 38, pp. 1-43.

2. Browning, K.A., and F.H. Ludlam, 1962: Air flow in convective storms. Quart. J.R. Met. Soc., 88, pp. 117-135.

3. Byers, H.R., and R.R. Braham, 1949: "The thunderstorm. Washington." D.C., U.S. Govt. Printing Office, 287 pp.

4. Chisholm, A.J., 1973: Alberta hailstorms, Part I: Radar case studies and airflow models. Meteor. Monogr., No. 36, pp. 1-36.

5. Čadež, M., 1964: "Weather in Yugoslavia", Papers, 5, Fac. of Math. and Nat. Sc., Beograd, 83 pp.

6. Ćurić, M., 1980: Dynamics of a cold air outflow from the base of the thunderstorm. A simple model . J. Rech. Atmos., 14, No. 3-4, pp. 493-498.

7. Ćurić, M., 1980: An indicator of precipitation enhancement due to cloud seeding aimed at reducing hail. Proc. Third WMO Sc. Conf. on Wea. Modif. Clermont-Ferrand, 21-25 July 1980, pp. 749-753.

8. Davis, W.M., 1894: "Elementary meteorology. Boston, Ginn and Co.", 366 pp.

9. Dennis, C.A., and all., 1970: Characteristics of hailstorms of Western South Dakota. J. Appl. Meteor., 9, pp. 127-135.

1o. Goff, R.C., 1976: Vertical structure of thunderstorm outflows. Mon. Wea. Rev., 104, pp. 1429-1440.

11. Howard, L., 1803: On the modifications of clouds. Phil.Mag.,

16, 97, 344; 17, 5.

12.Nelson, S.P., and R.R. Braham, 1975: Detailed observational
   study of a weak echo region . Pure and App. Geoph., 113,
   pp. 735-746.

13. Suckstroff, G.A., 1939: Die Ergebnisse der Untersuchungen
    an tropishen Gewittern. Gerl. Beitr. Z. Geophys., 55,
    pp. 138-185.

COMPARISON OF RADAR OBSERVATIONS OF A DEVASTATING HAILSTORM AND A CLOUDBURST AT JAN SMUTS AIRPORT

G Held

National Physical Research Laboratory, CSIR, Pretoria, South Africa.

Two similar-looking thunderstorms, which occurred on two different days in the vicinity of Johannesburg's International Airport, were monitored by an S-band radar. The one produced the highest rain- fall rate recorded at that site in 25 years, i.e. 174.0 mm/h during 20 minutes followed by 55.2 mm/h for 30 minutes, but almost no hail was reported. The other storm developed into one of the worst hailstorms ever observed in this area, hailing almost con- tinuously for 1½ hours within a 10 x 10 km area. The largest hail- stones were > 5 cm in diameter, causing many millions of dollars damage. The two storms differed remarkably little in their struc- ture as well as in their behaviour. The comparison illustrated clearly the difficulties which are encountered in discriminating between a severe hailstorm and a heavy cloudburst.

1. INTRODUCTION

Damage to crops caused by hailstorms in South Africa amounts to an average of 60 to 70 million dollars annually (1). Therefore it is of great economic importance to study and fully understand thunderstorms and the formation of hailstones, with the ultimate aim of possibly preventing or at least reducing the damage caused by hailstorms on the South African Highveld, where damaging storms are most frequent. In 1962 the Atmospheric Sciences Division of the CSIR initiated a programme to study the hail fall-out pattern on the ground, based on a large team of voluntary hail observers. This was supplemented by radar observations in 1971. Figure 1 shows the position of the S-band radar and the 40 x 70 km hail- observing area with Pretoria in the northeast and the densely populated Witwatersrand (a conglomerate of municipal areas,

273

E. M. Agee and T. Asai (eds.), Cloud Dynamics, 273–284.
Copyright © 1982 by D. Reidel Publishing Company.

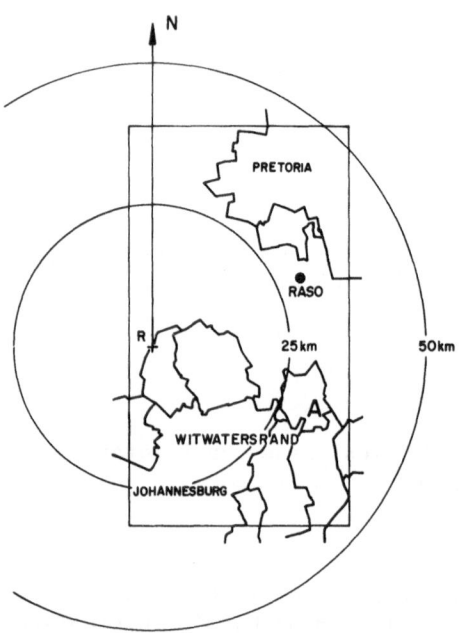

Figure 1. The hail-observing network; the radar station (R),
the radiosonde station (RASO) and the position of Jan Smuts
Airport (A) are indicated.

incorporating Johannesburg) in the south. Some 3000 voluntary
hail observers have been enrolled to report details of hailfalls.
The hail-observing network which is located on the continental
plateau, about 1500 m above MSL, and the hail climatology are
described extensively in (2), (3), (4) and (5).

        During the afternoon of 15 January 1980 a storm which had
developed just south of the airport eventually became one of the
worst hailstorms ever observed in this area, causing many millions
of dollars damage to aircraft, motor cars and property. This particular storm produced hail almost continuously for 1½ hours over
a 50 to 60 km² urban area and the largest hailstones were reported
to have been bigger than hens' eggs.

        About 13 months earlier, a similar-looking storm had
developed in the same area, which produced the highest rainfall
rate recorded there in 25 years, i.e. 174 mm/h during a 20-minute
period, but almost no hail was reported (6).

        It is not intended to go into the details of the case
studies here but rather to concentrate on the differences and
similarities of the two storms.

2. THE HAILSTORM ON 15 JANUARY 1980

Figure 2 shows a typical picture of this storm, more or less at the time when it began to produce the very large hailstones. The area on which hail fell at this particular time (17h18 - all times in South African Standard Time) is hatched and coincides very well with the area enclosed by the 52.5 dBZ contour, except for its southern portion (the calibration of the radar system had to be adjusted; the accuracy is estimated to be ± 2 dB). Vertical cross-sections through this storm, shown below in Figure 2, indicate a 50-60° tilt of the echo core towards west and southwest, as well as the storm's penetration through the tropopause. The echo tops reached up to 14 to 15 km AGL.

The storm which occurred concurrently with others was in fact a multicellular complex. The very first echo was detected about 6 km southeast of the airport at 15h28 between 4.5 and 8 km AGL (Storm I). Its life cycle is shown in the form of vertical

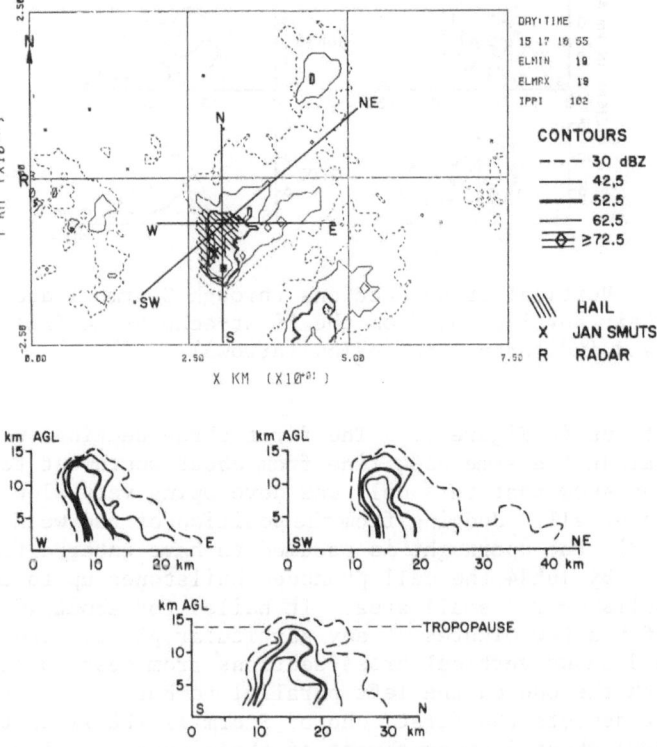

Figure 2.   15 January 1980, 17h18: Computer-generated plot of a PPI at 2° elevation, showing the area where it hailed, and vertical cross-sections along the indicated base lines.

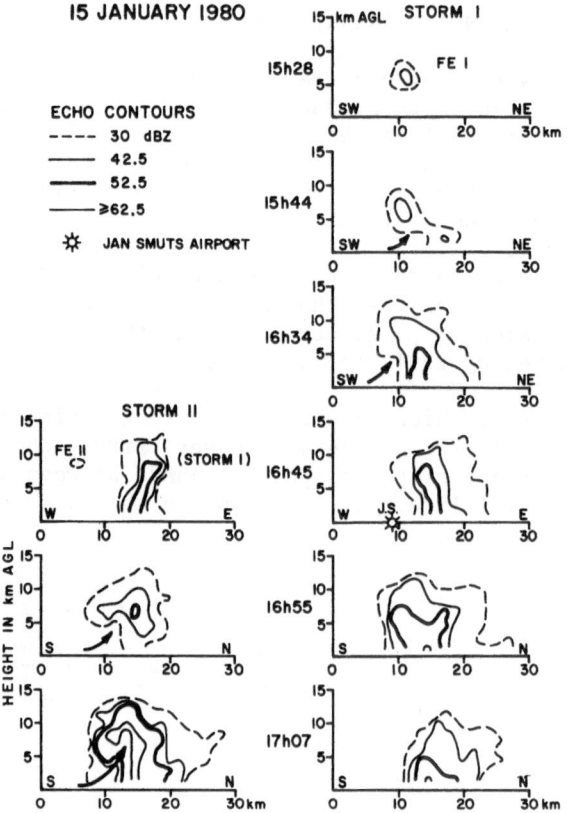

Figure 3.   Vertical cross-sections through Storms I and II,
showing their development from the first-echo stage (FE) to
maturity and indicating the inflow (arrows).

cross-sections in Figure 3.   The first three sections were constructed along the same base line from about southwest to northeast. They show that this cell was developing very slowly and
not moving at all. Judging from the position of the weak echo
region (WER), the updraught is assumed to have entered from the
southwest. By 16h34 the cell produced hailstones up to the size
of golf balls over a small area. It hailed for about 40 minutes,
but only for a few minutes at any particular place. The next row
in Figure 3 shows vertical cross-sections from west to east at
16h45, with the one on the left parallel to but 4 km south of the
other. It depicts the first echo of Storm II (FE II of the *Jan
Smuts Storm*) about 5 km southwest of the airport. Unlike Storm I,
this one developed very rapidly and within 10 minutes had reached
the ground and extended up to 13 km AGL, drawing its supply of
moist air in from the south. At 17h07, i.e. 22 minutes after it

had been detected, its echo core had reached a reflectivity of
≥ 62.5 dBZ with a very strong updraught from the south being indicated by the well-pronounced WER and the area of high reflectivity suspended aloft. Both north-south cross-sections through
Storms I and II at 16h55 and 17h07 are in the same position and
exactly parallel at a distance of 6.5 km. Although the two storms
developed towards each other and eventually merged into one complex, they never lost their identity throughout their unusually
long life span of 1½ to 2 hours. This is emphasised in Figure 4,
which shows the positions of the first echoes of Storms I and II
(FE I and II, cross-hatched) as well as the areas where the reflectivity exceeded 52 dBZ at any time during the given period (dotted).
No systematic movement of the echo cores or trend was obvious
within the dotted areas.

    A large storm to the southeast of the airport which
propagated northwards deflected to the northwest and completely
merged with Storm I at 17h34 (Figure 4). Similarly, storms in the
west converged slowly towards Storm II. Most other storms in the
vicinity behaved in the same way. This is indicative of a strong
local convergence zone around the airport and would also explain
the stationary character of Storms I and II.

    The hailfall pattern of both storms is shown in Figure 5.
The areas on which hail fell on the ground are outlined at different times. In some places near the airport it hailed for 60
to 90 minutes with very brief interruptions. The area on which

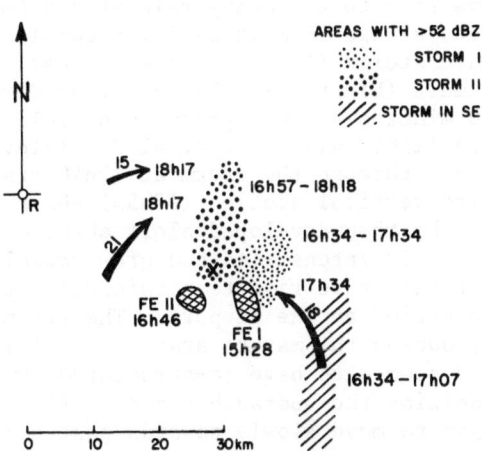

Figure 4.   15 January 1980: Position of first echoes (FEI and II)
relative to Jan Smuts Airport (X) and areas where the radar
reflectivity was > 52 dBZ during any time of the specified time
interval. Convergence of other storms and their speeds (km/h)
are indicated by arrows.

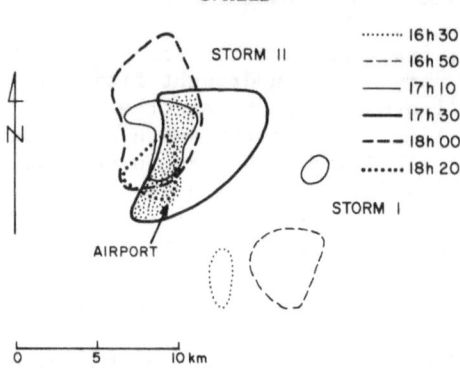

Figure 5. Isochrones of hailfalls from Storms I and II on
15 January 1980. The dotted area shows where the largest hailstones were > 5 cm in diameter.

hailstones > 5 cm in diameter fell is dotted. It was almost
15 km² large and included the airport where enormous damage was
caused to the aircraft. The average point hailfall duration of
Storm II was 32.1 minutes over an area of 50 - 60 km² and two-
thirds of the reports from this area indicated that the largest
hailstones were > 3 cm in diameter.

3. THE CLOUDBURST ON 22 DECEMBER 1978

The storm that gave rise to the heavy rain at Jan Smuts Airport
was part of a large, elongated multicellular complex, which consisted of three main storms (A, B, C) more or less in a line from
northwest to southeast (Figure 6). The heaviest downpour of 174.0
mm/h lasted for 20 minutes and was produced by Cell 1 of Storm B,
accompanied by very light hail. A typical PPI (elevation 2°) and
two vertical sections through the storm at 17h40 are shown in
Figure 6. The third vertical section (17h53) shows the early stage
of Cell 2 which had by then developed aloft about 4 km north-northwest of the airport. It intensified and grew rapidly on its
northern flank. It then gave rise to a rainfall rate of 55.2 mm/h
during a 30-minute period at the airport. The storm remained more
or less stationary during the mature stage of Cell 1, but the development of Cell 2 seemed to have re-structured the whole storm,
possibly by reorganizing the updraughts and airflow pattern within
it. The storm began to move slowly towards east-northeast at
18h03. The maximum reflectivity of both cells was ⩾ 67 dBZ. It
is noteworthy that the other storms of the complex all moved in
various directions towards east to southeast at 30 - 40 km/h
(Figure 6), while Storm B remained more or less stationary during
its most severe stage, thus giving rise to the very high rainfall
rates during extended periods. This would be indicative of a

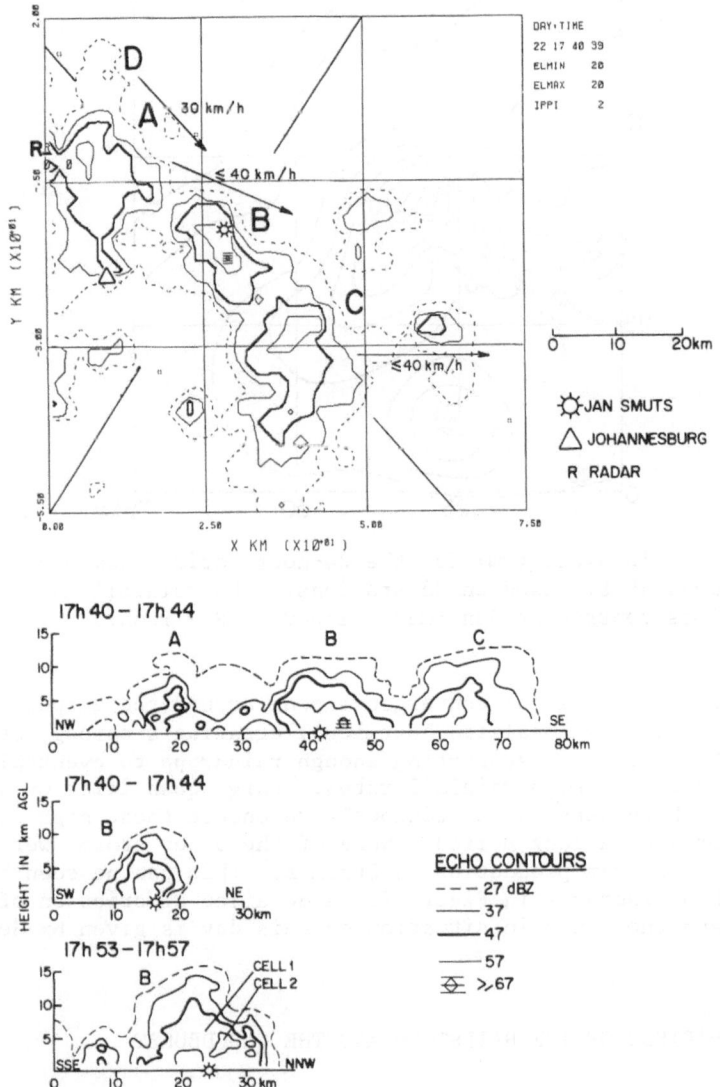

Figure 6. Radar echoes on 22 December 1978: Computer-generated plot of a PPI (elevation 2°) at 17h40 showing Storms A, B, C and D (top). Vertical cross-sections through the storms from northwest to southeast (top) and through the Jan Smuts Storm from southwest to northeast (centre) at 17h40 and at 17h53 from south-southeast to north-northwest showing the newly-developed Cell 2 (bottom). (The approximate position of Jan Smuts Airport is indicated in the cross-sections and the ends of their base lines are shown in the PPI at 17h40. For heights above MSL add 1.5 km.)

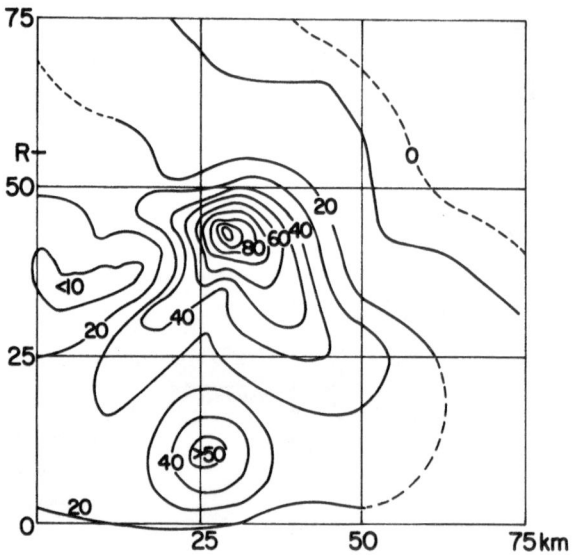

Figure 7.   Isohyets (mm) for the 24-hour period from 08h00 on
22 December 1978, based on 35 stations. The rainfall maximum
(102 mm) is centred at Jan Smuts Airport; R = radar.

small, but strong local convergence zone around the airport (possibly caused by a localized heat low), creating a strong, steady
updraught capable of supporting enough raindrops to eventually
produce the very high rainfall rates. Large quantities of water
must have been supplied continuously to ensure these high rainfall
rates for such a long period. None of the other storms were as
efficient in rain production as Storm B. This can be seen from
the 24-hour isohyets in Figure 7. A detailed description of the
storms and the synoptic situation on this day is given by Held
in (6).

4.   COMPARISON OF THE HAILSTORM AND THE CLOUDBURST

An attempt was made to quantify the differences and similarities
of both storms in order to present them in a tabulated form which
does not need much discussion (Table 1). However, there are a few
points which need qualification. For instance, the volume of both
storms was of the same magnitude, but if one looks at the volume
of high reflectivities integrated in time, then the hailstorm was
severer than the cloudburst, i.e. the top of the 52.5 dBZ radar
reflectivity contour remained above 10 km during a 60-minute
period (with peaks extending up to 12.5 km), while the top of the
47 dBZ contour of the cloudburst did not exceed 10 km but nevertheless it remained above 8 km for a 40-minute period.

Table 1. Comparison of a severe hailstorm (15 January 1980)
and a heavy cloudburst (22 December 1978).

15 January 1980   22 December 1978

	GENERAL DIFFERENCES	
Hailfall: 90 min		Cloudbursts: 20 + 30 min
Largest hailstones: diam. >5 cm		Rainfall rates: 174 55 mm/h
55.0 mm	*Total precipitation from storms* (Jan Smuts Airport)	88.2 mm
	RADAR PARAMETERS	
>72 dBZ	*Max. reflectivity*	⩾67 dBZ
±15 km	*Echo tops (AGL)*	±15 km
10 - 12.5 km/60 min	*52.5 dBZ contour*	
	47 dBZ contour	8 - 10 km/40 min
	STORM STRUCTURE	
Multicellular	*Cells*	Multicellular
S & W-NW flank	*Discrete development*	N of old cells
	Contin. development	Western flank
From S & SW, later from W	*Inflow*	(from SSW & SW)
Fluctuating	*Updraught*	Fairly steady
Tilted	*Echo core*	No tilt
During early stages	*WER*	None
1½ - 2 hours	*Lifetime of cells*	±1 hour
Stationary	*Movement*	Stationary for ½ hour
	ENVIRONMENT	
-6.5	*Showalter stab. index*	-3.2
Strong directional	*Wind shear*	Speed only

In the case of the cloudburst the air is thought to have
entered the storm from the south-southwest and northwest. A more
definite statement can be made about the hailstorm, since it displayed a significant tilt of the echo core and at times a WER.
During the early stages the inflow was from the south and southwest. Later, air was entering all along the western flank of the
storm, where there was a strong reflectivity gradient. The fact
that the high-reflectivity area aloft was quite variable in time,
position and volume, leads to the conclusion that there must have
been a very strong, but fluctuating, possibly pulsating updraught.
This would then also explain the various bursts of differently
sized hailstones, with the very large hailstones falling about
halfway through the mature stage of the storm.

The upper air sounding at 14h00 (1200 GMT) indicated an
unstable atmosphere on both days, but the Showalter index was
considerably more negative on the day of the hailstorm. The vertical wind shear certainly also contributed towards the severity
of the hailstorm. As can be seen from Figure 8, there was a
strong directional wind shear from the low (L = below cloud base)
to the middle level (M ≈ 5-6 km AGL) of the atmosphere from where
it veered to the higher levels (H ≈ 10 km AGL) on the day of the
hailstorm. The wind speeds did not differ significantly from one
level to another. However, on the day of the cloudburst the wind
speed almost doubled from the middle to the high levels, with very
little variation in direction. But over and above this there must
have been other differences, e.g. in the internal airflow,
to account for such different end products of these two storms.

There were also a number of *similarities* in environmental
conditions under which the storms occurred as well as in their

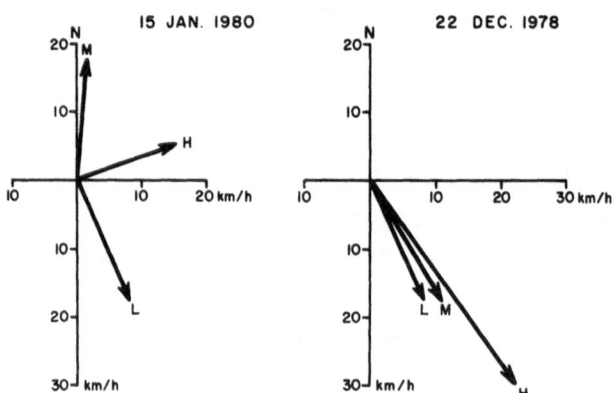

Figure 8.   The upper winds at 3 different levels. (L = below
cloud base, M ~ 5-6 km AGL, H ≈ 10 km AGL.)

behaviour: On both days the thermodynamic conditions were quite favourable for the development of severe storms; from the sparse network of surface wind recorders and the behaviour of surrounding storms a zone of fairly strong local convergence near the airport was inferred; both storms were multicellular and behaved quite differently from others on the same day, i.e. they were considerably more severe and stationary during most of their lifetime; the storm areas and echo cores had more or less similar magnitudes and both storms penetrated the tropopause which happens only in exceptional cases on the South African plateau. In terms of graphs which relate the height of the 40 dBZ radar reflectivity contour to the probability of hail on the ground, both storms had extremely high probabilities for producing large hailstones (5). Even the cloudburst had a probability of > 95 per cent to produce hailstones > 1 cm in diameter on the ground. Yet only very isolated reports of light falls of hailstones ≤ 1 cm in diameter were reported.

5. CONCLUSIONS

Two similar-looking thunderstorms of which one produced an extremely heavy cloudburst and the other hailstones > 5 cm in diameter were analysed in great detail, but remarkably few differences were found: Both were multicellular complexes with continuous and discrete development; they remained more or less stationary for most of their lifetime, which would explain the long point durations of the severe hailfall and the extended periods of high rainfall rates (174 mm/h during 20 minutes, followed by 55.2 mm/h for 30 minutes) yielding a large rainfall total of the cloudburst (88.2 mm). The three-dimensional structure of the radar echoes, however, differed somewhat, since only the hailstorm showed a significant tilt of the echo core and occasionally a well-pronounced weak echo region (WER), which together with a steep reflectivity gradient and unusually high reflectivities aloft marked the position of a strong updraught on its western flank. This updraught was thought to have pulsated in the hailstorm while it appeared to have been more steady in the cloudburst. Also, the Pretoria upper air sounding indicated somewhat greater instability on the day of the severe hailstorm than on 22 December 1978.

The comparison of these storms illustrates clearly the difficulties which are encountered in discriminating between a severe hailstorm and a heavy cloudburst. Perhaps parameters other than the conventional ones should be sought additionally (e.g. the electrical structure and behaviour of storms) to differentiate between storms. Airflow measurements in and around the storms, e.g. by a multiple Doppler radar system and a surface meso-network, would have also provided some valuable information.

ACKNOWLEDGEMENTS

The author would like to thank his colleagues who maintain and operate the CSIR radar, as well as all the hail observers who contribute diligently to the study of hailstorms on the South African Highveld.

REFERENCES

(1)   Carte, A.E.: 1977, S.A. Journal of Science 73, pp. 327-330.

(2)   Held, G.: 1973, J.de Rech. Atmos. 7, pp. 185-197.

(3)   Held, G.: 1974, Pure and Applied Geophysics 112, pp. 765-776.

(4)   Carte, A.E. and Held, G.: 1978, J. Appl. Meteor. 3, pp. 365-373.

(5)   Held, G.: 1978, J. Appl. Meteor. 17, pp. 755-762.

(6)   Held, G.: 1980, Meteorol. Rdsch. 33, pp. 37-42.

USE OF THE RADAR DIFFERENTIAL REFLECTIVITY RADAR TECHNIQUE FOR OBSERVING CONVECTIVE SYSTEMS

T. A. Seliga, K. Aydin[1], C. P. Cato and V. N. Bringi[2]

Atmospheric Sciences Program and Department of
Electrical Engineering, The Ohio State University,
Columbus, Ohio 43210, U.S.A.

The differential reflectivity (Z_{DR}) radar technique is a new
observational tool for measuring important parameters of convective storm systems. The measurement depends on two fundamental
naturally occurring properties of hydrometeors -- their non-
sphericity and their tendency toward common alignment, usually
along the vertical axis. Evolution and structure of storms are
particularly susceptible to Z_{DR} measurements which accurately
estimate drop size distributions, rainfall rate and water content,
and provide a detection capability of hydrometeor phase. Several
examples of radar observations in strong convective storms
obtained during the SESAME 1979 field program are presented.
The results are particularly well-suited for statistical
descriptions of storm behavior.

1. INTRODUCTION

The first measurements of the differential reflectivity (Z_{DR})
radar signal [1, 2, 3, 4] have clearly indicated its potential
and illustrate that its role in meteorological research, particularly in the study of convective storms, is very great.
Potential applications include: studies of precipitation rates,
water content and drop size distributions; hydrometeor phase
detection; rainout and washout processes; cloud-scale climatology;
effects of natural terrain and anthropogenic activities on storm
development and outputs; mesoscale organization; characterization
and evolution of storms; hydrology; forecasting; cloud modeling;
weather modification; etc. This paper examines this potential
by reviewing the technique and demonstrating how Z_{DR} measurements
might influence such problem areas. Examples of data obtained

E. M. Agee and T. Asai (eds.), Cloud Dynamics, 285–300

during the SESAME 1979 field program in Oklahoma are used to support these assertions.

2. REVIEW OF THEORY

The differential reflectivity (Z_{DR}) radar technique, introduced by Seliga and Bringi [5], provides a methodology of obtaining both quantitative and qualitative information about the hydrometeors which are responsible for producing meteorological radar signals. In rainfall, raindrops flatten into nearly oblate spheroidal shapes, exhibit greater deformation with increasing size and tend toward a common alignment with their axes of revolution vertical. Consequently, the ratio of reflectivity factors ($Z_{H,V}$) at horizontal and vertical polarizations

$$Z_{DR} = 10 \log (Z_H/Z_V) \quad dB \tag{1}$$

is a measure of the median volume diameter, D_o, of the usually applicable exponential raindrop size distribution,

$$N(D) = N_o \exp(-3.67 \, D/D_o) \quad 0 < D \leq D_m, \quad m^{-3} cm^{-1} \tag{2}$$

where D is the equivolumic diameter of the nonspherical raindrops, N_o is the distribution's intensity or magnitude and D_m is the maximum drop size diameter. Combined measurements of (Z_{DR}, $Z_{H,V}$), therefore, provide estimates of (N_o, D_o), since

$$Z_{H,V} = \frac{\lambda^4}{\pi^5 |K|^2} \int_0^{D_m} \sigma_{H,V} \, N(D) \, dD \tag{3}$$

λ is the radar signal wavelength; $K = (\varepsilon_r - 1)/(\varepsilon_r + 2)$, ε_r being the relative dielectric constant of water; and $\sigma_{H,V}$ are the radar cross sections of the oblate spheroidal drops at horizontal and vertical polarizations. $\sigma_{H,V}$ are functions of drop size according to the axial ratios of the drops [6, 7]. In turn, the two drop size distribution parameters (N_o, D_o) can be used to compute estimates of rainfall rate (R) or liquid water content (M). For example,

$$R = 0.6\pi \int_0^{D_m} D^3 v(D) \, N(D) \, dD \quad mm \, h^{-1} \tag{4}$$

$$M = \frac{\pi}{6} \int_0^{D_m} D^3 N(D) \, dD \quad g \, m^{-3} \tag{5}$$

where v(D) is the terminal velocity (m s^{-1}) [8]. By eliminating

N_O and D_O using Eqs. (1), (3) and (4), R can be directly related to Z_H and Z_{DR} by the approximate relationship

$$Z_H = 147R(Z_{DR} + 0.474)^{2.454} \quad 0.57 \leq Z_{DR} \leq 4.8dB \qquad (5)$$

where it is assumed that $\lambda = 10cm$, $D_m = 1cm$, R is in mmh^{-1} and Z_H in mm^6m^{-3}.

In addition to the quantitative utility of the Z_{DR} technique, qualitative information about the presence of ice particles such as hail and/or ice crystals as well as hydrometeor phase change may also be obtained from the Z_{DR} radar signal. For example, measurements of Z_{DR} as a function of height clearly showed that $Z_{DR} \approx 0$ dB above the melting layer, presumably corresponding to the presence of randomly oriented ice crystals [2]. Z_{DR} also increased with decreasing height through the melting layer, reaching a maximum value before returning to a value corresponding to the raindrop size distribution existing below the layer. The detection of hail of certain shapes and sizes also appears possible. In particular, at $\lambda = 10cm$, moderate sized oblate-spheroidal hail, falling with preferred orientation, is capable of producing $Z_{DR} \gtrsim 4dB$; nearly spherical hail would produce a $Z_{DR} \approx -0dB$; and conical shaped hail would produce $Z_{DR} \approx -1dB$ [9, 10]. It is obvious that such inferences are not unique. However, when combined with other radar observations such as reflectivity, Doppler velocity and dual-wavelength signals, they should be possible with good reliability.

3. RAINFALL MEASUREMENTS

To date there have been several opportunities to compare radar estimates of rainfall with ground-based and in-situ measurements. These experiments utilized raingages, disdrometers and airborne spectrometer probes. The results are summarized below.

3.1. Raingage Network Comparisons

Seliga et al. [3] compared radar measurements of average rainfall rate over two separate 550 km^2 areas with raingage measurements in the same areas over time intervals of about 1 h duration in Chicago during the summer of 1978. The (Z_{DR}, Z_H) measurements were obtained with the CHILL radar, operated by the Illinois State Water Survey (ISWS), in conjunction with measurements from the Chicago Hydrometeorological Area Project's (CHAP) raingage network. Although the average rainfall rates were relatively light over the two areas (2.80 and 1.20 mm h^{-1}), excellent agreement was found between the radar Z_{DR}-derived rates and the gage values. Over one of the areas the Z_{DR} method gave 86% of the average of 26 gages while the other gave 117% of the average of

27 gages. Subdivision of the two larger 550 km^2 areas into 14
subareas provided a basis for added statistical comparison of
the results. The absolute average difference between the gage
and radar estimates was 22% using the Z_{DR}-derived values; this
compared with a 42% difference obtained when using a calibrated
Z-R relationship.

3.2. Disdrometer Rainfall Rate Comparisons

During the same field experiments in Chicago an opportunity for
heavy rainfall rate comparisons occurred when a storm progressed
directly toward the radar from the West [4]. A ground-based
Joss-Waldvogel type disdrometer was operating at the radar site
from which rapid response rain rates were derived. Under the
assumption of steady-state, range-dependent radar-derived rainfall
rates could be compared with time-dependent disdrometer values.
The radar measurements supported the steady-state hypothesis
which in turn led to the comparative results shown in Fig. 1.

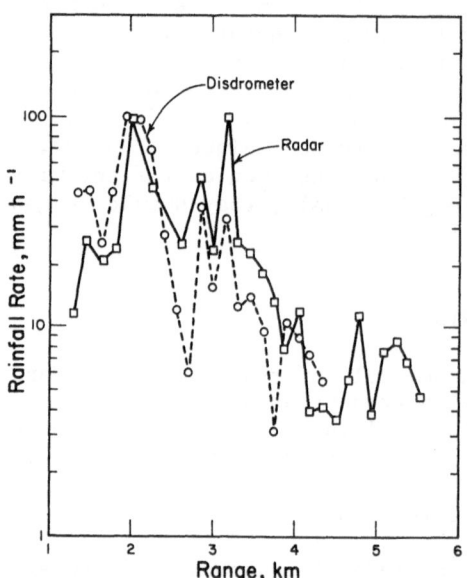

Figure 1. Comparison of disdrometer- and radar-derived
rainfall rates [4].

Obviously, the agreement is excellent with the radar describing
not only the variability of the rainfall very well but also the
intensity.

3.3. Disdrometer Differential Reflectivity Comparisons

A joint research program between The Ohio State University and
the Rutherford-Appleton Laboratory, U. K., produced simultaneous
radar and disdrometer measurements at the same location in England
using the Chilbolton radar system [11, 12]. The drop size distributions derived from the disdrometer were used to calibrate
Z_{DR} based on equilibrium drop shapes [6]. Comparison of these
with simultaneous radar observations of Z_{DR} are shown in Fig. 2.
Very good agreement occurred, particularly considering the possible sources of errors such as the different sampling volumes
(e.g., radar sampling volume > 10^4 x disdrometer sampling volume),
effects of wind shear, disdrometer limitations in detecting small
droplets, etc.

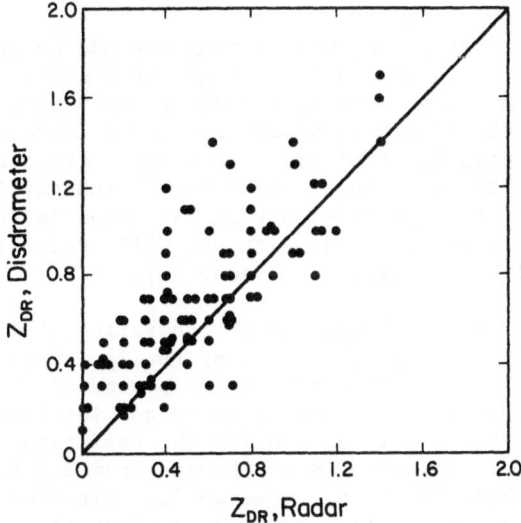

Figure 2. Comparison of disdrometer-derived
and radar-measured Z_{DR}'s [11, 12].

3.4. Aircraft Median Volume Diameter Comparison

Stickel and Seliga [13] reported on comparison of an airborne
measurement of median volume diameter (D_0) with ground-based
radar observations of the circular depolarization ratio (CDR)
which is directly related to Z_{DR} and thus to D_0. The average
CDR value measured by the radar was -21.2dB which corresponds to
a Z_{DR} of about 1.5dB which in turn gives a $D_0 \doteq$ 1.4 mm. The
airborne data yielded an integrated value for D_0 of 1.31 mm
during the 3 min observation time of the combined radar-aircraft
correlation measurements -- a very favorable comparison.

3.5. Rainfall Summary

All of the quantitative estimates of rainfall rates, thus far
obtained with the Z_{DR} radar technique, have compared very favor-
ably with other measurements as indicated above. Although much
more additional data and verification are needed, it is clear
that the dual polarization Z_{DR} method has great potential for
improving significantly the remote measurement of rainfall and
its characterization in terms of the parameters (N_o, D_o) of its
drop size distribution. This potential is illustrated further
in section 6 by examining data obtained during the SESAME 1979
field program in Oklahoma with the CHILL radar.

4. HYDROMETEOR PHASE

Since the dual polarization Z_{DR} measurement depends on both the
non-spherical shape and common alignment of the hydrometeors in
the radar scattering volume, inferences about the presence of
non-raindrop hydrometeors are possible from Z_{DR} measurements
whenever the Z_{DR} signal is atypical of raindrop scattering or
behaves uncharacteristically in space or time. Indeed, this
atypical behavior of the Z_{DR} radar signal was noted in the first
measurements obtained in the U. S. with the CHILL radar [1, 14]
and in the U. K. with the Chilbolton radar [2].

The CHILL data were taken in Oklahoma in the spring of 1977 and,
in addition to confirming the existence of the Z_{DR} signal and its
range of variation in rain, illustrated the potential of Z_{DR} for
the detection of other types of hydrometeors and in-cloud electric
fields. Fig. 3 indicates this by showing the variations in Z_H
and Z_{DR} with range as elevation angle changes from 1.3 to 2.1 to
$4.1°$. At $1.3°$ Z_{DR} behavior is as expected for rain ($Z_{DR} < 5dB$).
However, as the elevation angle increases, causing the radar beam
to pass through greater heights, Z_{DR} changes dramatically. At
$2.1°$ the fourth storm cell at a height of around 2.8 km gives Z_{DR}
values varying slightly around 0dB over a range of around 3 km
compared to 2–3dB at the same range (75 km) when the beam was at
$1.3°$ elevation angle. The $4.1°$ beam exhibits additional changes
in Z_{DR}: moderate negative values in the first cell; strongly
negative values in the second cell; positive values in the third
cell; and values near zero in the fourth cell. A number of
hypotheses, including the presence of ice crystals, hail and/or
graupel of different shapes, sizes and orientations, and the
influences of electric fields and motion on these, may explain
these results. Nevertheless, the data show that the in-cloud
hydrometeor characteristics are very different than below-cloud
and that this information can be used to describe cloud behavior
and evolution. The U. K. results [2] also support the use of
Z_{DR} for hydrometeor phase detection; Hall et al.'s data, when

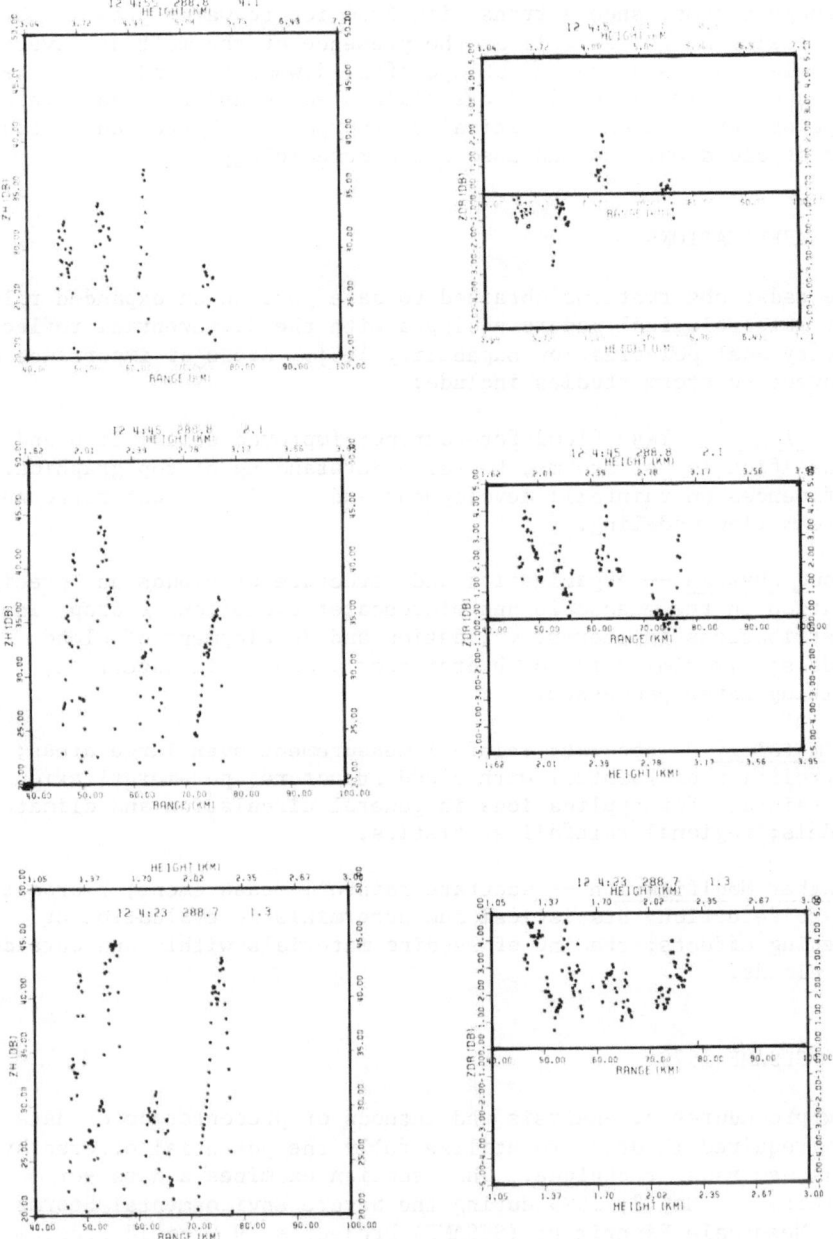

Figure 3.   Variations of Z_H(dBz) and Z_{DR}(dB) vs. range for three rays at elevation angles of 1.3°, 2.1° and 4.1°.

presented in color or gray-scale graphics of vertical scans
through storms, show a transition from ice to water phase with
decreasing height as well as the presence of the melting layer.
In selected cases regions of updrafts, downdrafts and the presence
of super cooled water droplets could also be ascertained. This
type of information is critically important to better understanding of cloud physics and associated meteorology.

5.   APPLICATIONS

The radar observations obtained to date portend an expanded role
for meteorological radars equipped with the differential reflectivity dual polarization capability [15]. Areas of importance to
convective storm studies include:

Hydrology -- flash flood forecasting; improved description and
classification of storms; better understanding of topographical
influences on rainfall; development and testing of watershed and
stream flow modeling.

Cloud Physics -- organization and structure of clouds and precipitation on the mesoscale and microscale; evolution of drop size
distributions in storms; validation and development of cloud
models; discrimination of hydrometeor phase; hail detection;
melting layer processes.

Climatology -- accurate rainfall measurement over large areas;
correlation of rainfall with cloud structure; parameterization
of rainfall for applications in general circulation and climate
models; regional rainfall statistics.

Weather Modification -- accurate rainfall measurement; hydrometeor
phase detection; statistical and deterministic evaluation of
seeding effects; tracing of seeding materials within and outside
of clouds.

6.   SESAME 1979

New procedures of analysis and methods of presentation of data
are required in order to utilize fully the potential offered by
this new radar technique. This section examines a data set
obtained on May 2, 1979 during the Severe Environmental Storms
and Mesoscale Experiment (SESAME) Project's 1979 Field Program
conducted in Oklahoma and surrounding states. The emphasis here
will focus on Z_{DR} data interpretation and presentation rather
than on meteorological commentary. The consequences for the
latter should be apparent.

6.1. Example of Single Ray Analysis

During the evening of May 2, 1979, an isolated thunderstorm
developed southwest of the CHILL radar site in Anadarko, Oklahoma
as a remnant of a squall line which developed in the early evening
along a northeast-southwest line and travelled generally in a
southeasterly direction [16, 17]. This storm was reported to
have produced wind damage in Lawton and Duncan, Oklahoma.

Fig. 4 is a composite picture of the radar measurements (Z_H, Z_{DR})
for a single ray taken from this storm and their interpretation
in terms of D_o, N_o and R. A plot of R, calculated from the
Marshall-Palmer Z-R relationship $Z = 200R^{1 \cdot 6}$, is also shown for
comparison. In this particular region of the storm the rainfall
is fairly uniform in size with D_o generally varying around 0.2 cm
and reaching minimum and maximum values of around 0.12 and 0.36 cm,
respectively. The N_o Marshall-Palmer value of 80,000 $cm^{-1}m^{-3}$
is indicated by the straight line on the N_o plot. It is interesting to note that in the region where N_o is near this value the
two radar estimates of R agree very well as expected. Below
40 km range, however, where N_o < 80,000 the discrepancy between
R's is as large as a factor of ten, demonstrating the importance
of using two radar observables to estimate R instead of a single
observable such as reflectivity factor Z. These results also
illustrate radar's ability to deduce the temporal and spatial
variability of storm behavior. Several methods of examining
these storm properties are presented in the next section.

6.2. Squall Line Properties

The squall line of May 2, 1979, noted above and shown in Fig. 5,
provided an opportunity for Z_{DR} measurements encompassing nearly
all aspects of severe convective storm activity. The Z(dBz)
contour plot shows the location and extent of the line relative
to the location of the CHILL radar. Regions of the storm designated A-E coincide with values of reflectivity factor equal to
33 dBz and are the general locations of major thunderstorms
embedded within the squall line. The results presented here
were taken from the PPI scan used to produce Fig. 5. Note that
only five such scans were available over a 30 min observational
period because of the slow-switching capability employed by the
CHILL radar during SESAME [18, 1] and because of other project
priorities. During the Z_{DR} measurements the squall line moved
steadily as a whole toward the southeast at a speed of approximately 10 - 15 m s^{-1}. The storm also grew rapidly in size and
intensity. Comparison of the first and last PPI scans indicated
around a 200% increase in the land area covered by the observed
portion of the storm (from 300 to 1,000 km^2) and that this growth
was primarily due to a lateral expansion of the squall line.

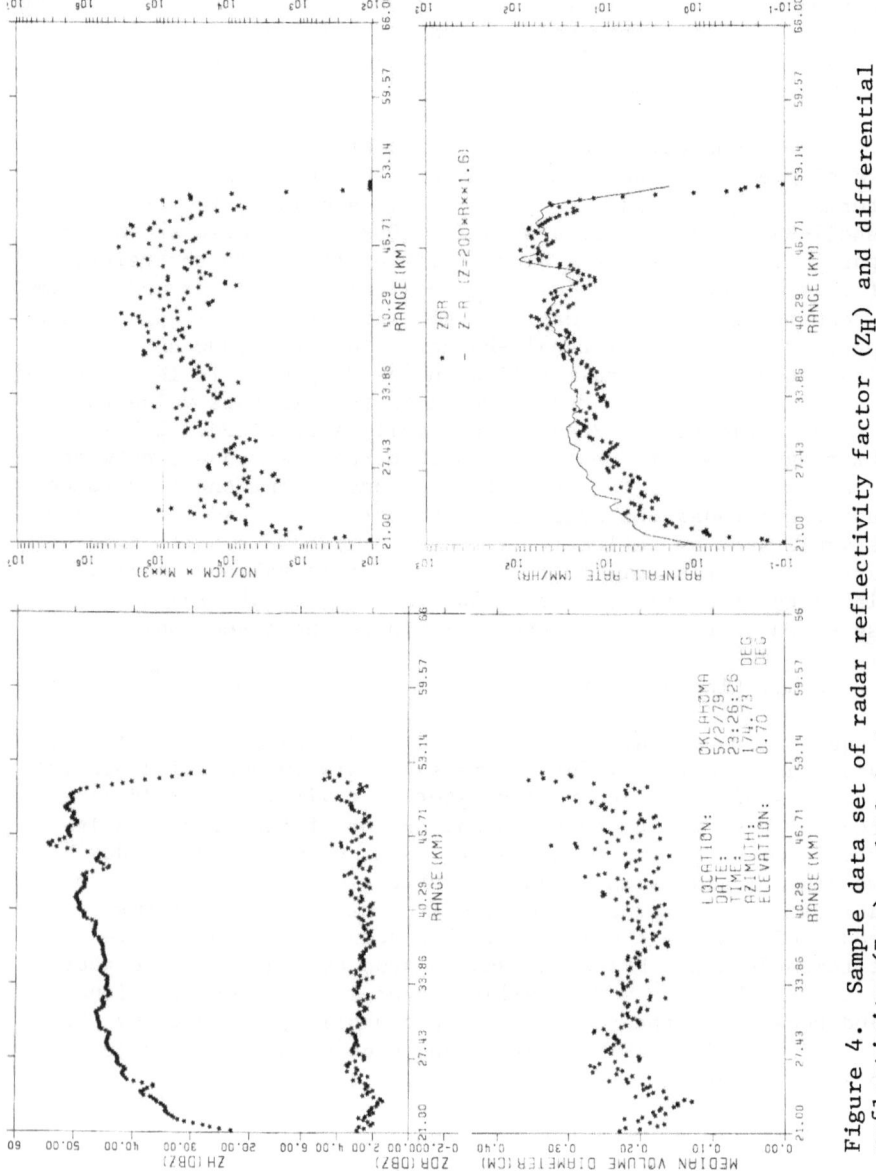

Figure 4. Sample data set of radar reflectivity factor (Z_H) and differential reflectivity (Z_{DR}) and their interpretation in terms of the exponential drop size distribution parameters (N_o, D_o) and rainfall rate (R).

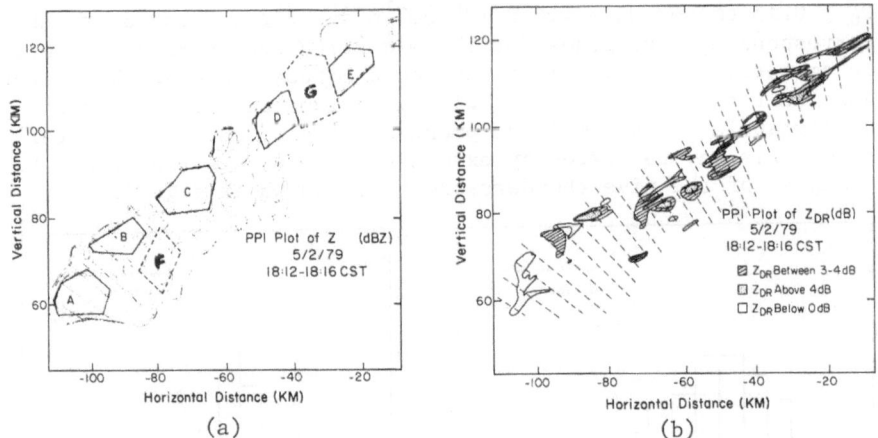

(a)   (b)

Figure 5. (a) PPI plot of reflectivity factor Z_H and
(b) regions of selected differential reflectivity Z_{DR}
values at 18:12 CST, May 2, 1979. Major thunderstorm
cells are marked A through E. Note the scale differences along the axes.

In order to illustrate the potential utility of Z_{DR} radar measurements for investigating storm behavior, relative frequency histograms and scatter diagrams of the measurements of Z_{DR} and Z_H are
useful along with the spatial and temporal information on the
location and occurrence of events of interest. Fig. 6 gives the
relative frequency histograms of reflectivity factor Z_H and Z_{DR}
obtained from the entire scan. The corresponding mean values of
Z_H and Z_{DR} for these data are 32dBz and 1.75dB, respectively.
Excluding values of $Z_{DR} \leq$ 0dB yielded means of approximately

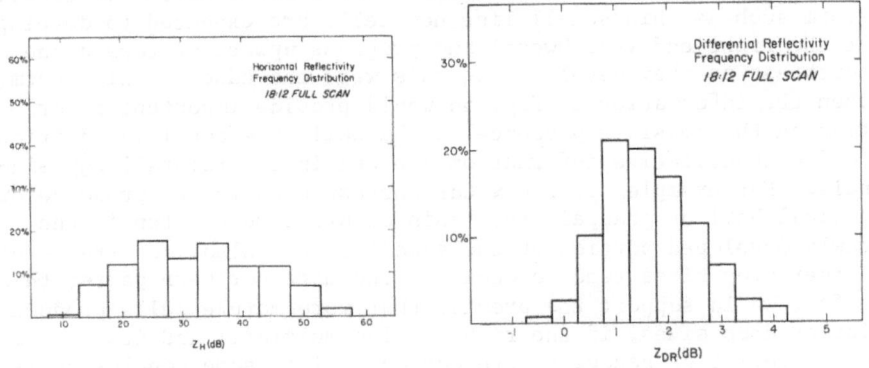

Figure 6. Relative frequency histograms of Z_H
($Z_H \leq$ 10dBz) and Z_{DR} for the entire squall line
scan at 18:12.

$\overline{D_O}$ = 0.15 cm and $\overline{N_O}$ = 4.6 x 10^8 $cm^{-1}m^{-3}$. Figs. 7 and 8 are the corresponding histograms for these derived drop size distribution parameters. Such distributions could be used to describe a storm and their changes in time or with location could further the understanding of storm development and evolution. This latter hypothesis may be tested by examining the Z_{DR} distributions, for example, of the five thunderstorm cells shown in Fig. 5.

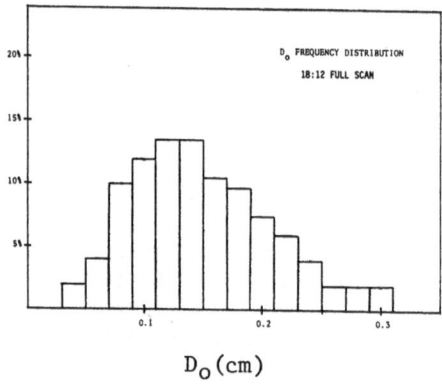

D_O (cm) N_O $(cm^{-1}-m^{-3})$

Figure 7. Frequency distribu-   Figure 8. Frequency distribution of D_O for data from the   tion of N_O for data from the
full scan at 18:12 CST.   full scan at 18:12 CST.

Figure 5b displays the spatial locations where relatively large, large and negative values of Z_{DR} occurred. It is seen that these regions of Z_{DR} occur in the general vicinity of the major cells. Regions of minus Z_{DR} occur predominantly in the southwest portion while most of the large values (>4dB) appear mostly near the upper, northeast end of the storm. For a right-moving multicell storm such as this squall line new cells are expected to develop at the right end (southwest) and progress upward or toward the northeast as they develop. If this were the case in this storm, then the information in Fig. 5b would provide important information on the possible presence of the hail or graupel in addition to the quantitative information present in the rainfall Z_{DR} signals. For example, if $Z_{DR} \lesssim$ 0dB corresponded to the presence of conical hail or graupel, then this occurred more often in the newly developed portion of the squall line. Also, occurrence of larger drop sizes tend to occur in the more northern parts; this effect would support the premise that more mature cells produce larger drop sizes, if the right-moving multicell and its presumed characteristics behavior were correct. This same conclusion is supported by examination of the frequency distributions of Z_{DR} in cells A-E which are shown in Fig. 9. Assuming that the right-most cells (A, B) of the forward moving storm are the youngest, a correlation appears to be present between the occurrence of low

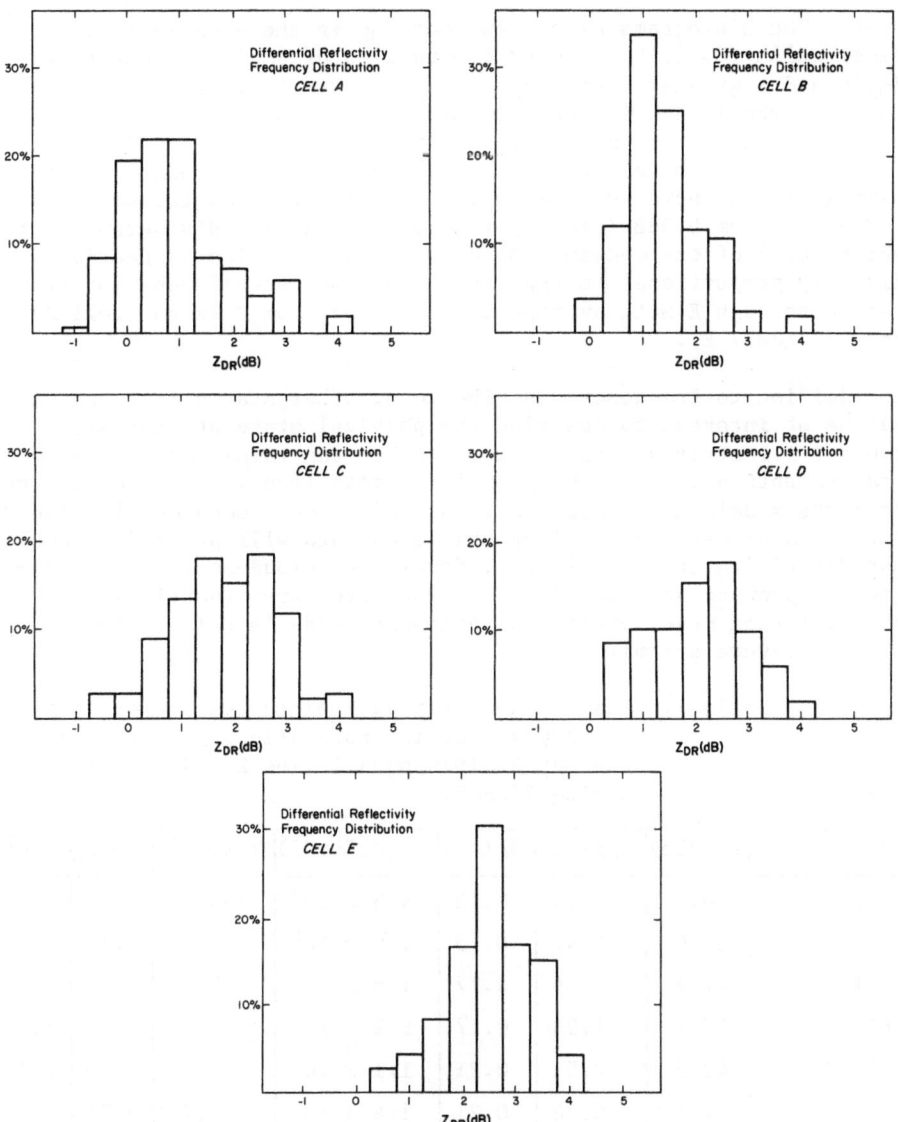

Figure 9.   Histograms of Z_{DR} for thunderstorm
cells A–E shown in Figure 5.

and negative Z_{DR} values and the newness of the cells. These
phenomena indicate the probable presence of graupel and/or
conical hail in these cells which would suggest cloud processes
and dynamics favorable to such hydrometeor development occurs
more readily in newly developed thunderstorms than in more mature

ones. The histograms also show that Z_{DR} in the progressively
older cells (D, E) have lower frequencies of small and negative
Z_{DR}'s and a greater frequency of $Z_{DR} \gtrsim 4dB$. A plausible explanation for the latter is that the older cells in the northeast
portion of the storm may contain large, oblate spheroidal hail.
The sequence from cell A-E shows a tendency for the width of the
histograms to increase from cell A to cell E; also, the mean Z_{DR}
increases from 0.9dB (cell A) to 2.5dB (cell E). Histograms (not
shown here) of the median drop diameter (D_0), which is nearly
directly proportional to Z_{DR}, show the same general behavior from
cell A through E with average D_0's ranging from 0.12 cm (cell A)
to 0.21 (cell E).

In addition to frequency distributions, other statistical measures
may be of interest to describe the physical state of storms.
Table 1 illustrates this by listing the mean values of the radar
measurements and their physical interpretations which were derived
from the squall line data base. As this radar technique is placed
into greater use, it is clear that such data will prove invaluable
for describing storm evolution, developing cloud-scale climatolo-
gies, improving understanding of cloud processes (particularly
precipitation development) and extending radar's role in forecasting severe storms.

Table 1. Mean values of radar-related parameters and
their physical interpretation. Derived from
the May 2, 1979 squall line Z_{DR} data set
during SESAME.

	$\overline{Z}_H$ (dBz)	$\overline{Z}_{DR}$ (dB)	$\overline{D}_0$ (cm)	$\overline{N}_0$ (cm^{-1}m^{-3})	$\overline{R}$ (mmh^{-1})	$\overline{\log N_0}$	$\overline{\log R}$
Cell A	46.0	0.93	0.12	9.3×10^9	760	6.69	2.1
Cell B	38.0	1.42	0.13	1.2×10^8	35	5.15	1.0
Cell C	47.7	1.86	0.17	1.5×10^8	250	5.45	1.8
Cell D	42.7	1.94	0.17	9.2×10^6	84	4.96	1.2
Cell E	43.5	2.51	0.21	1.1×10^7	81	4.30	1.1
Sector F	24.3	0.38	0.21	1.4×10^5	0.7	3.91	-0.32
Sector G	34.5	1.93	0.17	5.5×10^5	3.9	4.02	0.38
Squall Line	32.3	1.75	0.154	4.5×10^8	64	4.26	0.34

7. CONCLUSIONS

This paper reviewed the theoretical foundation of the differential
reflectivity radar technique and presented evidence supporting
both its quantitative and qualitative uses in the study of

convective storms. Although the data base available to date is highly limited because of the newness of the technique and the lack of adequate numbers of properly instrumented radars, future research and applications employing Z_{DR} measurements appear unlimited. The reason for this expectation is that the Z_{DR} signal adds a new dimension to radar observations -- one which significantly improves the radar's capability for interpreting reflectivity fields. This second dimension is essential since rainfall is at best a two parameter problem, size and number density, and because hydrometeor phase is at the very least a two-state question, water or ice. These fundamental facts have limited the meteorologist's ability to interpret reflectivity factor data by themselves. As illustrated here, utilizing reflectivity factor measurements at both horizontal and vertical polarizations as is done with the Z_{DR} radar technique should significantly alleviate this deficiency in the future.

ACKNOWLEDGMENTS

This research was supported by the Atmospheric Research Section, National Science Foundation, under Grant Nos. ATM-7908666 and ATM-8003376. Additional support for collaboration with the Rutherford and Appleton Laboratory (RAL) was provided under NATO Research Grant No. RG 054.80. Both of the disdrometer results cited herein were obtained with an instrument provided by the Air Force Geophysics Laboratory. The collaboration and assistance of E. A. Mueller of the Illinois State Water Survey relative to operation of the CHILL radar and of M. P. M. Hall, S. M. Cherry and J. S. F. Goddard of RAL relative to operation of the Chilbolton radar are gratefully acknowledged. Their contributions to the dual polarization Z_{DR} research have been truly outstanding.

---

[1]Dr. Aydin is currently on leave from the Electrical Engineering Department, Middle East Technical University, Ankara, Turkey.

[2]Dr. Bringi is presently with the Department of Electrical Engineering, Colorado State University, Fort Collins, Colorado.

REFERENCES

1. T. A. Seliga, V. N. Bringi and H. H. Al-Khatib, 1979: IEEE Trans. Geos. Elect., GE-17, pp. 240-244.
2. M. P. M. Hall, S. M. Cherry, J. W. F. Goddard and G. R. Kennedy, 1980: Nature, 285, pp. 195-198.
3. T. A. Seliga, V. N. Bringi and H. H. Al-Khatib, 1981: J. Appl. Meteor., 20, pp. 1362-1368.

4. V. N. Bringi, T. A. Seliga and E. A. Mueller, 1982: IEEE
    Trans. Geos. Elect. (in print).

5. T. A. Seliga and V. N. Bringi, 1976: J. Appl. Meteor., 15,
    pp. 69–76.

6. H. R. Pruppacher and K. V. Beard, 1970: Quart. J. Roy.
    Meteor. Soc., 96, pp. 247–256.

7. A. W. Green, 1975: J. Appl. Meteor., 14, pp. 1578–1583.

8. R. Gunn and G. D. Kinzer, 1949: J. Meteor., 6, pp. 243–248.

9. V. N. Bringi and T. A. Seliga, 1977: Ann. des Telecomm., 32,
    (11–12), pp. 392–397.

10. T. A. Seliga and V. N. Bringi, 1978: Radio Sci., 13(2),
    pp. 271–275.

11. T. A. Seliga and V. N. Bringi, 1981: "The Differential
    Reflectivity Dual Polarization Method of Rainfall Measure-
    ments," in *Precipitation Measurements from Space*, Workshop
    Report, D. A. Atlas and O. W. Thiele, eds., NASA Goddard
    Space Flight Center, Oct. 1981, pp. D98–D104.

12. J. W. F. Goddard, S. M. Cherry and V. N. Bringi, 1982: J.
    Appl. Meteor. (accepted for publication).

13. P. G. Stickel and T. A. Seliga, 1981: "Cloud Liquid Water
    Content Comparisons in Rain Using Radar Differential Reflec-
    tivity Measurements and Aircraft Measurements," Preprints
    AMS 20th Conf. on Radar Meteor., Nov. 30 – Dec. 3, 1981,
    Boston, Mass., pp. 567–571.

14. H. H. Al-Khatib, T. A. Seliga and V. N. Bringi, 1979: Differ-
    ential Reflectivity and Its Use in the Radar Measurement of
    Rainfall, Atmos. Sci. Prog. Rept. No. AS-S-106, Ohio State
    Univ., Columbus, Ohio, April 1979.

15. T. A. Seliga, 1980: Nature, 285, pp. 191–192.

16. G. M. Heymsfield, S. Schotz and R. Blackmer, 1981: "Structure
    of Thunderstorms Along a Squall Line on May 2, 1979," Pre-
    prints AMS 20th Conf. on Radar Meteor., Nov. 30 – Dec. 3, 1981,
    Boston, Mass., pp. 44–51.

17. T. A. Seliga, J. R. Peterson and V. N. Bringi, 1981: "Hydro-
    meteor Characteristics in the May 2, 1979 Squall Line in
    Central Oklahoma as Obtained from Radar Differential Reflec-
    tivity Measurements During SESAME," Preprints AMS 20th Conf.
    on Radar Meteor., Nov. 30 – Dec. 3, 1981, Boston, Mass.,
    pp. 561–566.

18. E. A. Mueller and E. J. Silha, 1978: "Unique Features of the
    CHILL Radar," Preprints 18th AMS Conf. on Radar Meteor.,
    Atlanta, Georgia, 28–31 March, pp. 381–382.

COMPLEX RADAR METHODS OF HAIL CLOUD STRUCTURE, EVOLU-
TION DYNAMICS AND MICROSTRUCTURE

M.T.Abshaev

High Mountain Geophysical Institute,
Nalchik, USSR

Methods and equipment for complex radar investigation
of structure, evolution dynamics and microstructure
of hail clouds and updrafts inside them using multi-
wave active and passive, coherent and incoherent ra-
diolocation and rocket sounding are discribed. Classification of hailstorm processes in the North Caucasus
according to their structure and evolution dynamics
is given. Results of investigation of hail cores in
hailstorm processes of various types and regularities
of hail formation process distribution in space are
considered. Recommendations for hailstorm modification
with an account of hailstorm's structure and evolution
dynamics are presented.

For using of macro - and microphysical characteristics of hailstorms with the purpose of their classification and improvement of the modification method
on them, the procedure of complex hailstorm study which
provides three-dimentional pattern of structure and
hail cloud cell structure, structure of individual convective cells, their evolution dynamics, hail core lo-
calization, obtaining of air flow structure, cloud microstructure and the type of falling precipitation depending on thermodynamic instability, water content and
wind regime in the atmosphere has been developed by the
laboratory of radar meteorology of High-Mountain Geophysical Institute.
    The difficulty of the realization of such comp-
lex investigations is in organization of simultineous

E. M. Agee and T. Asai (eds.), Cloud Dynamics, 301–313.
Copyright © 1982 by D. Reidel Publishing Company.

observations by active, passive, coherent and incoherent multiwavelength radar,and also by combining the total obtained information and its interpretation.

The complex of research equipment includes:
-highpotential radar hail detector MRL-5 (3,2 and 10cm) with devices of multicontour iso-echo, hail core selection, turbulence zone determination, range correction system and photo-registration equipment/1/;
-three-wavelength (2,0; 3,2 and 10cm) experimental radar with devices of multicontour iso-echo, hail core selection, range correction system, descrete setback of selector pulses, pulse signal registration and radial distribution of echo power at three wavelengths, power difference of echo at each two pairs of wavelengths and attenuation coefficient at 2,0 and3,2cm/2/;
-pulse-coherent decimetre range radar, combined with incoherent pulse RLS of 0,86cm range and equiped with multichannel optical spectrum analizer of Doppler frequenceis with control display, providing during the observations reflection velocity spectrum of hydrometeor vertical movement in height-range coordinates, and also by devices of descrete setback of selector pulse registration of echo power and Doppler characteristics on self-recorder and magnetic tape at every 210 height /6/;
-pulse-coherent centimetre range wavelength radars, providing the obtaining of horisontal flow structure in horisontal and slope sounding;
-rocket sounding apparatus, including radar "Meteorit" and rocket sonde, ensuring in regime of automatic following the obtaining of vertical and horisontal flow velocities, temperature and three components of electric field on the flight track in cloud on the parachute /2,3/;
- radar-radiometric station (RRS) having high potential radar channel of 10cm range and two radiometric channels (3,2 and 10cm) with combined diagrams of direction, equipped with registering device of cloud thermal radio-radiation intensity /4/;
-control computer "Electronics K-200", used during observations for processing the information, more bulky in computing plan;
- apparatus of obtaining and microfilming of thin hailstone sections to study their structure and hail embryo nature /8/.

For good coordination the apparatus complex was compactly installed taking into account the cooperation of some groups of observers.

Mutual noise of RLS was excluded by their discrimination on frequency and polarization and also

using narrow band preselectors. For the synchroniza-
tion of observations the operational contact between
stations envisaged. The control of complex experiments
was carried out by radar MRL-5, providing duty observations of cloud formation, evolution and displace-
ment. Observations and information registering on
MRL-5 and TRS (three-wavelength radar station) were
conducted in the range of 100km, on RRS – 50km, on
Doppler RLS of horisontal scanning – 30km, in rocket
soundings – 8km and in vertical Doppler sounding during hail cloud passage above observational station.
        For obtaining of three-dimentional pattern
of structure and echo cell structure of clouds with
MRL-5, periodical (in 5-7min) photoregistration of
echo pattern on PPI and RHI at different elevation
angles and azimuths is carried out. Cloud picture photography on PPI and RHI is taken in isocontours of
radar reflectivity at 10cm range with fluorescence
against this background the hail cores and turbulence
zones. Microstructure investigations were carried out
by experimental three-wavelength RLS. In this case
spatial-time distribution of microstructure parameters
is calculated from radar reflectivity distribution and
attenuation coefficient at wavwlengths of 2,0; 3,2
and 10cm according to the following formulas:
- in the case of hailfalls

$$d_{3h} = 2,0 \left(\eta_{3,2} / \eta_{10} \right)^{-0,27} ; \qquad (1)$$

$$N_h = 10^6 \, \eta_{3,2}^{1,6} \, \eta_{10}^{-0,6} ; \qquad (2)$$

$$q_h = 1,5 \cdot 10^6 \, \eta_{3,2}^{0,7} \, \eta_{10}^{0,3} ; \qquad (3)$$

$$J_h = 1,1 \cdot 10^8 \, \eta_{3,2}^{0,6} \, \eta_{10}^{0,4} ; \qquad (4)$$

- in the case of rain:

$$d_{32} = A_1 \left(K_{\lambda_i} / \eta_{\lambda_j} \right)^{\alpha} ; \qquad (5)$$

$$N_z = A_2 \, K_{\lambda_i}^{d_1} \, \eta_{\lambda_j}^{d_2} ; \qquad (6)$$

$$q_z = A_3 \, K_{\lambda_i}^{d_3} \, \eta_{\lambda_j}^{d_4} ; \qquad (7)$$

$$J_z = A_4 \, K_{\lambda_i}^{d_5} \, \eta_{\lambda_j}^{d_6} , \qquad (8)$$

where d_3, N, q and J - meancubical diameter of hydrometeor spectrum, their concentration, liquid water
content and precipitation intensity, respectively
(indices "h " and "z " are rain and hail);
$A_1,...,A_4, d_1,...,d_6$ - constants, depending on used wavelengths.
        Measurements of integral liquid water content
and total water content of hail clouds are carried
out by simultaneous measurements of cloud own thermal
radio radiation intensity and parameters of radial
distribution of radar reflectivity. The calculation
of integral liquid water content of thunderstorm cloud
radial column (Qkg/m²), limited by acceptance angle
on RLS data is performed according to /1/ and the
expression:

$$Q = 0,46 \; \lambda^{2,85} \left(\frac{T_b}{T_k}\right)^{1,03} \left(\frac{R_m}{\Delta R}\right)^{2,06} \tag{9}$$

where T_b and T_k - radiobright and kinetic cloud temperature respectively;
ΔR - cloud thickness extension (km) with echo power
$P_2 > P_m/2$;
R_m - cloud extension from near boundary to the area
of maximum radar echo (km).
        Total water content of the whole cloud volume
Q_Σ is calculated by the formulae:

$$Q_\Sigma = \sum_{i=1}^{n} h_i \int_{A_1}^{A_2} Q(A) dA, \tag{10}$$

where n - number of cloud layers, at which the measurements of tangential profile T_b/T_k and calculation of
tangential profiles Q(A) are carried out;
h_i -thickness of cloud layers;
A_1 and A_2 - asimuths of left and right cloud echo
boundary.
        Investigation of air flow structure in cumu-
lonimbus is performed by rockets and Doppler cloud
sounding. Tracking and reception of data on cloud
temperature and electric field intensity on rocket
trajectory and its coordinates is performed by radar
"Meteorit" in regime of automatic tracking and automatic registration of received information. Doppler
observations of vertical flows are performed in vertical sounding regime during the cloud passage over
the polygon.
        Velocity of vertical flows W is calculated
by the expression:

$$W = \bar{V}_D - (67 - 6,7 \; lg \frac{K_{0,8}}{\eta})\left(\frac{\rho_o}{\rho_H}\right)^{0,5}, \tag{11}$$

where $\bar{V}_D$ - mean Doppler velocity of hydrometeors;
ρ_o and ρ_H - air density on the sea level and height H, respectively.

Velocity of vertical flow during rocket sounding is calculated by the expression:

$$\pm W = V_c - V_{nc} , \qquad (12)$$

where V_c and V_{nc} - velocity of rocket descending with parachute in cloud and cloudless atmosphere, respectively; W with "minus" denotes the updraft and with "plus" - the downdraft.

When detecting hail cores, a group for collecting of hailfall spectrum was sent to the area of hailfall to analize hail embryo origin (frozen drop, graupel) and to determine hailstone and embryo size and shape distribution.

The whole cloud information was processed jointly and was represented as type figures and tables, i.e. spatial-time echo structure, precipitation bands, height-time diagram of air flow distribution and cloud microstructure against the background of echo structure in PPI and RHI as reflectivity isolines in 10cm wave range (Fig.2), with hodograph and so on.

Limits of measured parameters and measurement errors are in Table 1.

With the help of described method in 1975-78 the study of 150 hailstorms of different intensity was carried out. Classification of these storms according to their cell structure and evolution dynamics was performed, characteristic features and evolution dynamics of singlecell, multicell and supercell storms, location of hail cores and severe streams of updrafts, height and temperature conditions of hail generation, time of hail growth, rules of hail formation space propagation and other macro- and microphysical characteristics of hailstorms were found out.

In Fig.1 there is an example of height-time diagram of radar reflectivity distribution η_{10}, maximum diameter d_{max} and hydrometeor concentration N, liquid water content (ice content) q and velocities of vertical flows W in supercell hail cloud in quasisteady stage.

Maximum overheating of cloudness in comparison with the environment (5-7°) were found in the regions of the strongest updrafts.

In Table 2 characteristic features of structure and evolution dynamics of main hail cloud types, aerosynoptic conditions of their evolution and repea-

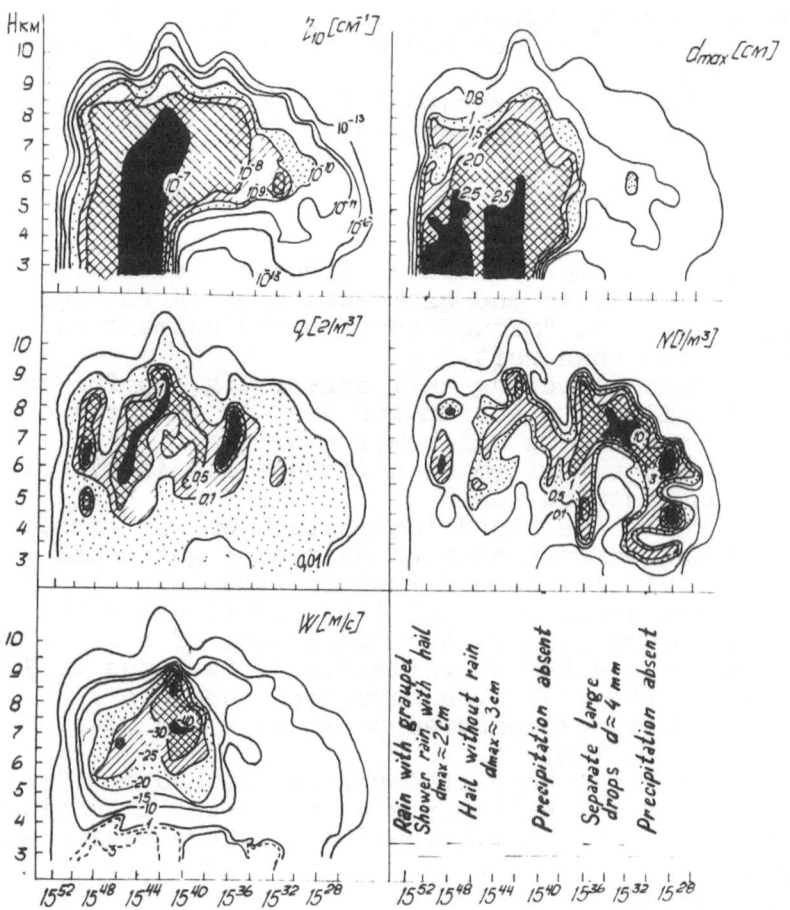

Fig.1 Height-time diagram of echo structure in isolines $\eta_{10}, d_{max}, q, \mathcal{N}$ and W in supercell hail cloud, June 10, 1976.

tability are presented.

On the basis of obtained results, improved schemes of hail cloud seeding by crystalizing agents with the purpose of interrupting and prevention of hail damage were worked out (Fig.2 and 3).

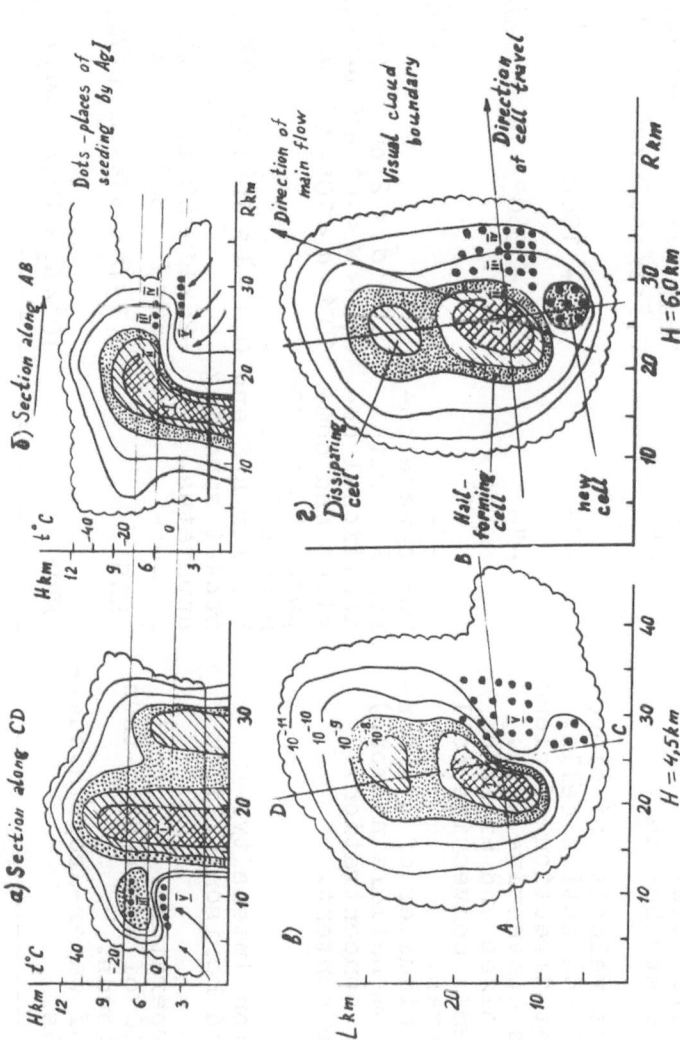

Fig.2 Typical echo structure of multicell storm and the scheme of its seeding with crystalizing agents. a) vertical cloud section scheme along CD line; b) vertical cloud section scheme along AB line; c) horisontal section scheme of multicell storm at the heights: 4,5(t=-5°C) and 6,0 km(t=-12°C); I-hail,localezation region; II-hail growth region; III-hail generation region; IV-"formation" region; V-weak echo region. Dots are the places of seeding.

Table 1. Complex of measured hail cloud parameters, measurement limits and errors.

NN	Measured parameters	apparatus	limits	errors
1.	Space radioecho structure in reflectivity isolines on PPI and RHI at $\lambda = 10$cm.	MRL-5 with additional devices.	6 reflectivity levels	$(\Delta\eta/\eta)_{max} = \pm 3dbz$
2.	Hail core structure.	-"-	$d_{max} > 1,0$cm	$\Delta L_{mean} \approx 0,15$km
3.	Turbulence distribution.	-"-	$0,5 < \Delta V < 4,0$m/s	$\Delta V \leqslant 0,5$m/s
4.	Displacement velocity of cloud systems and convective cells.	-"-	$0 - 100$km/h	$\dfrac{\Delta V}{V} \leqslant 20\%$
5.	Displacement direction of cloud systems and convective cell.	-"-	$0 - 360°$	$\Delta A \leqslant 10\%$
6.	Transverse sizes and areas of cloud systems, convective cells and hail cores.	-"-		$\Delta L/L \leqslant 10\%$
7.	Mean cubic diameter of hydrometeor spectrum(hail&rain drops)	Three-wavelength RLS with additional devices and computer "Electronics K-200".	$0,01 < \bar{d}_3 < 2$ cm	$(\Delta d_3/d)_{max} \leqslant 70\%$
8.	Hydrometeor concentration.		$0,01 < N < 10^4 \, m^{-3}$	$(\Delta N/N)_{max} \leqslant 200\%$
9.	LWC and ice content.		$10^{-4} < q < 10^2 \, g/m^3$	$(\Delta q/q)_{max} \leqslant 60\%$
10.	Precipitation intensity.	Doppler RLS and rocket sounding apparatus.	$0,1 < J < 10^3 \, mm/h$	$(\Delta J/J)_{max} \leqslant 60\%$
11.	Vertical and horisontal flow velocity.	-"-	$-25 < W < 60$ m/s	$\Delta W_{max} \leqslant -2m/s$
12.	Incloud temperature.		$-60 < T < 30°C$	$\Delta T_{max} \leqslant \pm 1,5°C$
13.	Integral LWC of water-drop thunderstorm clouds.	RRS	$0,1 < Q < 30kg/m$	$(\Delta Q/Q)_{max} \leqslant 90\%$
14.	Total LWC of water-drop thunderstorm clouds.	RRS	$10^4 < Q_z < 10^8 kg$	$(\Delta Q_z/Q_z)_{max} \leqslant 100\%$

Table 2. Characteristic features of single, multi and supercell hailstorms in the North Caucasus.

NN: Features and parameters	Types of hailstorms		
	singlecell	multicell ordered	supercell
1. Cloudiness cellular structure.	Some(many) spatially isolated or non isolated non-interacting cells.	Some interacting cells, boundaring with each other.	1,2 or 3 rather dispersed in space and non-interacting cells.
2. Convective cell structure.	Axisymmetric.	Asymmetric.	Asymmetric.
3. Evolution dynamics of cloud system.	Randomized generation of new and dissipation of old convective cells in space and time.	Periodical renewing of storm by generation of new and dissipation of old convective cells.	Quasicontinuous renewing of front part and disspation of back cell part.
4. Evolution dynamics of convective cells.	Development-10-20 min and dissipation-15-20 min without quasistationary stage.	Development-10-20 min, quasistationary state-10-30 min and dissipation-10-30 min.	Development-15-20min quasistationary state 30-250min, dissipation - 20-30min.
5. Generation rules of new and dissipation of old convective cells.	Random in space and time.	Periodical(in10-20 min) generation on the right flank of new and dissipation of old cells on the left flank.	Possible generation of new supercell in 1-2 hours later and dissipation 40-70 km south-east from previous.
6. Hail formation space propagation.	Descrete.	Descrete-continuous.	Continuous.
7. Direction of cell	Randomized	Right from main	Right from main

Table 2 (Continued)

NN	: displacement.	: flow on 5-40°.	: flow on 10-90°.
8. Velocity of cell displacement V km/hr.	$0 \div 20$	$10 \div 70 \approx (\frac{1}{2} \div 1)V_n$	$20 \div 40 \approx \frac{1}{2} V_n$
9. Maximum echo height, H_m km.	8-12	10-14	11-16
10. Maximum radar reflectivity at λ=10cm	$5 \cdot 10^{-9} - 10^{-7}$ / 35-50	$5 \cdot 10^{-9} - 5 \cdot 10^{-7}$ / 35-55	$5 \cdot 10^{-8} - 2 \cdot 10^{-6}$ / 45-60
11. Maximum size in hailfall spectrum d_{max} cm.	1-3	2-5	3-8
12. Area of hailfall S km^2.	0,5-10	5-400	60-1800
13. Hail path sizes L km width length	0,3-3 / 0,5-4	0,5-10 / 3-60	3-20 / 15-120
14. Synoptic situation.	Intermass development, small-size fields of low and high pressure.	Passage of main and secondary cold front.	Passage of cold fronts and occlusion fronts as cold ones.
15. Direction of main flow.	Undetermined.	South-west, west, north-west.	South-west, sometimes west.
16. Velocity of main flow V_n km/hr	30	30-100	50-80
17. Wind shear γ sec^{-1}	10^{-4}	$10^{-4} - 10^{-3}$	$5 \cdot 10^{-4} - 5 \cdot 10^{-3}$
18. Convective instability.	Moderate and strong.	Moderate and strong.	Moderate and strong.

NN:	Liquid water content of atmosphere.	Weak and moderate.	Moderate and increased.	Increased.
19.	content of atmosphere.			
20.	Temperature at condensation level, °C.	+2 ÷ +16	- 5 ÷ +28	+7 ÷ +17
21.	Relative air humidity at level 1,5-5,5km,%	35-70	35-90	50-80
22.	Repeatability,%	20	30	10

Note: 40% of hailstorms have non-ordered structure, not included in above pointed three types, or intermediate characteristics.

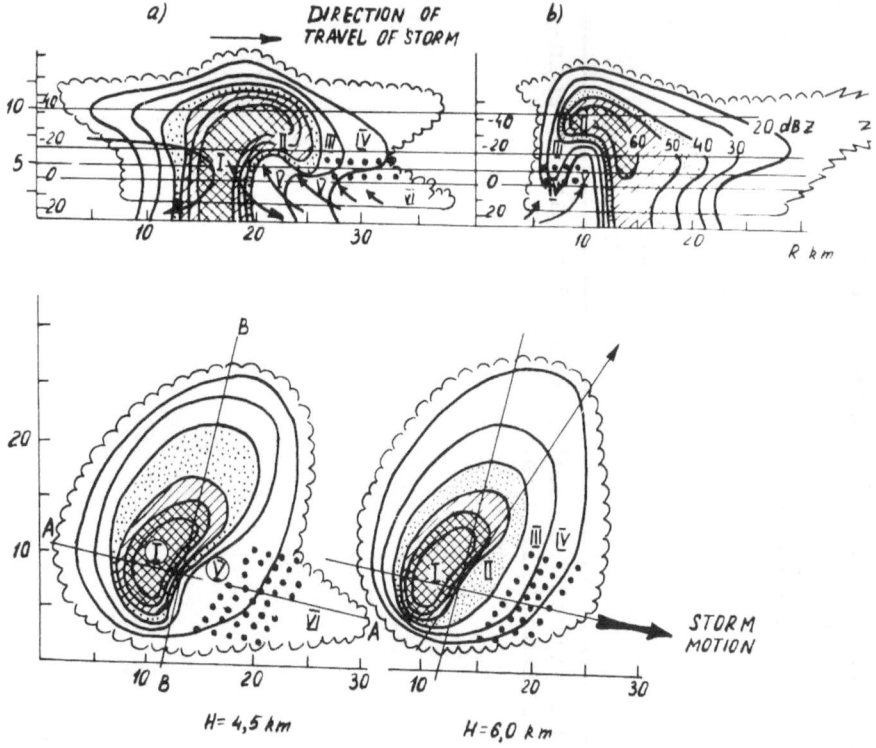

Fig. 3 Typical echo structure of supercell storm and the scheme of its seeding with the purpose of hail formation interrupting: a) vertical section through the weak echo region along AB line in the plane of cloud movement direction; b) vertical section through the weak echo region along CD line in the plane normal to cloud movement direction; c) horisontal section at height level 4 km; d) horisontal section at height level 5,5 km; e) horisontal section at height level 7 km.
Signs are the same as in Fig.2.

REFERENCES

1. Abshaev, M.T., "Radar radiometric measurement method of integral liquid water content of cumulonimbus". Trudy of the V All Union Conf. on Radar
   Meteorology, 1978,M.
2. Abshaev, M.T., Juboev, M.M., "Rocket observations
   of air flow structure in cumulonimbus". Trudy VGI,
   1976, 33, pp. 57-66.
3. Abshaev, M.T., Imyanitov, E.M., Mashukov, Kh.M.
   "Device for electric field sounding of thunderstorm and hail clouds". Trudy VGI, 1976, 31,
   pp.15-31.
4. Abshaev, M.T., Karmov, Kh.N. "Active-passive method of hail core detection in cumulonimbus".
   Trudy VGI, 1976, 33, pp.43-56.
5. Abshaev, M.T., Dadali,Yu.A., Malbahova,N.M., Pash-
   kevich, M.Yu., Pravosudov,A.V., Chekun, V.N.
   "On space-time distribution of microstructure and
   water content parameters in cumulonimbus", Trudy
   VGI, 1976, 33, pp.67-80.
6. Abshaev, M.T., Belyavsky,A.V., Tkhamokov, B.Kh.,
   Vlaznev,V.A. "Methods and equipment for Doppler
   vertical flow measurements in hail clouds". Trudy
   VGI, 1976, 31, pp.41-55.
7. Abshaev, M.T., Burdakov, F.E., Vasilyev,G.V.,
   Gornostayev, N.V., Shevela, G.F. "Special radar
   MRL-5 for hail detection and storm warning and
   its meteorological efficiency", Trudy VGI, 1976,
   33, pp. 3-30.
8. Tlisov,M.I., Khorguani, V.G. "Study of hail embryos in wind tunnel", Trudy VGI, 1975, 29, pp.122-139.

A DIAGNOSTIC STUDY OF THE TORNADIC STORM BASED ON DUAL-DOPPLER
WIND MEASUREMENTS

Yeong-jer Lin and Robert Pasken

Department of Earth and Atmospheric Sciences

Saint Louis University, St. Louis, Missouri, USA

Dual-Doppler data for a tornadic storm were used to derive the
three-dimensional wind field at twelve analysis levels within the
storm. A grid spacing of 1 km covering the domain of 26(km) x 26
(km) x 12(km) was used. The three wind components were then used
as input to the three momentum equations for recovering the perturbation fields of pressure, density and virtual temperature
within the storm. The solution obtained was smoothed, using a
nine-point smoother horizontally and a three-point smoother vertically. Results obtained show that the recovered thermodynamic
fields are in good agreement with the storm's updraft-downdraft
structure.

1. INTRODUCTION

    The advancement of Doppler radar technique has indeed added
additional knowledge of the kinematics of severe thunderstorms.
Although multi-Doppler radars can provide three-dimensional winds
within a convective storm, they cannot detect thermodynamic variables, e.g., temperature and pressure fluctuations, directly.
These thermodynamic variables are of vital importance for understanding the structure and dynamics of a severe convective storm.
It is generally known that the fluctuations (from their environmental means) of temperature and pressure inside the storm are in
the order of few degrees and few millibars, respectively. These
quantities cannot be measured accurately by the conventional upper air technique.

    Recently, several feasibility studies were conducted to re-
cover thermodynamic variables from the detailed wind field inter-

315

E. M. Agee and T. Asai (eds.), Cloud Dynamics, 315–328.
Copyright © 1982 by D. Reidel Publishing Company.

nal to the storm; for example, see studies by Hane and Scott (1978), Gal-Chen (1978) and Hane et al. (1981). All these studies employed the model-simulated data instead of real storm data. Gal-Chen (1978) proposed an algorithm whereby the combined use of the governing equations and the observed wind will uniquely determine the denisty and pressure perturbations. The algorithm has been shown to have two important features: 1) it involves the use of only the momentum equations without resorting to any thermodynamic parameterizations, and 2) there is no need to make any artificial assumptions about the boundary conditions. This method was applied to a three-dimensional numerical cloud model by Hane et al. (1981). It is essential not only to determine the importance of physical processes represented in the method, but also to assess the impact of errors due to observation by multiple Doppler radars on the derived thermodynamic variables. A total of ten cases were investigated by the above authors for testing the accuracy of the retrieval method and for sensitivity studies. The results obtained indicated that Gal-Chen's method was feasible for recovering perturbation pressures, densities and virtual temperatures within a severe convective storm. The results also showed that when local time derivative and frictional terms were omitted, the solution deteriorated considerably, but still useful if general temperature patterns are desired. The retrieval fields were sensitive to errors added to the input velocity fields. However, if the retrieved fields are properly filtered, the results are still usable in a qualitative manner.

The structure and internal dynamics of the storm in the sheared and veered environment were studied by many researchers; for example, see studies by Newton and Newton (1959), Lin and Chang (1977), Bonesteele and Lin (1978) and Lin and Whiton (1980). These studies found that the unique storm-environment interaction produces three-dimensional perturbation pressure gradient forces (hereafter referred to as PPGF) around the main updraft core. These forces play an important role in protecting the updraft core by forcing the environmental air to flow around it. As a result, the impact of lateral entrainment on the main updraft is greatly minimized thereby prolonging the storm's lifetime.

The main purpose of this study is to further understand certain physical properties of the tornadic storm using real dual-Doppler storm data as input to the basic equations. The domain of investigation had a volume of 26 x 26 x 12(km)^3 with twelve analysis levels in the vertical. Following the procedures outlined by Gal-Chen (1978), a set of internally consistent data including three wind components, perturbation fields of pressure, density and temperature at a grid spacing of 1 km was obtained. This data set was then used to study the structure, dynamics and thermodynamics of the storm during the tornado occurrence period.

2. DATA AND COMPUTATIONAL PROCEDURES

The dual-Doppler data used in this study were associated with the Oklahoma City storm. This storm produced three tornadoes, the Oklahoma City, Spencer and Luther tornadoes. The last tornado occurred at 1412 CST and dissipated just south of Luther, Oklahoma at 1428 CST. The data, received from the National Severe Storms Laboratory (NSSL), contain only the radial velocities (V_r), the standard deviations of V_r, positions of the target, reflectivities (Z) and automatic gain control values. These data were meticulously processed and analyzed following the procedures introduced by NSSL scientists (see Brown et al., 1981) for a single scan through the storm at 1421 CST. All observations within the storm volume were adjusted for mean storm motion (230° at 15 m s^{-1}) to a common reference time near the mid-point of the data collection, and only observations with reflectivity > 20 dBZ were considered. Following the studies by Armijo (1969) and Brandes (1977,1978), two horizontal wind components (u,v) <u>relative to the storm</u> in a Cartesian coordinate system can be derived from two radial velocities along the radar beams. The Doppler radars were located at Norman and Cimarron Airfield. The latter was about 42 km NW of Norman.

Since errors within the wind field, used as input data to perturbation pressure recovery, can seriously affect the results, a careful consideration of the errors in the wind field is necessary. Ray et al. (1980) found that the horizontal wind fields deduced from two to four radars are nearly the same for all analysis schemes used. Errors in the dual-Doppler case are typically < 3 m s^{-1} and the improvement obtained by additional radars is small when compared to a mean wind speed of 15 m s^{-1} interior to the storm. Based on the variance analysis study by Doviak et al. (1976), Brandes (1978) estimated that errors in horizontal winds for the 8 June 1974 storm case are roughly 10 degrees in direction and 1 m s^{-1} in speed. Using the formula derived by Lhermitte and Miller (1970), we also conducted a series of error analyses. Under worst case conditions, errors in the wind field were 5 m s^{-1} in vertical velocity and one standard deviation in radial velocity. Assuming a one standard deviation error in V_r, the calculated errors in u and v are estimated to be 3 m s^{-1} which are comparable with that reported by Ray et al. (1980). Thus it is reasonable to assume that dual-Doppler data for the 8 June 1974 tornadic storm contain roughly 10% random errors in horizontal wind fields.

Vertical velocities (w) were computed by integrating the anelastic continuity equation downward with a variational adjustment (see Ray et al, 1980). Once the three-dimensional field of u, v and w wind components was obtained, values of perturbation pressure (P') at each analysis level were recovered by solving the horizontal pressure equation in the form of

$$\frac{\partial^2 \boldsymbol{P'}}{\partial x^2} + \frac{\partial^2 \boldsymbol{P'}}{\partial y^2} = \frac{\partial F}{\partial x} + \frac{\partial G}{\partial y} \qquad (1)$$

$$F = -\frac{\delta}{\delta t}(\bar{\rho}_a u) - \frac{\partial}{\partial x}(\bar{\rho}_a uu) - \frac{\partial}{\partial y}(\bar{\rho}_a uv) - \frac{\partial}{\partial z}(\bar{\rho}_a uw)$$

$$+\bar{\rho}_a 2\Omega \ (v \ sin\phi - w \ cos \ \phi) + \mathit{f}_x$$

$$G = -\frac{\delta}{\delta t}(\bar{\rho}_a \ v) - \frac{\partial}{\partial x}(\bar{\rho}_a uv) - \frac{\partial}{\partial y}(\bar{\rho}_a vv) - \frac{\partial}{\partial z}(\bar{\rho}_a vw)$$

$$-\bar{\rho}_a 2\Omega u \ sin\phi + \mathit{f}_y$$

Where f_x and f_y are frictional components toward the x and y directions, respectively, $\bar{\rho}_a$ is the basic density of the adiabatic hydrostatic atmosphere, and other symbols have their conventional meanings.

A unique solution for equation (1) exists only if the horizontal average of P', denoted by $<P'>$, is removed from P' in (1). Equation (1) was solved by sequential relaxation with Neumann boundary conditions. Finally, density fluctuation fields were computed from the vertical momentum equation:

$$(\boldsymbol{P'} - <\boldsymbol{P'}>)g = -\frac{\partial}{\partial z}(P' - <P'>) + (H - <H>) \qquad (2)$$

$$where \ H = -\frac{\delta}{\delta t}(\bar{\rho}_a w) - \frac{\partial}{\partial x}(\bar{\rho}_a uw) - \frac{\partial}{\partial y}(\bar{\rho}_a vw)$$

$$-\frac{\partial}{\partial z}(\bar{\rho}_a ww) - \bar{\rho}_a 2 \Omega u \ cos\phi + \mathit{f}_z$$

With the aid of the perturbation equation state, fields of perturbation virtual temperature (T_v') were obtained. Both P' and T_v' fields recovered were subject to horizontal and vertical smoothing using a nine-point and three-point smoother, respectively. According to the studies by Gal-Chen (1978) and Hane et al. (1981), such smoothing is necessary in order to reduce the impact of input errors on the thermodynamic retrieval solution. Although the perturbation pressure fields required little smoothing, the perturbation virtual temperature fields were very distorted until both the horizontal and vertical smoothing was applied.

3. DISCUSSION OF RESULTS

In this study only one set of dual-Doppler data at 1421 CST was considered. Hence, the tendency terms in the momentum equations cannot be evaluated. For simplicity, frictional terms were also omitted for thermodynamic retrieval. These two assumptions together with the uncertainty inherent in horizontal wind fields do not warrant any quantitative assessment. Instead, the results should be interpreted in a qualitative manner. In the follow-up experiments, these two terms will be included in the retrieval method as suggested by Gal-Chen (1978) and Hane et al. (1981).

The storm-relative winds at the lowest level (0.3 km) are displayed in Figure 1. In this figure, dashed lines are contours of constant reflectivities in dBZ. The heavy dashed line represents the gust front which separates the warm moist inflow from the cold outflow. A distinct wind shift along the gust front is apparent. Note that the main vortex, denoted by •, is located in the weak echo region (WER). Field of horizontal convergence (C) and (D) at the same level is shown in Figure 2. It is evident, a strong convergence occurs in front of the gust front, while a strong divergence due to the effect of cold downdraft occurs behind it. The vortex center is found in the region with strong convergence. Figure 3 depicts the recovered p' field at 0.3 km level. A large perturbation pressure excess of 2 mb occurs behind the gust front. This high is consistent with the so-called meso-high observed by Fujita (1963) underneath the cold downdraft. The downdraft, originating at mid-levels on the left rear flank of the main updraft, produces pronounced cold outflow at this level behind the gust front. On the other hand, a weak pressure deficit (≈ -0.5 mb) occurs in front of the gust front. As a result, significant PPGF are found across the gust front. According to Bonesteele and Lin (1978), such forces act to 1) sustain the mesocyclone, 2) accelerate moist, low-level inflow toward the WER, 3) induce upward motion at low levels, and 4) decelerate the downdraft near the surface as it is converted into horizontal flow behind the gust front.

In the mid-troposphere, the main updraft core tends to behave like an obstacle causing the environmental flow to divert around it (see Figures 4-6). The SW winds entering from the storm's west side are forced to flow around the updraft core with refelctivities > 40 dBZ increasing speeds on the right and left flanks and decreasing speeds at the leading and trailing edges to include the wake region. The direction of the diverted winds is observed almost parallel to reflectivity contours. The corresponding p' fields computed at 6.3, 7.3 and 8.3 km levels are illustrated in Figures 7-9, respectively. It is seen that the main pressure deficit up to -1.5 mb corresponds well to the position of the main vortex core which has a cyclonic rotation. Two

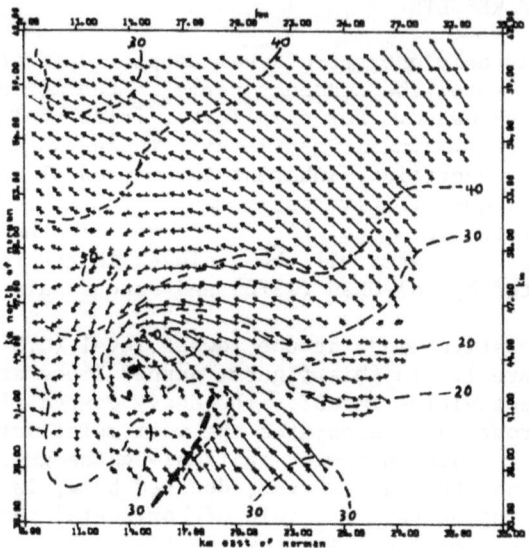

Figure   1.   Storm-relative winds and radar reflectivity contours
              (dashed lines) in dBZ at 0.3 km level.   A heavy
              dashed line indicates the gust front.

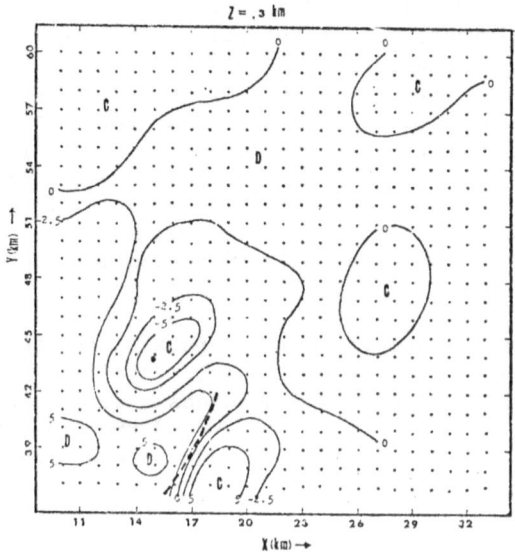

Figure   2.   Field of convergence (C) and divergence (D) in 10^{-3}
              s^{-1} at 0.3 km level.

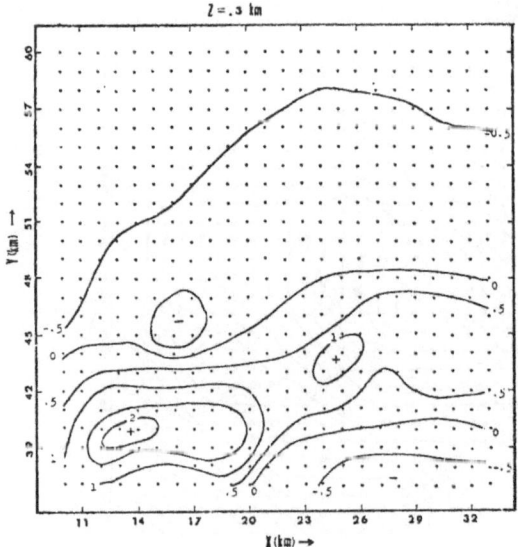

Figure 3. Field of perturbation pressure in mb at the level of 0.3 km.

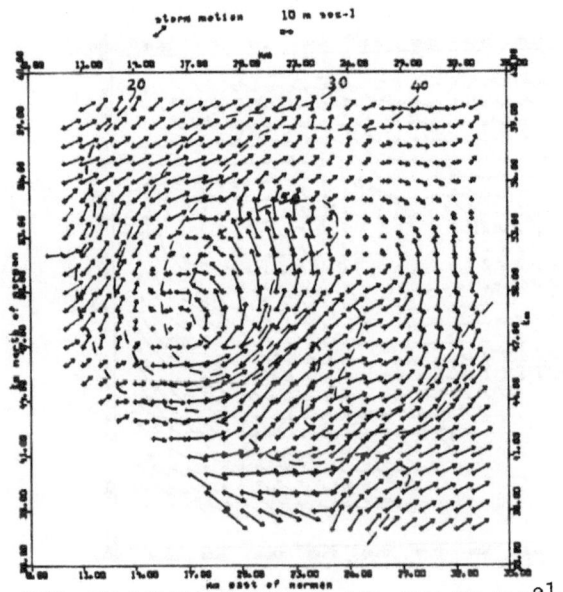

Figure 4. Same as Figure 1 except for 0.3 km level.

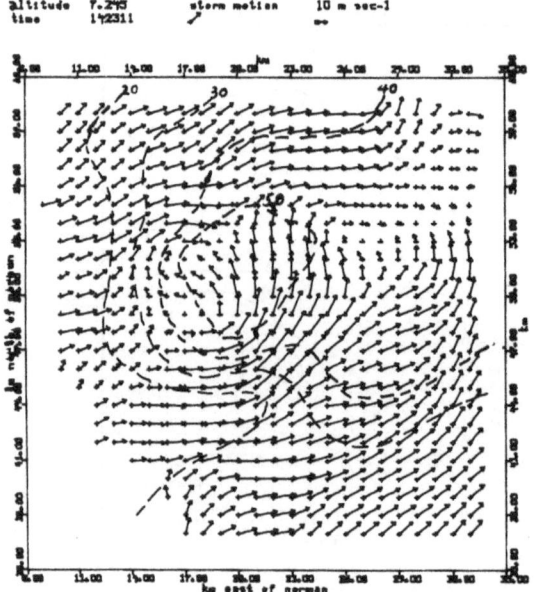

Figure 5. Same as Figure 1 except for 7.3 km level.

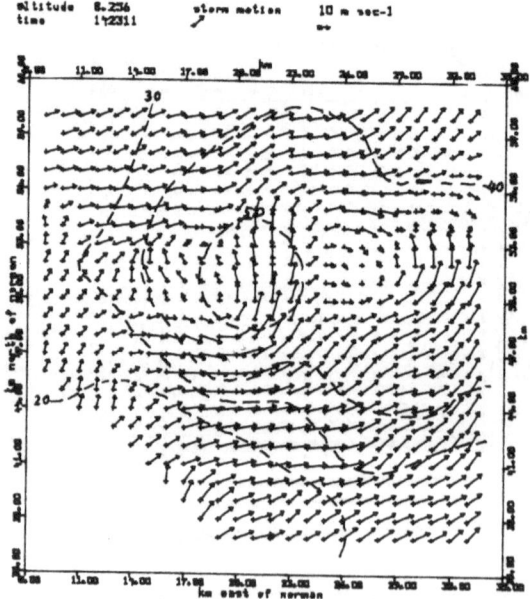

Figure 6. Same as Figure 1 except for 8.3 km level.

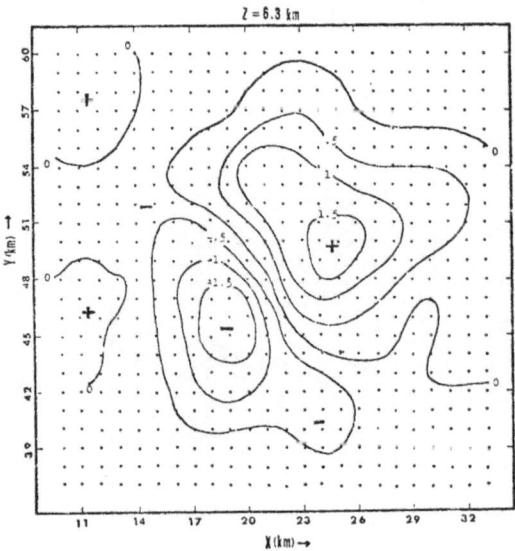

Figure 7. Field of perturbation pressure in mb at the level of 6.3 km.

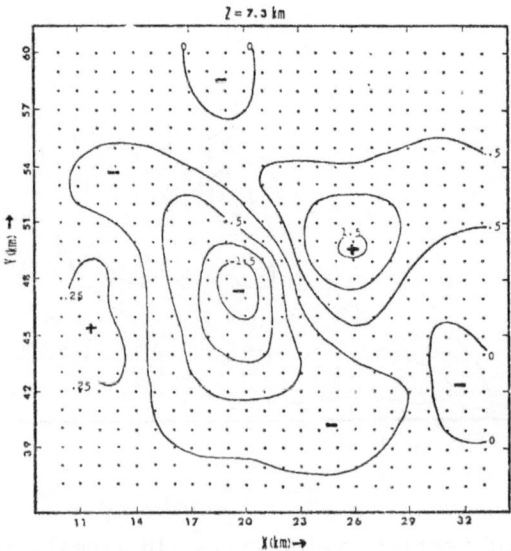

Figure 8. Field of perturbation pressure in mb at the level of 7.3 km.

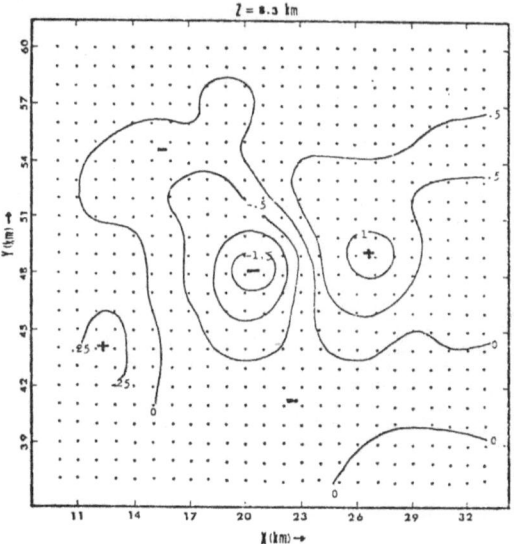

Figure   9.   Field of perturbation pressure in mb at the level of
              8.3 km.

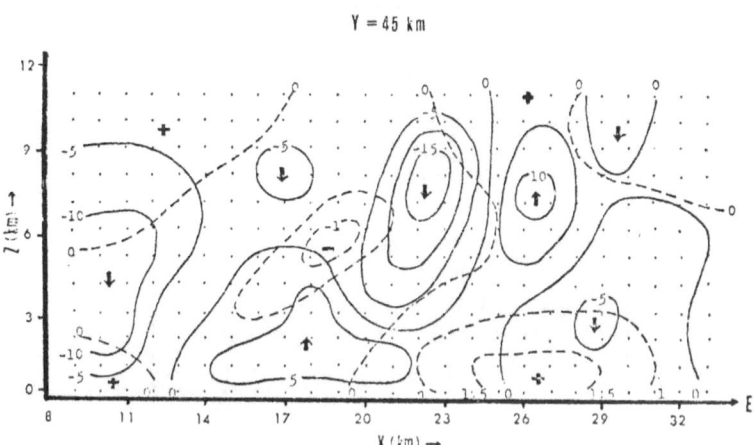

Figure 10.   Values of vertical velocity (solid lines), in m s^{-1},
             and perturbation pressure (dashed lines), in mb,
             along the east-west cross-section at 45 km north of
             Norman, Oklahoma.

secondary pressure deficits are found on the right and left flanks associated with the areas with strong diverted winds shown in Figures 4-6. On the other hand, perturbation pressure excesses occur at the leading and trailing edges of the updraft with much stronger values in the wake region. Such pressure differentials around the main updraft in turn produce the significant horizontal PPGF in the middle troposphere where the barrier flow concept is most applicable. These forces tend to accelerate the incoming air away from the main updraft and then decelerates it toward the wake region. As a consquence, the impact of lateral entrainment on the main updraft is greatly minimized thereby allowing the storm to maintain its vigorous convection for many hours. A similar pattern of P' field is also found at 9.3 km level (not shown). The overall pattern of P' fields shown in Figures 7-9 is in good agreement with that reported by Lin and Chang (1977) and Lin and Whiton (1980) in which the model-generated data were used as input.

Figures 10 and 11 display the computed values of w, P' and T_V' along the east-west cross-section at 45 km north of Norman. The downdraft shown on the left side of Figure 10 corresponds to the position of cold downdraft behind the gust front. This intense downdraft, originating from the mid-troposphere aloft where the air is relatively dry and cold, carries a strong downward momentum (w $\sim$ -10 m s^{-1}) and causes cooling (T_V' $\sim$ -2°C) in the lower layers near the surface (see Figure 11). It also produces an excessive pressure (see Figure 10) near the ground due to downward transport of momentum and mass. These results are in good qualitative agreement with observation reported by Newton (1963). In other regions, the fields of P' and T_V' are found to be associated with the updraft-downdraft structure. Generally, warmer regions correspond to updrafts, while colder regions match well with downdrafts. The magnitude range of T_V' is about ±6°C, which is comparable to that calculated from the parcel theory. On the other hand, pressure deficits are generally associated with the high relfectivity regions, while pressure excesses correspond to the areas with strong horizontal outflow underneath the downdraft.

4. CONCLUDING REMARKS

This study employs a three-dimensional wind field derived from dual-Doppler data to recover the perturbation pressure and temperature fields at each analysis level. A total of twelve levels from 0.3 to 11.3 km were used. Results obtained show that:

1) The three-dimensional PPGF play an important role in affecting the structure and dynamics of the tornadic storm.

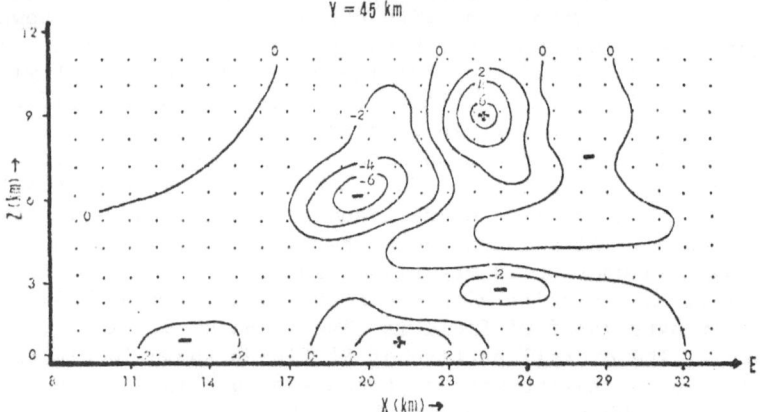

Figure 11. Same as Figure 10 except for T_v' values in $^{\circ}C$.

2) The horizontal PPGF across the gust front are found respon-
    sible for maintaining updraft-downdraft interaction and vig-
    orous circulation in the lower layers.

3) In the middle troposphere, the three-dimensional PPGF cause
    incoming environmental winds either to flow around the main
    updraft or to move downward forming the sloping downdraft.
    As a result, the storm's main updraft core is well protected
    from being entrained thereby prolonging its lifetime.

4) The perturbation temperature field obtained from the thermo-
    dynamic retrieval method is found to be in good qualitative
    agreement with the updraft-downdraft structure.

Generally, warmer regions correspond to updrafts and colder regions match with downdrafts. The magnitude range of T_v' appears
to be consistent with that calculated from the parcel theory.

ACKNOWLEDGMENTS

    The authors wish to thank the NSSL for providing the dual-
Doppler data and technical assistance. Thanks also go to many
scientists, especially R. Brown, E. Brandes and C. Hane of NSSL
and T. Gal-Chen of NSSA for their advice and constructive sug-
gestions throughout the course of this study. This material was
based upon work supported by the Division of Atmospheric Sciences,
National Science Foundation, under Grant ATM-7823138-01.

REFERENCES

Armijo, L. : 1969,"A theory for the determination of wind and precipctation velocities with Doppler radars", J. Atmos. Sci., 26, pp. 570-573.

Brandes, E. A.: 1977, "Gust front evolution and tornado genesis as reviewed by Doppler radar", J. Appl. Meteor., 16, pp. 332-338.

Bonesteele, R. G., and Lin, Y. J.: 1978, "A study of updraft-downdraft interaction based on perturbation pressure and single-Doppler radar data", Mon. Wea. Rev., 106, PP.62-68.

Brown, R. A., Safford, C. R., Nelson, S. P., Burgess, D. W., Bumgarner, W. C., Weible, M. L., and Fortner, L. C. : 1981, "Multiple Doppler radar analysis of severe thunderstorms: Designing a general analysis system, NOAA Tech.Memo.ERL-NSSI.-92.

Doviak, R. J., Ray, P. S., Strauch, R. G., and Miller, L. J.: 1976, "Error estimation in wind fields derived from dual-Doppler radar measurements", J. Appl. Meteor., 15, pp. 868-878.

Fujita, T. : 1963, "Analytical mesometeorology: A review", Meteor. Monogr., No. 27, pp. 77-125.

Gal-Chen, T. : 1978, "A method for initialization for the anelastic equations: Implications for matching models", Mon. Wea. Rev., 106, pp. 587-606.

Hane, C. E., and Scott, B. C.: 1978, "The temperature and pressure perturbations within convective clouds derived from detailed air motion information: Preliminary testing", Mon. Wea. Rev., 106, pp. 654-661.

Hane, C. E., Wilhelmson, R. B., and Gal-Chen, T.: 1981, "Retrieval of thermodynamic variables within deep convective clouds: Experiments in three dimenisons", Mon. Wea. Rev., 109, pp.564-576.

Lhermitte, R., and Miller, L. J.: 1970, "Doppler radar methodology for the observation of convective storms". Preprints, 14th Radar Meteor. Conf., Amer. Meteor. Soc., pp. 133-138.

Lin, Y. J., and Chang, P. T.: 1977, "Some effects of the shearing and veering environmental wind on the internal dynamics and structure of a rotating supercell thunderstorm", Mon. Wea. Rev., 105, pp. 987-997.

Lin, Y. J., and Whiton, R.: 1980, "A numerical study of thunderstorm-environment interactions determined from perturbation pressure gradient forces, Part I: Storm movement", Papers in

Meteorological Research, J. of Meteor. Soc. of ROC, 3, pp. 1-16.

Newton, C. W. : 1963, "Dynamics of severe convective storms",
Meteor. Monogr., No. 27, pp. 33-58.

Newton, C. W., and Newton, H. R. : 1959,"Dynamical interaction
between large convective clouds and environment wind vertical
shear", J. Meteor., 16, pp. 483-496.

Ray, P. S., Ziegler, C. L., Bumgarner, W., and Serafin, R. J.:
1980, "Single- and multiple Doppler radar observations of
tornadic storms", Mon. Wea. Rev., 108, pp. 1607-1625.

AN EXPERIMENTAL INVESTIGATION OF THE FACTORS GOVERNING
THE DYNAMIC STRUCTURE AND INTENSITY OF ATMOSPHERIC
VORTICES

Christopher R. Church

Miami University, Oxford, Ohio 45056

Experiments have been performed on convergence-
driven vortices in a large vortex generator. Flow
visualization demonstrated the development of intense
laminar single-celled vortical cores at low values of
swirl ratio, S, which evolved into broader turbulent
two-celled structures as S was increased. Vortex
systems containing up to six multiple subsidiary
vortices developed at high S. It was established that
the swirl ratio was the parameter which primarily
determined the dynamic structure of vortices. The
magnitudes of peak velocity and surface pressure
associated with single-celled vortices increased
strongly with increasing S. The magnitudes of the
corresponding quantities in two-celled vortices showed
relatively little variation over a wide range of S.
It was found that for all vortices, the mean axial
velocity component in the apparatus, corresponding
to the mean vertical velocity in a thunderstorm updraft,
was the principal scaling parameter for velocity and
pressure.

1. INTRODUCTION

Intense atmospheric vortices may occur as tornadoes,
waterspouts, dust devils or fire whirls. When rendered
visible by whatever wind-blown material or condensate
is available, the most frequently observed form is of
an erect axisymmetric column; one might be tempted to
form a rather simple physical picture of the dynamics

E. M. Agee and T. Asai (eds.), Cloud Dynamics, 329–345.
Copyright © 1982 by D. Reidel Publishing Company.

of such a simple form.   More comprehensive examinations
of natural vortices, however, have revealed a variety
of dynamic details which are far from simple and which
point to many different dynamic forms for tornado-like
vortices.   For example, although we might intuitively
picture the vortex core as a narrow suction tube in
which the vertical velocity field is everywhere upward,
Sinclair's (1972) field measurements on dust devils
showed downward motion along the axis of the core,
surrounded by a field of upward motion.   Also, Hoecker's
(1960) photogrammetric analysis of the Dallas,Texas,
tornado of April 2, 1957 showed an upward axial
velocity diminishing with height, becoming zero at the
tip of the condensation funnel, with presumed downward
motion within the funnel.   The multiple vortex phenomenon is a more graphic example of complex forms, in
which varying numbers of intense subsidiary vortices
are in rotation about the axis of the parent tornado
vortex.   A particularly large number of this type of
tornado ocurred on April 3, 1974 in the midwestern
states.   Photographs of the Xenia, Ohio tornado on
that day (Fujita, 1976) show a pair of condensation
funnels embedded in the generally clear core of the
parent circulation.   The Parker, Indiana tornado
(Agee et.al.,1975) displayed four multiple vortices.

These and other observations present an array of
information, puzzling at first, but which ultimately
have order and conformity.   Some success in establishing
this order has been achieved through recent laboratory
modelling of vortices.   One objective of this article
is to discuss the techniques employed and principal
results of these endeavors.   Another important aspect
of research on atmospheric vortices is to establish
physical relationships which may be of practical
significance, such as might lead, for example, to an
improved predictive capability in severe storm
surveillance operations.   Storm surveillance depends
heavily on remote probing techniques.   Doviak et.al.
(1979) have discussed the various types of information
which can be obtained from Doppler radar observations
of severe thunderstorms.   These provide good resolution
on the mesoscale, but are much more limited in observing
events on the scale of the tornado.   Although such
techniques provide some velocity information, they do
not provide much detailed information about the dynamic
structure, size and intensity of the vortex and nothing
directly about the pressure field.   These factors
however, are surely related to the kinematics of the
tornado mesocyclone.   From a comprehensive investigation

of the relationships between the velocity and pressure
fields and the background flow in laboratory vortices,
it may be possible to provide indications of how remote
observations of tornado mesocyclones can be used to
provide more detailed information on the scale of the
tornado vortex. Experimental results which bear on
this topic are discussed in this article.

2. MODELLING CONCEPTS

    When compared with the difficulties encountered
in making detailed measurements on intense natural
vortices, laboratory modelling of them seems particularly attractive. Problems of prediction and safety
aspects do not arise with laboratory vortices, and
they can be created and sustained in one of a variety
of configurations, and probed in detail at will.
However, whereas unimpeachable measurements of velocity
and pressure can be obtained for natural vortices,
the results obtained in the laboratory depend not only
upon accurate measurement but also upon the validity
of the modelling technique, i.e. the design of the
vortex simulator. All vortex simulators possess
rigid walls which have no counterpart in nature.
This results in the development of spurious secondary
flows which can exert a dominant influence on the
vortices under study. In his evaluation of vortex
simulators, Davies-Jones (1976) concluded that, so
far, the most appropriate approach to tornado modelling
is the Ward (1972) type apparatus. This apparatus
replicates the most basic flow features of a situation
wherein tornado production is the result of low-level
radial convergence of angular momentum into a rotating
updraft. Observations of tornadic thunderstorms have
established that intense tornadoes are usually produced
in the intense updraft of a rotating thunderstorm,
a region identified as the tornado mesocyclone. A
schematic version of the Ward tornado simulator is
shown in Figure 1. this apparatus allows for independent
control of the updraft velocity and angular momentum,
and also the geometry of the inflow/updraft region.
Its physical arrangement is such that dynamic and
geometric similarity to a tornado mesocyclone is
attained. Examinations of the streamlines sketched in
Figure 1 shows that a secondary circulating flow exists
in the lower corner of the convection region, which
has a negligible influence on the vortices which form
along the centerline. A more detailed discussion of
the physical basis for the Ward-type simulator is

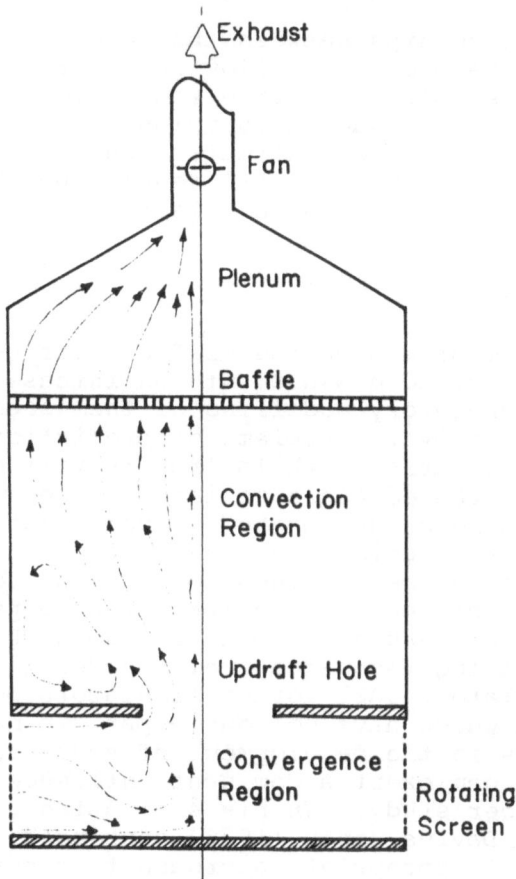

Figure 1.   Conceptual sketch of the Ward-type
vortex simulator

contained in Church et.al.(1979).

Lewellen (1962) and Davies-Jones (1973) showed
that a complete set of governing equations in terms
of non-dimensional circulation and stream function
may be written for incompressible, axisymmetric vortex
flow in a closed cylinder in terms of three dimensionless parameters: radial Reynolds number Re_r, swirl
ratio S, and a geometric aspect ratio a. These may
be defined as

$$Re_r = \frac{Q}{\nu} \tag{1}$$

$$S = \frac{r_o \Gamma}{2\,Qh} \tag{2}$$

$$a = \frac{h}{r_o} \qquad\qquad (3)$$

Where Q is the volume flow rate per unit axial length, ν the kinematic viscosity, r_o the radius of the updraft region, h the depth of the inflow layer and Γ the circulation at r_o. The aspect ratio defined above is an internal geometric parameter; external geometric parameters formed from dimensionless combinations of characteristic radial and axial dimensions of the apparatus (and mesocyclone) may also be significant in characterizing the flow. These are: r_s/r_o, r_w/r_o, and l/h, where r_s is the radius of the convergence region, r_w the radius of the convection region and l the depth of the convection region.

The results to be presented in this article were obtained in a Ward-type simulator which was developed at Purdue University (Church et.al., 1979). Accurate simulation of tornadoes requires that the swirl ratio, Reynolds number and aspect ratio be the same for the model as for the natural event. The range of values of these quantities likely to occur in nature are given in Table 1, and for comparison the range of values of the analogous quantities obtainable in the Purdue simulator are summarized in the right-hand column of this table. A comparison of the tabulated data shows that similarity is attainable in the simulator with respect to geometry and swirl ratio, but not with respect to Reynolds number. The range of Reynolds numbers was computed both for the laboratory and the atmosphere using the same value of $1.5 \times 10^{-5} m^2 s^{-1}$ for molecular kinematic viscosity. This results in the atmospheric values being some six orders of magnitude greater than the laboratory values. It can be argued however that eddy, rather than molecular, viscous effects are relevant in atmospheric flows. Agee (1975), for example, showed that the application of classical relationships to certain types of meso-scale atmospheric instabilities yielded numbers which were reasonable estimates of eddy, not molecular, viscosities. Based on such results it may be suggested that the flow into a tornado mesocyclone can be char-acterized by an effective Reynolds number, obtained by substituting a likely value of eddy viscosity for the molecular value. This would effect a much closer correspondence with respect to Reynolds number simil-arity. Experience has shown, furthermore, that the core radius and key core transition points are relatively insensitive to Reynolds number, provided it

Table 1. A comparison of typical values of dimensional and dimensionless parameters for actual rotating thunderstorm - tornado cyclone systems with those of the laboratory vortex simulator discussed in this article

QUANTITY	Likely Atmospheric Range	Range of Attainable values
r_s	5 - 10 km	152 cm(fixed)
r_0	1 - 3 km	20.3-79.0 cm
r_w	3 - 6 km	142 cm(fixed)
h	0.5 - 2 km	17.0-61.0 cm
l	5 - 16 km	133 cm(fixed)
Qh	$10^8-10^9 m^3 s^{-1}$	0.24-2.03 $m^3 s^{-1}$
Γ	$2.5 \times 10^4 - 2.5 \times 10^5 m^2 s^{-1}$	0.16 - 16.7 $m^2 s^{-1}$

Dimensionless grouping	Likely Atmospheric Range	Range of Attainable values
r_s/r_0	2 - 5	1.9-7.5
r_w/r_0	1.5 - 4	1.8-7.0
$a = h/r_0$	0.2 - 1	0.2-3.0
l/h	5 -16	2.2-7.8
S	0.05- 2	0.01-27.5
Re_r	$10^9 -10^{11}$	$2.57 \times 10^4 - 7.8 \times 10^5$

is sufficiently large, but are strong functions of the swirl ratio. Two other parameters which were not taken into account in the flow analysis are the surface roughness and the level of background turbulence in the converging flow. These influence the evolution of the vortex core and the core transition points. In order to minimize the effects of these parameters, the surface of the Purdue simulator was made as smooth as possible and antiturbulence screens were used to maintain the turbulence of the inflow at a very low level. Flow visualization was achieved by injecting smoke into the surface layers of the convergence zone. Velocity measurements were made using hot-film sensors,and surface pressure measurements were obtained using a variable reluctance differential pressure transducer connected to a surface-mounted static port.

3.   RESULTS

The core flow has been examined both qualitatively and quantitatively for a variety of different aspect ratios, Reynolds numbers and swirl ratios. It has been found that the parameter which most strongly influenced the dynamic structure of the core was the swirl ratio. As the swirl ratio is increased from zero, the simulator was found to produce a progression of ever more complex swirling flows. Visualized by the introduction of kerosene fog, observations of these flows have provided much insight into vortex dynamics and have aided in interpreting both the laboratory measurements and observations of actual tornado events.

3.1.   Evolution of the Vortex Core

Experimentally the range of core flows was observed by maintaining a constant geometry and Reynolds number and controlling the inflowing angular momentum (and hence the swirl ratio) by adjustments to the speed of the rotating screen. The magnitudes of swirl ratio were measured using a hot film velocity probe to determine the mean radial inflow angle at some large radius in the convergence region. The tangent of this angle divided by twice the aspect ratio is equal to the swirl ratio. Starting with very low values of swirl ratio the following sequence of events was observed.

Formation. For very small values of swirl ratio, no vortex core is found at the surface. This is a consequence of separation in the inflow on the lower surface. As the swirl ratio is increased a developing core zone approaches the surface. It was determined experimentally that for a critical swirl ratio of $S \sim 0.1$ the core makes contact with the surface.

Intensification and Breakdown. For $S \sim 0.1$ the central core appears as a smooth laminar cylinder extending upwards from the lower surface for the full height of the experimental volume. The vertical velocity has a maximum on the central axis, so that the core is said to be single-celled. A laminar core of this type responds to an increase in swirl by contracting in radius.

In the upper portion of the convection zone, the core structure changes character rapidly as the baffle is approached. This change in structure is conjectured to occur due to two inter-related effects: vortex breakdown, and development of downflow through the baffle into the center of the core. Vortex breakdown

is accompanied by an abrupt increase in core radius.
A stagnation point is found on axis, at the leading
edge of a partially closed circulating bubble of fluid
contained within the expanded core. The flow around
the leading edge of the breakdown bubble maintains
its laminar appearance and then further downstream
becomes turbulent, resembling a wake. A cylindrical
shear zone develops within this region of wake-like
flow. The centerline flow is upward but undergoing
deceleration. Further downstream an organized downflow exists. It is conjectured that a second free
stagnation point separates the region of decelerating
upflow from the downflow coming from the baffle. At
this point the vortex undergoes transition to a two-
celled structure.

Two-celled Structure. As the swirl parameter is
increased still further, so that $S \rightarrow 0.5$, the leading
breakdown point penetrates to the surface. A slight
further increase in swirl parameter results in a general
radial expansion of the core, leaving the vortex with
a calm inner sub-core and an axial downflow that penetrates to the surface. The vortex has then become two-
celled over its full length and resembles a turbulent
cylindrical column. As the swirl ratio $\rightarrow 1.0$, the
core expands to almost completely fill the updraft
hole. Downstream of the updraft hole, for large swirl
ratios ($S > 1.0$) the upflow gives way to a slowly
rotating turbulent plume containing subsidiary vortex
features.

Large Scale Instabilities. Organized large scale
instabilities develop due to the destabilization of
the strongly sheared annullar flow downstream of the
breakdown. The lowest order disturbance takes the
form of a helical, spiraling roll vortex surrounding
the upflow. As spin-up continues, the shear zone
associated with leading breakdown appears to thin out
so that higher order modes may be excited. A second
pattern in the form of two inter-twining spiral vortices
emerges. With further spin, patterns of three and four
subsidiary vortices appear in a similar manner. These
vortices are confined to the region between the two
breakdowns. With increased swirl the second breakdown
moves towards the surface, and the shear zone downstream of this feature also becomes unstable. The
tendency towards instability is further enhanced by
the presence of a weakly rotating downdraft in the
core. Visually, one sees the development of a system
of subsidiary vortices similar to that following the

first breakdown except that they now tend to be larger
and to extend further in the experimental volume. The
largest number of subsidiary vortices that has been
maintained experimentally is six.

3.2   Vortex Transition and Similarity Parameters

The preceding section has discussed several
aspects of the complicated flow associated with vortices
and it is evident several distinct vortex configurations,
separated from one another by points of transition, can
be identified. Obvious points are the transition from
laminar to turbulent flow associated with the leading
breakdown, the transition from a vortex containing a
single large helical disturbance to one containing two
large subsidiary vortices associated with the second
breakdown, the transition from two to three subsidiary
vortices, and so on. A series of experiments was
conducted to determine how the critical values of
swirl ratio associated with these transitions depended
on internal aspect ratio and Reynolds number. Although
the patterns exhibited some hysteresis, i.e. two
different patterns existing under identical swirl
conditions, it was possible to obtain values of swirl
ratio for which both the higher and lower vortex
modes appeared to be equally preferred. The results
are shown in Figure 2. The swirl ratio values are
plotted versus Reynolds number for the different
aspect ratios. The data suggest no more than a weak
aspect ratio dependence and a single smooth curve
has been drawn for each vortex transition. A variation
with Reynolds number is evident. In general, the
critical swirl ratio is a decreasing function of
Reynolds number and an asymptotic limit is approached
at high values. Physically, this behavior may be
explained as follows: The fluid which establishes
the core flow reaches the core via the boundary layer.
Interaction with the surface results in a decrease in
the tangential velocity component while maintaining
the converging radial flow. This effect becomes pro-
gressivly less important as the Reynolds number is
increased, because the boundary layer is thinner and
a smaller fraction of the angular momentum is lost
through surface drag. The curve asymptotically
approaches a value of swirl ratio corresponding to a
free slip condition.

3.3. Velocity Measurements and the Background Flow

Simultaneous measurements of all three components

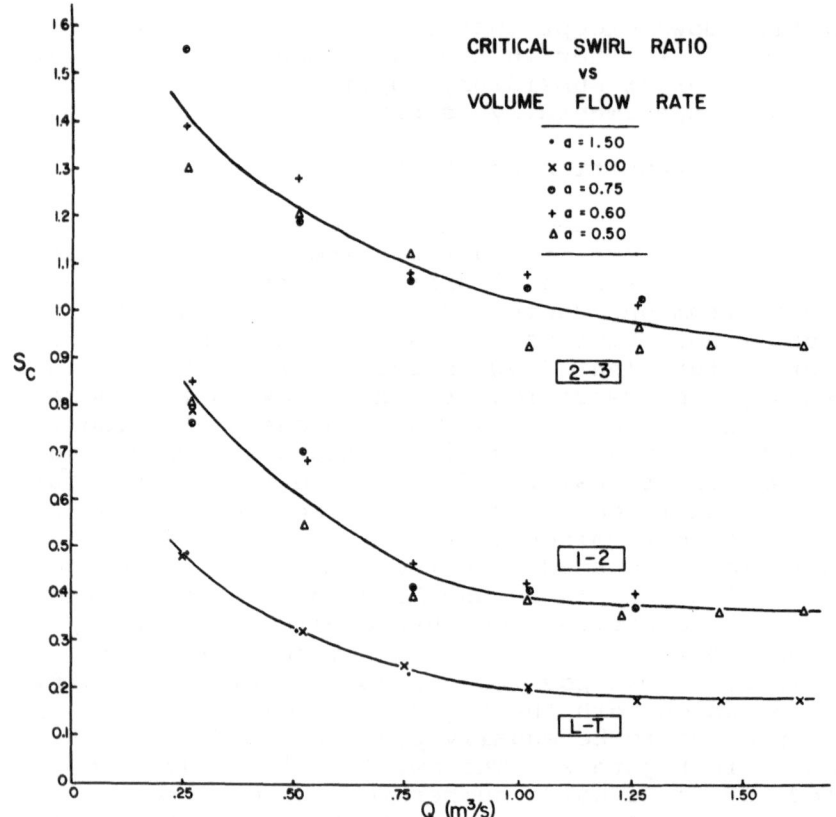

Figure 2.   Critical swirl ratio as a function
of radial Reynolds number, for the
transitions between various vortex
types.   L-T indicates the transition
between a laminar and a turbulent
core at the height of the updraft
hole.   1-2 represents the transition from a single spiralling
roll vortex to a configuration
containing two subsidiary vortices.
2-3 represents the transition
from two to three subsidiary
vortices.

of velocity and their spatial distribution in a
variety of vortex flows represents a formidable experimental challenge.   Although several papers have dealt
with certain aspects of velocity measurements in
vortices, a detailed examination of even these results
lies outside the scope of the present article.   Here
we focus instead on how the magnitudes of core

velocities are related to the background flow parameters. Baker and Church (1979) conducted a simple
experiment to investigate this by placing a single
hot-film sensor with its axis aligned in a radial
direction at a height 0.5 h. The sensor was moved to
a position of maximum velocity where an average
peak value was obtained. This procedure was repeated
for a wide range of swirl ratios and Reynolds numbers.
The results are shown in Figure 3.

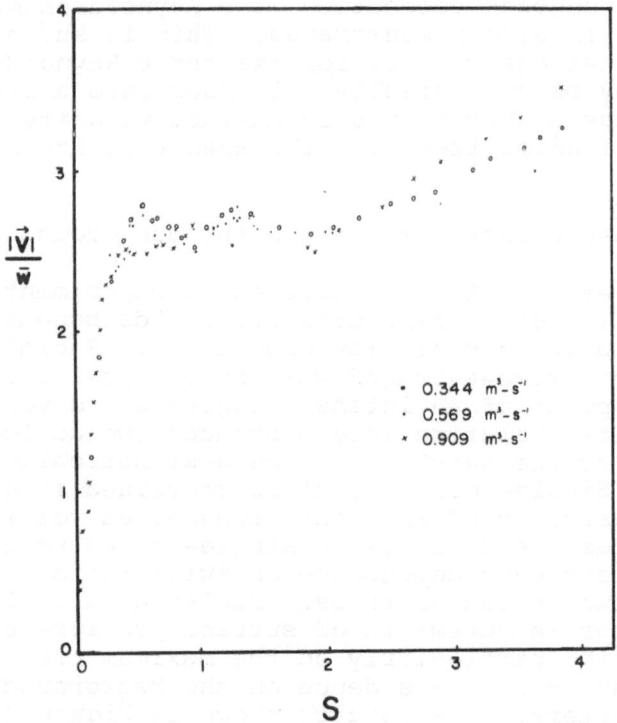

Figure 3. Dimensionless mean total core
           velocity as a function of swirl
           ratio, for different volume flow
           rates, corresponding to radial
           Reynolds numbers in the range
           0.6 to 1.6 x 10^5.

The velocity data have been nondimensionalized with
$\overline{w}$, the mean updraft velocity defined by $Qh = \pi r_o^2 \overline{w}$.
It was found that for laminar vortices $(0.1 < S < 0.4)$
the dimensionless peak velocity increases fairly

linearly to a maximum value of about 2.6. For turbulent
vortices in the range 0.4<S<2.0 the dimensionless
velocity remains quite close to this value. It
should be noted that the locally intense effects
associated with multiple vortices in the turbulent
cores were to a large extent removed by the averaging
process, and only slight peaks associated with these
subsidiary features can be discerned in the data. For
turbulent vortices above S = 2.0, the dimensionless
velocity increases gradually with increasing swirl
ratio. As well as illustrating the swirl ratio
dependence on peak velocities, the Reynolds number
dependence is also demonstrated. This is shown by
the fact that the results for the three Reynolds
numbers may be conveniently collapsed into a single
curve. Thus a characteristic maximum velocity in the
core can be normalized with the mean velocity in the
updraft.

3.4. Pressure Measurements and the Background Flow

Snow et.al. (1980) conducted an experimental
investigation of surface pressure fields beneath vortex
flows for a variety of flow conditions. Radial profiles of the time-averaged wall static pressure showed
the development of an intense single-celled vortical
core and its evolution into a broader two-celled
structure as the swirl ratio was systematically increased. Single-celled vortices contained larger
radial pressure gradients than two-celled vortices,
and the pressure deficits in single-celled vortices
exhibited stronger dependence on swirl ratio than
those in two-celled vortices. Pauley et.al. (1982)
made further measurements of surface pressure deficits,
concentrating particularly on the maximum pressure
deficits and their dependence on the background
flow parameters. The results shown in Figure 4 illustrate the complicated manner in which the maximum
pressure deficit varies with swirl ratio. The behavior
is summarized thus: at low swirl a pressure deficit
builds up and diminishes as S→0.25. This pressure
feature is apparently associated with flow separation
phenomena in the inflow. Beyond S = 0.25 the laminar
core intensifies and the pressure deficit increases
precipitously. Transition to a turbulent vortex
occurs at the surface for S = 0.45. Further increases
in swirl ratio result in an expanded core and smaller
pressure deficits. For 0.85<S<1.6 the pressure
deficits are associated with cores containing a pair
of subsidiary vortices. Beyond S = 1.6 other multiple

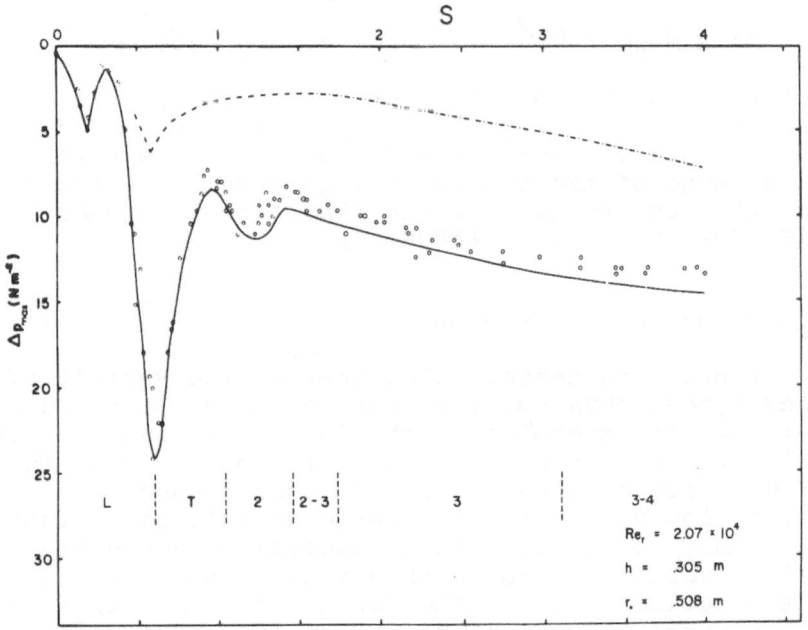

Figure 4. Maximum surface pressure deficit
as a function of swirl ratio for
an aspect ratio of 0.6 and radial
Reynolds number of 2.07×10^4.

vortex cores evolve and the maximum pressure deficits
increase slowly.

In order to investigate how the magnitudes of the
pressure deficits depended on the background flow,
particular attention was paid to the maximum pressure
deficit for different flow geometries and Reynolds
numbers. The measured pressure deficits increased
markedly as the updraft radius was decreased and the
aspect ratio increased, forcing stronger convergence.
The pressure drops were found to be approximately
proportional to the square of the radial Reynolds
number. Since Re_r is proportional to the volume flow
rate, this relationship suggested a non-dimensionalization with respect to a function of a through-flow
velocity. Non-dimensionalization with respect to
the mean vertical velocity at the updraft hole gave

the most satisfying results, i.e.,

$$p^* \equiv \Delta p_{max} / \rho \, \bar{w}^2 \qquad\qquad (4)$$

where Δp_{max} is the maximum pressure deficit, $\bar{w}$ is the mean vertical velocity at the updraft hole, and ρ is the air density. While the dimensional data varied over a range of two orders of magnitude, the dimensionless data tend to fall within the narrow range $\Delta p^* = 40 \pm 8$, particularly at large Re_r.

4.   SUMMARY AND DISCUSSION

It has been demonstrated that a wide variety of vortex types, physically resembling those observed in nature can be generated in the laboratory and that the swirl ratio is the key parameter for determining the configuration of a particular vortex. Above some critical low value of swirl ratio an intense laminar core forms, to be followed at successively higher swirl by vortex breakdown phenomena, the evolution of the two-celled core and the formation of multiple subsidiary vortices within the two-celled structure. When one compares the laboratory observations with the often unusual and varied eyewitness accounts of atmospheric vortex phenomena, it becomes possible to make sense of apparent contradictions in the latter, and to understand the order of rank in atmospheric vortex flows. The degree of visual correspondence between laboratory and natural vortices has been explored in some detail by Church and Snow (1979), where selected photographs of unusual or significant tornado features were displayed alongside photographs of corresponding laboratory vortex features. These exhibit a striking similarity, and this in itself is a validation of the Ward apparatus as a means of studying tornado-like flows. Of several vortex generators to have been developed, the Ward apparatus is the only one which has produced the full range of vortex phenomena found in nature.

Relatively little attention has been paid to determining the magnitudes of swirl ratio associated with natural vortices, and it would be instructive to have such estimates for comparison with the laboratory values. From mesoscale analysis Barnes (1978) was able to derive an estimate of the swirl ratio associated with the Mustang, Oklahoma tornado of 30 April, 1970. A value of S = 0.4 $\pm$ 0.2 was obtained, which would correspond to a single intense funnel close to breakdown, using the laboratory results for comparison. Although the uncertainty in this value is quite large, the order of magnitude is reassuring. More estimates of this type are needed.

Measurements of maximum velocities and pressure deficits in laminar vortices have shown that these quantities are strongly dependent on swirl ratio. In turbulent vortices the core velocities and pressure deficits are much less sensitive to changes in swirl ratio. For both types of vortices a characteristic axial velocity, the mean updraft velocity in the apparatus, has been found to be the most satisfactory parameter for non-dimensionalizing the velocity and pressure data. It follows that estimates of character-istic velocities and pressures in a particular vortex can be provided if S and $\bar{w}$ are known. It is tempting to apply these results to natural vortices. Thus if it is possible to determine an effective swirl ratio and mean updraft velocity in a tornado mesocyclone, it is conceivable that these estimates could provide details about the intensity and configuration of the tornado vortex. A great deal of work needs to be done before this can be realized, however: techniques need to be developed for routinely determining swirl ratio and updraft velocities in severe storms; the numerical coefficients derived from studies of laboratory vortices cannot yet be applied with confidence to atmospheric flows, and companion measurements on intense natural vortices are needed. Finally not all parameters which may influence the intensity of vortex, e.g. surface roughness, external geometry, have been fully investi-gated in the laboratory. Additional insight into these areas may result in an operationally useful technique for tornado surveillance.

Acknowledgements. The research reported in this article represents the results of efforts by many co-workers. In particular the author would like to thank Drs. E.M.Agee, J.T.Snow and G.L.Baker, Ms.B.J. Barnhart and Mr. R.L.Pauley for their valued contributions. Ms. Marcia Olcott typed the manuscript. This research was supported by the Atmospheric Sciences Section of the National Science Foundation.

5. REFERENCES

Agee,E.M.: 1975, J.Atmos.Sci. 32, pp. 642-646."Some inferences of eddy viscosity associated with instabilities in the atmosphere."

Agee, E.M., Church, C.R., Morris C. and Snow, J.T.:
1975, Mon.Wea.Rev. 103, pp. 318-333. "Some synoptic
aspects and dynamic features of vortices associated
with the tornado outbreak of 3 April 1974."

Baker, G.L., and Church, C.R.: 1979, J.Atmos.Sci.36,
pp. 2413-2424. "Measurements of core radii and peak
velocities in modelled atmospheric vortices."

Barnes, S.L.: 1978, Mon.Wea.Rev. 106, pp. 685-696.
"Oklahoma thunderstorms on 29-30 April 1970. Part II:
radar-observed merger of twin hook echoes."

Church, C.R., Snow, J.T., Baker, G.L. and Agee, E.M.:
1979, J.Atmos.Sci.36, pp. 1755-1776. "Characteristics
of tornado-like vortices as a function of swirl ratio"
a laboratory investigation."

Church, C.R. and Snow, J.T.: 1979, J.Rech.Atmos.13,
pp. 111-133. "The dynamics of natural tornadoes as
inferred from laboratory simulations."

Davies-Jones, R.P.: 1973, J.Atmos.Sci. 30, pp.1427-
1430. "The dependence of core radius of swirl ratio
in a tornado simulator."

Davies-Jones, R.P.: 1976, Proc,Symp. on Tornadoes,
Texas Tech. University, Lubbock, TX. pp. 151-173.
"Laboratory simulations of tornadoes."

Doviak, R.J., Zrnic, D.S. and Sirmans, D.S.: 1979,
Proc.IEEE 67, pp. 1522-1533. "Doppler weather radar."

Fujita, T.T.: 1976, Bull.Amer.Meteor.Soc. 57, pp.401-
412. "Graphic examples of tornadoes."

Hoecker, W.H.: 1960, Mon.Wea.Rev.88, pp. 167-180.
"Wind speed and air flow patterns in the Dallas tornado
of April 2, 1957."

Lewellen, W.S.: 1962, J.Fluid Mech. 14, pp. 420-432.
"A solution for three-dimensional vortex flows with
strong circulation."

Pauley, R.L., Church, C.R. and Snow, J.T.: 1982,
J.Atmos.Sci. 39, to be published. "Measurements of
maximum surface pressure deficits in modelled atmospheric vortices."

Sinclair, P.C.: 1973, J.Atmos.Sci. 30, pp. 1599-1619.
"The lower structure of dust devils."

Snow, J.T., Church, C.R. and Barnhart, B.J.: 1980,
J.Atmos.Sci.38, pp. 1013-1026. "An investigation of
the surface pressure fields beneath simulated tornado
cyclones."

Ward, N.B.: 1972, J.Atmos.Sci. 29, pp. 1194-1204.
"The exploration of certain features of tornado
dynamics using a laboratory model."

DETECTION OF CONVECTIVE STORMS BASED ON PENETRATIVE CLOUD TOP
FROM SATELLITE INFRARED AND RAWINSONDE DATA, AND GRAVITY WAVES
FROM DOPPLER SOUNDER

R. J. Hung

The University of Alabama in Huntsville, Alabama, U.S.A.

R. E. Smith

NASA/Marshall Space Flight Center, Alabama, U.S.A.

ABSTRACT

   The Louisiana tornado on March 24, 1976, and Arkansas
tornadoes on April 11, 1976, were investigated by using GOES
digital infrared data, rawinsonde observations, Doppler sounder
records, and radar summaries during the three-hour time period
immediately preceeding the touchdown of the tornadoes. Clouds
associated with the tornadoes were compared to clouds without
tornadoes using the observational data. It appears as if the
altitude to which the overshooting cloud top penetrates above
the tropopause height is the factor that controls the formation
of tornadoes. Gravity waves were first observed at ionospheric
heights when the overshooting cloud top began to penetrate the
tropopause.

INTRODUCTION

   Geosynchronous satellite visible and infrared observations
provide a powerful tool for studying severe convective storms,
such as thunderstorms, tornadoes, hail storms, hurricanes, etc.
(1 to 5). The infrared image provides an indication of the
equivalent blackbody temperature of the observed cloud tops. In
the Geosynchronous Operational Environmental Satellite (GOES)

347

E. M. Agee and T. Asai (eds.), Cloud Dynamics, 347–362.
Copyright © 1982 by D. Reidel Publishing Company.

infrared sensor, 256 different digital count values are assigned
to represent specific ranges of blackbody temperatures. By
referencing the temperature–height profiles from conventional
rawinsonde observations to the satellite infrared data sets at
different time periods the development of convective clouds
can be studied in detail from the formation of the cloud, the
initiation of the updraft motion, to the development of the
tornadic cloud.

Association of gravity waves and severe convective storms
has been studied extensively in the laboratory and in the field
during the past decade (6, 7, 8). Recently, gravity waves
associated with tornado activity (9, 10) and hurricanes (11)
have been observed. These observations were made with a high-
frequency CW Doppler array system in which radio wave receivers
located at a central site, NASA/Marshall Space Flight Center,
monitored signals transmitted from three independent remote
sites on three sets of frequencies and reflected off the ionosphere approximately halfway between the transmitter and
receiver sites. By using a ray tracing technique, Hung et al.
(12, 13) have shown that the enhanced convection–initiated gravity waves associated with tornadoes were generated by thunder-
heads embedded in a squall line and/or an isolated cloud with
intense convection. A comparison of the location of the computed
wave sources and the time of wave excitation with published
tornado touchdown data showed that the computed wave sources were
near the locations where tornadoes touched down more than one
hour after the waves were excited (9, 14).

Recently, Hung et al. (2) investigated the change of cloud
top temperature with respect to time for the clouds associated
with the source of gravity waves compared to the clouds which
were not associated with gravity wave generation. The study
of GOES infrared data during the time period between when the
gravity waves were being excited and the touchdown of the tornado indicated that clouds associated with tornado activity are
characterized by both a very low temperature at the cloud top,
which is equivalent to a higher penetration above the cirrus
canopy, and a very high growth rate of the cold region of the
cloud top, the signature of enhanced convection in the cloud.

In this article the life cycles of two isolated cloud systems, one in Louisiana on March 24, 1976, and the other in Arkansas on April 11, 1976, are used to illustrate how the visible
and infrared images observed from geosynchronous satellite can
be utilized to study severe storm development. A comparison of
the life cycles of the cloud systems associated with tornadic
storms and non-tornadic storms are made. There are some special
features of the cloud associated with the tornado. These special
features are the very low cloud top temperature of the overshooting

turret; a much higher growth rate of cloud top above the tropopause; and also a much larger area above the tropopause. The
comparison of the gravity wave observations and the GOES infrared
digital data shows that the gravity waves was excited when the
overshooting cloud top was growing rapidly.

SATELLITE IMAGE PROCESSING AND DATA ANALYSIS

An Image Data Processing Systems (IDAPS) was developed by
NASA/Marshall Space Flight Center to be used for the image processing requirements of the Skylab experiments. IDAPS can be
used to process high resolution photographs, both visible and
infrared, from satellites to study cloud top height variability,
temperature distribution, and growth and collapse rates of clouds.
GOES digital infrared data during the time period between three
hours before the touchdown of tornadoes and the tornado touchdown time for two severe convective storms on March 24, 1976,
and April 11, 1976, and the other non-tornado-associated clouds
in the entire United States were used in this study. The period
between satellite observations was 15 minutes for the case of
March 24, 1976, and was 30 minutes for the case of April 11, 1976.

In this study, a cumulative histogram is compiled starting
from the cold end of the temperature distribution. The number
of pixels (picture elements), N_i, with blackbody temperature
equal to or less than temperature T_i, is obtained. Physically
the number of pixels, N_i, is proportional to the area of the
cold cloud top with temperature $\leq T_i$. This also provides data
about the horizontal area of the cloud penetrating above certain
altitudes.

Growth rate of penetrative overshooting cloud top is a significant characteristic for the evolution of thunder clouds to
tornadic clouds. The following equation was used to calculate
the growth and expansion rate of the cloud:

$$\gamma_i = \frac{dN_i}{dt} \quad (\text{Pixels} - \text{sec}^{-1})$$

where γ_i denotes the growth rate of the cloud area with temperature $\leq T_i$; N_i, the number of pixels with temperature $\leq T_i$; and
t, the nominal time period between observations.

(A) Study of Severe Convective Clouds on March 24, 1976

GOES digital IR data for the entire United States during
2102-2347 GMT, March 24, 1976, were analyzed in this study. For
the huge cloud extending from Ohio, Illinois down to Arkansas,
Mississippi, Louisiana and Texas, a small portion of this cloud

located in southwestern Louisiana was the only cloud with a cloud
temperature $\leq$ -68.2°C.

The areal expansion and growth of the cold element of the
cloud top have been studied also. Comparison between IR digital
data of cloud top located at southwestern Louisiana and rawinsonde data from Lake Charles, Louisiana, shows that the cloud
top started to penetrate above the tropopause at about 2131 GMT.
Figure 1 shows the changes in the areas of the cloud top penetrating above the tropopause at different temperatures during
the 2131-2347 GMT time period for the cloud located in Louisiana. The figure illustrates the time-dependent increase (or
decrease) of the number of pixels for cloud top temperatures
$\leq$ -62.2°C, $\leq$ -66.2°C, $\leq$ -68.2°C and $\leq$ -69.2°C. The rawinsonde
observation from Lake Charles, Louisiana, at 2300 GMT, March 24,
1976, indicates that the tropopause height was 11.3 km with a
temperature of -58.3°C. This shows that the overshooting turrets
penetrated above the tropopause about 2.5 hours before the tornado touchdown. For the cloud area with temperature $\leq$ -62.2°C,
the expansion is very rapid between 2131 and 2202 GMT. The cloud
area expansion slows down in the next 30 minutes. A higher altitude cloud with a temperature $\leq$ -66.2°C is then visible. The
top of the cloud continued to rise until it reached a temperature of $\leq$ -68.2°C at 2247 GMT, and then it reached the coldest
temperature of $\leq$ -69.2°C at 2317 GMT. The growth of the cloud
top continued until 2332 GMT when cloud top areas with temperatures of both $\leq$ -68.2°C and $\leq$ -69.2°C started decreasing, implying that the cloud was collapsing. The tornado touched down
at 2400 GMT. The cloud collapsed about 30 minutes before the
tornado touched down, in agreement with the results on the May 29,
1977, storm (2). This result is also in good agreement with the
aircraft observations made by Fujita and his associates (15, 16).
The size of the cloud top penetration above tropopause just before the touchdown of the tornado, for this case, was 243 pixels.

By using the relationship established by Simpson (17), one
can learn much about the growth of cumulonimbus clouds by using
both infrared imagery data from satellites and vertical temperature distribution from regular rawinsonde observations. From
Lake Charles, Louisiana, 2300 GMT rawinsonde data, the tropopause height was determined to be 11.3 km with a temperature of
-58.3°C. Figure 2 shows the growth of cloud top height in comparison with the height of tropopause during the period 2131-
2347 GMT. It is shown in this figure that the highest altitude
of the turret top of the Louisiana cloud was more than 4.1 km
higher than the tropopause. Since the coldest overshooting top
temperature of the Louisiana cloud was about 11C° below the temperature of the tropopause, the density of the overshooting turret
is again much higher than the density of the surrounding air.

Figure 3 shows the growth·rates of the cold cloud areas with temperatures $\leq$ -62.2°C, $\leq$ -66.2°C, $\leq$ -68.2°C, and $\leq$ -69.2°C during the time period 2247-2347 GMT. The maximum growth rate of the area with temperature $\leq$ -62.2°C occurred during the period 2302-2317 GMT, which was about 43 to 58 minutes before the tornado touchdown. In other words, the maximum growth rate of the penetrative overshooting turret above the tropopause (essential for formation of tornadic storms) occurs approximately one hour or less before the tornado touchdown. On the other hand, the growth rate of the high-altitude clouds with temperatures $\leq$ -69.2°C became negative after 2332 GMT, implying that the higher altitude cloud began collapsing approximately 30 minutes before the tornado touchdown.

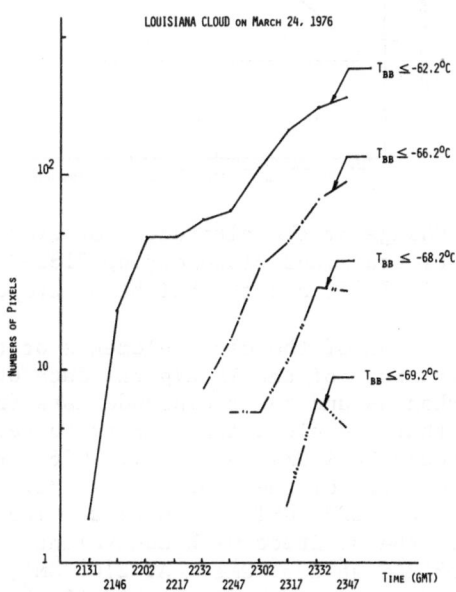

Figure 1.   Area expansion and vertical growth of penetrative overshooting cloud top in terms of time change of pixels with temperatures $\leq$ -62.2°C, $\leq$ -66.2°C, $\leq$ -68.2°C, and $\leq$ -69.2°C during 2131-2347 GMT, March 24, 1976, for cloud located in south-western Louisiana.

Similar analyses of both areal expansion rates and growth of cloud top heights during the same period were accomplished all over the United States. It was found that only clouds located over Idaho had a temperature 2C$^\circ$ warmer than the cloud top temperature over Louisiana; however, the growth rate of the Idaho cloud was less than that of the Louisiana cloud.

(B)   Study of Severe Convective Clouds on April 11, 1976

       GOES digital IR data for the entire United States during
2031-2231 GMT, April 11, 1976, were analyzed in this study. For
the large cloud of the squall line extending from Oklahoma through
Arkansas, Tennessee and North Carolina, only a small portion of
this cloud located in central and eastern Arkansas had a cloud
with a cloud temperature $\leq$ -60.2°C.

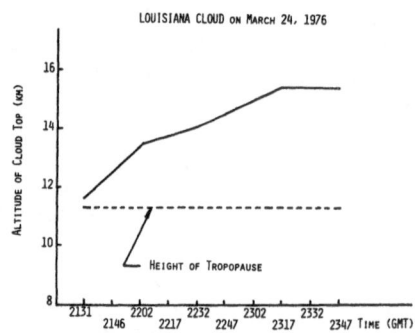

Figure 2.   Change of the altitudes for overshooting turret
             of Louisiana cloud during 2131-2347 GMT, March
             24, 1976, and the height of tropopause.

       The area expansion of the cloud elements of this cloud top
was studied.   Comparison of the IR digital data of this cloud
top located in Arkansas and the rawinsonde data from Nashville,
Tennessee, shows that the cloud top started to penetrate above
the tropopause around 2031 GMT, April 11, 1976.   Figure 4 shows
the changes in the areas of the cloud top at different temperatures during the 2031-2231 GMT time period.   The cloud top
started to grow to the altitude with equivalent blackbody temperature, $T_{BB} \leq$ -59.2°C at 2031 GMT.   At 2101 GMT, the cloud top
reached the altitude with a temperature $\leq$ -62.2°C; at 2131 GMT,
the cloud top reached a temperature $\leq$ -64.2°C; and at 2201 GMT,
the cloud finally reached the highest altitude with the lowest
temperature at -66.2°C.

       The rawinsonde data from Nashville, Tennessee were used to
relate the temperature of the penetrating turret to an altitude.
Based on the rawindonde data of 2300 GMT the same day, the temperature of the tropopause was around -59°C and the altitude of
the tropopause was around 11.2 km, while the lowest cloud top
temperature observed on the satellite IR imagery was -66.2 about
30 minutes before the touchdown of the tornadoes.   Our earlier
case studies of March 24, 1976 and May 29, 1977 using the 15
minute interval IR data show that the lowest cloud top temperature is observed about 15 minutes before the touchdown of the

tornado, just before the overshooting turret collapsed. In the present case, the lowest cloud top temperature could be even lower than -66.2°C because there was no 15 minute interval data available and thus there is no way to determine what the cloud top temperature was 15 minutes before the touchdown of the tornado. For the very same reason, the time of the collapse of the overshooting turret could not be determined precisely since it occurred about 15 minutes before the touchdown of the tornado (15, 16). However, it can be seen from Figure 4 that (1) the cloud top started to penetrate above the tropopause approximately two and one-half hours before the touchdown of the tornado, (2) the coldest overshooting top temperature was more than $7C^{\circ}$ below the temperature of the tropopause, and (3) the area of the cloud top above tropopause before the touchdown of the tornadoes covered 213 pixels.

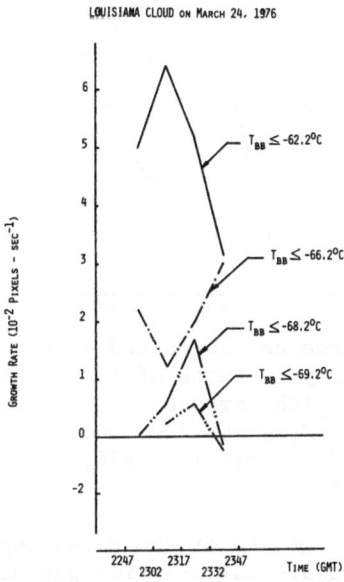

LOUISIANA CLOUD ON MARCH 24, 1976

Figure 3.   Growth/collapse rate of penetrative overshooting cloud top with temperatures $\leq -62.2^{\circ}$C, $\leq -66.2^{\circ}$C, $\leq -68.2^{\circ}$C, and $\leq -69.2^{\circ}$C during the time period 2247-2347 GMT, March 24, 1976, for cloud at southwestern Louisiana.

The relationship between the heights of overshooting tops of clouds above the tropopause and some physical parameters has been investigated by a group of scientists at NASA Goddard Space Flight Center. By using a downward pointing lidar on a high-altitude aircraft and by using sideviews of the thunderclouds photographed every 30 seconds from a second aircraft, Simpson

and her associates determined the heights of these clouds (17).
A detailed examination of the values of infrared equivalent blackbody temperature of the cloud tops obtained from the satellite,
and the lidar-determined heights for the overshooting tops above
the tropopause shows a high correlation between the heights and
temperatures, with the slope of the relation indicating an approximate adiabatic lapse rate (17).

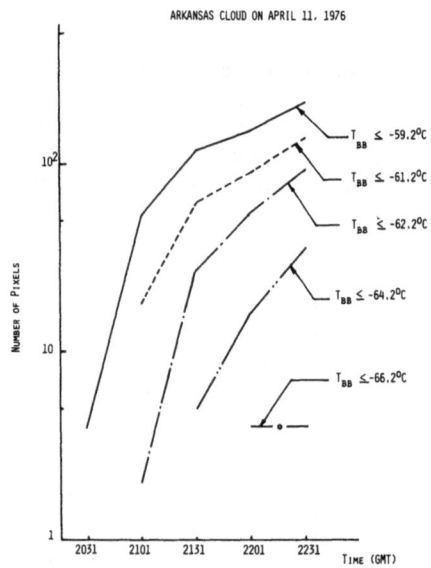

Figure 4.   Cold area expansion and vertical growth of the
             cloud top in terms of time change of number of
             pixels with temperatures $\leq$ –59.2°C, $\leq$ –61.2°C,
             $\leq$ –62.2°C, $\leq$ –64.2°C, and $\leq$ –66.2°C during 2031–
             2231 GMT, April 11, 1976, for cloud located in
             Arkansas.

Figure 5 shows the growth of the cloud top height above the
tropopause during the 2031-2231 GMT time period based on the relationship suggested by Simpson and her associates. This figure
shows that the highest altitude of the turret top of the Arkansas
cloud was more than 5.3 km higher than the tropopause altitude
approximately 30 minutes before the touchdown of the tornado.
Since the overshooting top temperature was more than 7C$^{\circ}$ below
the temperature of the surrounding air, the density of the overshooting turret is much higher than the density of the surrounding
air. The overshooting turret can only exist as long as it is
dynamically supported by intensive vertical convection; therefore,
as the intense vertical convection disappears, the overshooting
turret should collapse. In this case, the exact time that the
overshooting turret collapsed could not be determined due to use
of 30 minute interval satellite data.

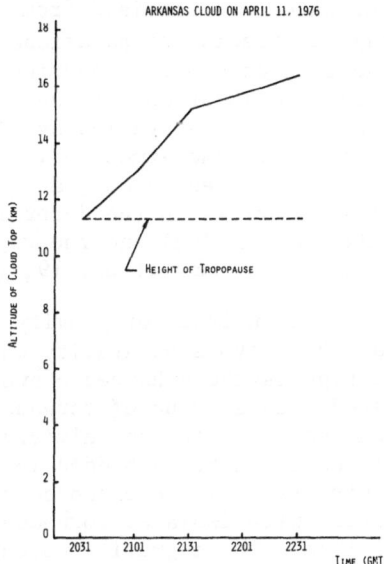

Figure 5. Change of the altitudes for overshooting turret
of Arkansas cloud during 2031-2231 GMT, April 11
1976, and the height of tropopause.

Figure 6 shows the growth rate of the cloud top areas above
the tropopause. The maximum growth rate of the area with temperature $\leq$ -59.2°C (with an altitude just above the tropopause) with
the value of 4.74 x 10^{-2} pixels - sec^{-1} occurred about 45 minutes
before the tornado touchdown. In other words, the maximum growth
rate of this overshooting turret above the tropopause (apparently
essential for the formation of tornadic storms) occurred approximately one hour or less before the tornado touchdown.

Similar analyses of both areal expansion rates and growth of
cloud top heights during the same time period were accomplished
throughout the United States. It was found that the Arkansas
cloud was the only cloud with a temperature lower than -60.2°C
combined with a high growth rate of the cloud top above the tropopause.

DOPPLER SOUNDER AND RADAR OBSERVATIONS

During time periods with severe weather activity, wave-like
disturbances are observed in the high frequency CW Doppler records.
The data from the Doppler sounder array are subjected to a power
spectral density analysis to obtain the wave periods of these
Doppler fluctuations while the direction of propagation and the

phase velocity of the waves are obtained from a cross correlation analysis (10). Group ray tracing computations using the best available data on the thermodynamic properties of the atmosphere are used in determining the locations of the sources of the waves. A detailed description of the observation system, the data pro- cessing techniques, wind data and atmospheric models used in the ray tracing computations is given in Hung and Smith (10) and Hung et al. (9). The probable errors in the determination of the azimuthal angle of the wave arrival and the ray tracing compu- tation have been discussed by Hung et al. (9, 12, 13, 14).

Based on our previous analysis of gravity waves associated with tornadic storms, three types of gravity waves have been detected. The first type is the enhanced convection-initiated gravity wave associated with a group of tornadoes (9, 13). The second type is the enhanced convection-initiated gravity wave associated with isolated tornadoes embedded in a squall line (12, 13, 14). The third type is the enhanced convection-initiated gravity waves associated with isolated tornadoes without the presence of a squall line. These gravity waves were excited under a wide variety of meteorological conditions ranging from the static conditions associated with air mass type convective storms to those associated with rapidly moving fronts, pre-frontal squall lines and isolated clouds with intense convection. In each instance, within the combined probable error bands of the detection system and analytical techniques, the wave sources were located at clouds with intense convection which eventually developed into tornadic storms.

(A) Severe Storm on March 24, 1976

During the time period of 2315-2415 GMT, March 24, 1976, Doppler records showed wave-like oscillations in the F-2 layer of the ionosphere. Figure 7 shows the oscillations in the high frequency, 5.734 MHz, CW transmissions.

Four gravity waves were observed and analyzed. The pro- pagation characteristics of these waves are listed in Table 1. The azimuthal angles of wave arrival were 42-46° and the hori- zontal phase speeds were 104-128 m/sec; therefore, it appears as if these waves could be from the same source. Ray tracing computations confirmed this.

Ray tracing computations show that the computed probable sources of waves were located at southwestern Louisiana, and the wave traveling times from the computed probable sources to the observation point were 112-136 minutes. The gravity waves were excited while the overshooting turret penetrated above the tropopause.

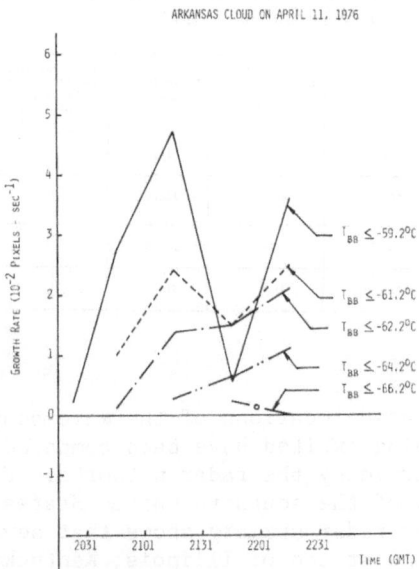

Figure 6. Growth rate of penetrative overshooting cloud top with temperatures $\leq -59.2^{\circ}C$, $\leq -61.2^{\circ}C$, $\leq -62.2^{\circ}C$, $\leq -64.2^{\circ}C$, and $\leq -66.2^{\circ}C$ during the time period 2031-2231 GMT, April 11, 1976 for cloud at Arkansas.

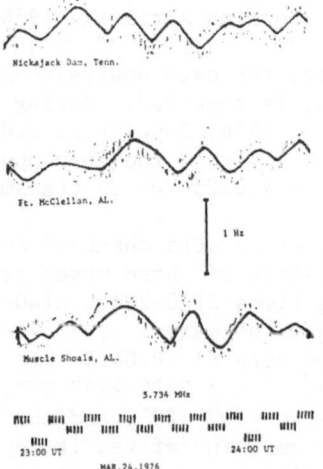

Figure 7. Doppler record on March 24, 1976, during the time period of 2255-2415 GMT, at the operating frequency 5.734 MHz.

Table 1.   Propagation Characteristics of the Observed
Gravity Waves Associated with Tornadic Storms
on March 24, 1976.

Data Sampling Time (GMT)	Wave Period (Min)	Horizontal Wavelength (km)	Azimuth Angle of Wave Arrival (Deg)	Horizontal Phase Speed (m/sec)	Wave Traveling Time From Source to Array		Location of Wave Source	Tornado Touchdown Location	Tornado Touchdown Time (GMT)
					Hr	Min			
2300 2400	20.00	153.60	46.0	128.0	1	52	30.38N 93.25W		
2315 2400	20.71	149.11	43.0	120.0	2	4	29.97N 93.35W	30.26N 92.42W	2400
2315 2415	20.24	134.80	42.0	111.0	2	14	29.95N 93.36W		
2330 2415	18.39	114.75	46.0	104.0	2	16	30.34N 92.99W		

The computed locations of the wave sources when the gravity
waves were being excited have been compared with the severe storm
activity observed by the radar networks. Figure 8 shows the
radar summary of the southern United States at 2235 GMT, March
24, 1976. The radar summary shows that severe thunderstorms occurred over the states of Illinois, Kentucky, Missouri, Tennessee,
Arkansas, Mississippi, Louisiana and the southern section of
Texas when the observed gravity waves were being excited. The
computed probable wave sources were located in the severe thunderstorm activity area where cloud top heights of 13.7 km were
observed by the radar when the observed gravity waves were being
excited.

(B)   Severe Storm on April 11, 1976

In this study, the data analysis of Doppler sounder records
on April 11, 1976, is reported. During the time period of 2140-
2240 GMT, April 11, 1976, Doppler records showed wave-like oscillations in the F-2 layer of the ionosphere. Figure 9 shows the
oscillations in the 4.0125 MHz CW transmissions.

Five gravity waves were observed and analyzed. The propagation characteristics of these waves are listed in Table 2. For
the data sampling times 2140-2210, 2140-2220, and 2145-2330 GMT,
the azimuthal angles of wave arrival were $95°-98°$, and the horizontal wavelengths were 81.98-84.56 km; while for the data sampling times 2210-2235 and 2210-2240 GMT, the azimuthal angles
of wave arrival were $110°-120°$, and the horizontal wavelengths
were 96.56-100.89 km. Therefore, it appears as if the gravity
waves detected during the data sampling times 2140-2210, 2140-
2220, and 2145-2230 GMT could be from one source; while the waves
detected during the data sampling times 2210-2235 and 2210-2240
GMT were from another source. Ray tracing computations

confirmed this.

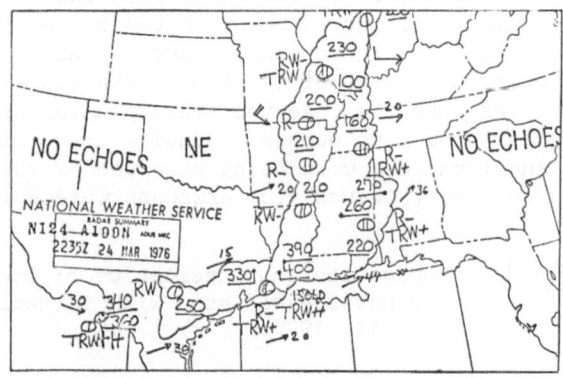

Figure 8.   Radar weather summary of the southern United
States at 2235 GMT, March 24, 1976.

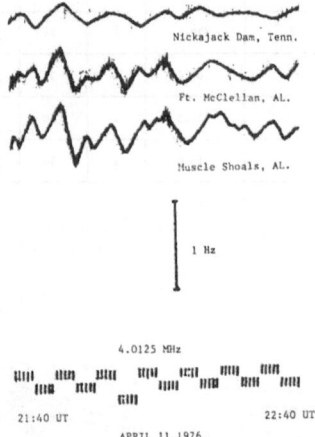

Figure 9.   Doppler record on April 11, 1976, during the
time period of 2140-2240 GMT, at the operating
frequency of 4.0125 MHz.

Ray tracing computation shows that the computed probable
sources of waves observed during 2140-2210, 2140-2220, and 2145-
2230 GMT time period were all located in the central eastern
region of Arkansas, in which the tornado touchdown was at 2255
GMT; while the computed probable sources of two gravity waves
detected during 2210-2235 and 2210-2240 GMT were located in the
northeastern corner of Arkansas near the Tennessee border in
which the tornado was touchdown at 2250 GMT.

The computed locations of the wave sources when the gravity

waves were being excited have been compared with the severe storm activity observed by the radar networks. Figure 10 shows the radar summary of the southern United States at 2035 GMT, on April 11, 1976. The radar summary shows that severe thunderstorms occurred over the states of Oklahoma, Arkansas, Tennessee, and North Carolina when the observed gravity waves were being excited. In particular, the line echoes extending from Oklahoma through Arkansas and Tennessee were being observed by the radar when the observed gravity waves from Arkansas were being excited.

Table 2. Propagation Characteristics of the Observed Gravity Waves Associated with Tornadic Storms on April 11, 1976.

Data Sampling Time (UT)	Wave Period (Min)	Horizontal Wavelength (km)	Azimuth Angle of Wave Arrival (Deg)	Horizontal Phase Speed (m/sec)	Wave Traveling Time From Source to Array		Location of Wave Source	Tornado Touchdown Location	Tornado Touchdown Time (UT)
					Hr	Min			
2140 2210	12.20	81.98	98	112.0	1	26	34.12N 91.12W	34.40N 92.14W	2255
2140 2220	12.44	82.48	95	110.5	1	29	33.94N 91.35W		
2145 2230	12.93	84.56	95	109.0	1	32	33.94N 91.58W		
2210 2235	12.93	96.59	110	124.5	1	26	35.00N 90.62W	35.35N 89.56W	2250
2210 2240	14.88	100.89	120	113.0	1	28	35.93N 90.86W		

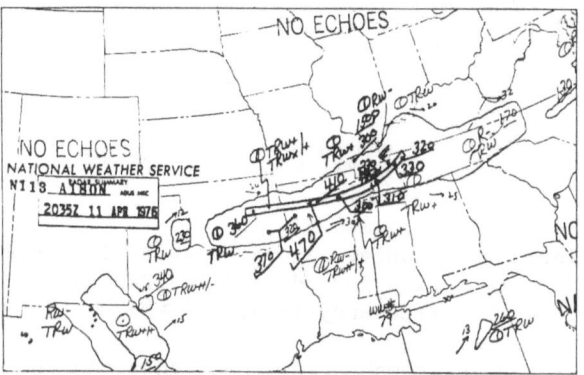

Figure 10. Radar weather summary of the southern United States at 2035 GMT, April 11, 1976.

DISCUSSIONS AND CONCLUSIONS

From the present analysis, the results obtained from the

combination of the cloud top temperature changes from the satellite infrared imagery, the rawinsonde data, radar summaries, and the Doppler sounder gravity-wave observations lead to the following:

(1) Tornado-associated clouds have overshooting turrets penetrating above the tropopause.

(2) The difference between the overshooting cloud top temperature and the tropopause temperature, a measure of how much the cloud has penetrated above the tropopause, rather than either the absolute temperature of the penetrative cloud or the height of the top of overshooting turret is important in the development of severe storms.

(3) The growth rate of the overshooting turret above the tropopause for severe storm-associated clouds is much greater than that for non-severe storm-associated clouds.

(4) The high density penetrative overshooting turret (temperature of the overshooting turret is much colder than the surrounding air temperature) above the tropopause collapses about 15 to 30 minutes before the tornado touchdown. However, the information on the collapse of the overshooting turret may be missed if the satellite observations are at 30 minute intervals.

(5) The life of a tornado-associated cloud, from the moment the overshooting turret penetrates above the tropopause to the touchdown of the tornado, is no more than 3 hours.

(6) Gravity waves are observed when there are severe convective storms. Ray tracing results show that the source of these gravity waves is located at the cloud with intensive convection at the time the overshooting turret of the cloud is penetrating above the tropopause.

This research suggests that the combination of satellite imagery, rawinsonde data, radar summaries, and gravity wave observations is useful for studying the evolution of tornadic clouds from thunderclouds. However, rapid scan, $\leq$ 15 minute intervals, satellite observation is necessary for the study of short-life mesoscale phenomena. Otherwise, important phenomena such as the time of the collapse of the overshooting turret before the tornado touchdown can be missed.

ACKNOWLEDGEMENTS

R. J. Hung appreciates the support of present study from the National Aeronautics and Space Administration through contract NAS8-33726.

REFERENCES

1. Purdom, J. F. W.:   1976, Mon. Wea. Rev. 105, PP. 1474-1483.
2. Hung, R. J., Phan, T., Lin, D. C., Smith, R. E., Jayroe, R. R., and West, G. S.:   1980, Mon. Wea. Rev. 108, PP. 456-464.
3. Adler, A. F., and Fenn, D. D.:   1979, J. Appl. Meteor. 18, PP. 502-517.
4. Gentry. R. C., Rodgers, E., Steranka, J., and Shenk, W. E.:   1980, Mon. Wea. Rev. 108, PP. 445-455.
5. Sikdar,D. N., Suomi, V. E., and Anderson, C. E.:   1970, Tellus, 22, PP. 521-532.
6. Gossard, E., and Sweezy, W. B.:   1974, J. Atmos. Sci. 31, PP. 1540-1548.
7. Hung, R. J., and Smith, R. E.:   1979, J. Geomag. Geoelect. 31, PP. 183-194.
8. Baker, D. M., and Davies, K.:   1969, J. Atmos. Terr. Phys. 31, PP. 1345-1352.
9. Hung, R. J., Phan, T., and Smith, R. E.:   1978, J. Atmos. Terr. Phys. 40, PP. 831-843.
10. Hung, R. J., and Smith R. E.:   1978, J. Appl. Meteor. 17, PP. 3-11.
11. Hung, R. J., and Kuo, J. P.:   1978, J. Geophys. 45, PP. 67-80.
12. Hung, R. J., Phan, T., and Smith, R. E.:   1979, J. Appl. Meteor. 18, PP. 460-466.
13. Hung, R. J., Phan, T., and Smith, R. E.:   1979, J. Geophys. Res. 84, PP. 1261-1268.
14. Hung, R. J., Phan, T., and Smith, R. E.:   1978, AIAA J. 16, PP. 763-766.
15. Fujita, T. T., and Byers, H. R.:   1977, Mon. Wea, Rev. 105, PP. 129-146.
16. Fujita, T. T., and Caracena, F.:   1977, Bull. Amer. Meteor. Soc. 58, PP. 1164-1181.
17. Simpson, J.:   1980, Progress Report for the Period October 1 1979 to October 1, 1980, NASA/GSFC, Greenbelt, MD, PP. 1-88.

PRECIPITATION IN CONVECTIVE STORMS: AN OBSERVATIONAL AND NUMERICAL STUDY

D.A. Bennetts and M.J. Bader

Meteorological Office
Bracknell, Berkshire
United Kingdom

SUMMARY

Observations of convective rainfall have been simulated in a numerical model with the aim of understanding some aspects of extreme rainfall. This paper reports on two case studies; the first shows a direct comparison between the observations and simulations and the second identifies a form of interaction between clouds which enhances surface rainfall.

1. INTRODUCTION

The concept of Probable Maximum Precipitation (PMP) is one which is invoked during the assessment of design limits for hydrological structures. It is an estimate of the maximum amount of precipitation likely to fall in a given catchment area in a specified period. For periods extending over a few hours such extreme rainfall inevitably results from convective storms (Ludlam 1980). At present PMP is calculated either by statistical analysis of past rainfall events, supplemented by empirical relationships, or by maximising precipitable water from observed storms (WMO 1973). In recent years considerable advances have been made in the understanding of convective processes (summarised in Ludlam 1980) and there is some hope that a more physically based approach may be fruitful. The present work makes a contribution to the development of such an approach by discussing a process which is found to influence rainfall in the convective cloud systems experienced in the United Kingdom. Evidence is presented that under some circumstances interaction between neighbouring cells leads to a

363

E. M. Agee and T. Asai (eds.), Cloud Dynamics, 363–377.
British Crown Copyright © 1982

merging process which may enhance the amount of surface rainfall compared with that expected from separate cells.

Observations obtained in the United Kingdom suggest that if two clouds of different sizes, having relative motion due to their different depths, come into close proximity, then, with a favourable combination of cell separation, orientation and vertical wind shear, it is possible for precipitation embryos from the larger to fall through the smaller. The accretive effects are enhanced, the evaporation reduced and larger surface rainfall rates result. A justification of this interpretation is offered following attempts to reproduce the phenomenon in a numerical model.

a.   Observational Systems

Precipitation amounts were obtained from the United Kingdom radar network (Browning 1979). In general the radars operate at 25 km^2 resolution but at one elevation ($1\frac{1}{2}^{\circ}$), the range-compensated data were available out to about 30 km with a horizontal resolution of 4 km^2. Data with this high resolution are particularly suitable for studying the rainfall characteristics of cumulonimbus clouds.

Rainfall rates measured by a radar at Camborne in the south west of the United Kingdom were chosen for the study because of the availability of radiosonde data from a nearby site. The radar was calibrated using rain gauges and the radiosonde information used to initiate the numerical model.

b.   The Numerical Model

The three-dimensional, pressure coordinate model of deep convection described by Miller and Pearce (1974) has been extended to simulate mid-latitude clouds by the inclusion of a parametrised ice phase. This is discussed in detail in Bennetts and Rawlins (1981). The model domain is 16 x 16 km in the horizontal with a 1 km grid resolution and 950 mb deep with a 50 mb resolution in the vertical.

2.   RESULTS OF THE CASE STUDY ON 23 JANUARY 1980

During the morning of the 23rd January, a westerly airstream covered the south west of the United Kingdom. The appropriate ascent (Figure 1) showed the air to be conditionally unstable and capable of sustaining convection over the sea which, in the observational area, had a surface temperature between 8 and 9°C. A warm front was approaching from the Atlantic and the resulting high level advection of warm air is prominent in the midday ascent (Figure 1). This progressively stabilised the atmosphere and the mean rainfall rate from the showers embedded in the airstream fell from 10 mm h^{-1} early in the period, to 5 mm h^{-1} at 0600 GMT. Shower activity ceased soon after midday.

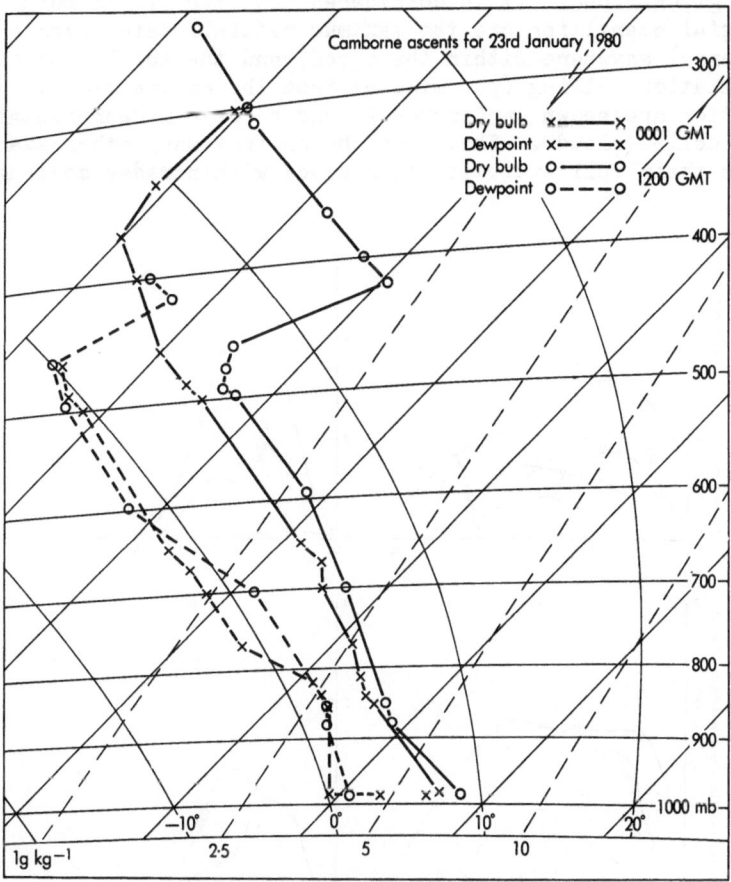

Figure 1. The 0000 and 1200GMT radiosonde ascents from Camborne on 23rd January 1980.

Study of the full record of radar observations from 0000 to 1200GMT showed that the strongest convective activity tended to occur in bands that moved eastwards through the area of interest. In the bands, the showers were heavier, and in general larger in area, than those which were isolated. The slow change in the environmental lapse rate was sufficient to account for the progressive decrease in intensity of precipitation, but the difference in intensity between isolated clouds and groups of clouds suggested that there might have been some organisation to the convection within the bands.

There are many ways in which to quantify the precipitation from a given cloud. It is considered that two of the more meaningful quantities are the maximum rainfall rate (over a 4 km^2 area) anywhere within the cloud, and the total mass of precipitation falling (per minute) from the entire cloud. These quantities are shown in Figures 2a and b for six representative single celled clouds. They were chosen from many other examples because their full evolution took place within radar coverage.

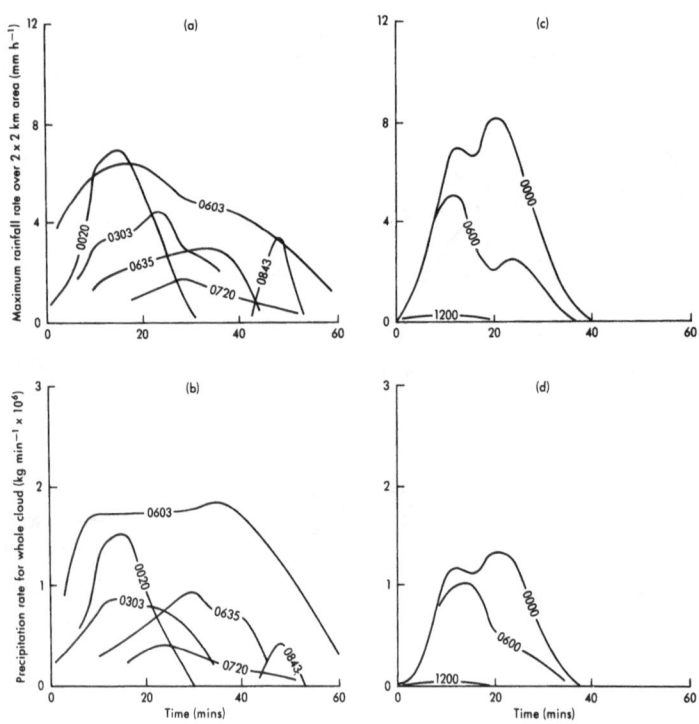

Figure 2. Maximum rainfall rates and total precipitation falling from observed and simulated clouds. There is no significance in the displaced starting times, the curves are spaced for clarity.

   The model was integrated using, as input, the radiosonde
information appropriate to midnight, midday and the mean of these
two soundings to represent an intermediate stage, labelled for
convenience 0600. A wind profile for 0600 was available from a
pilot balloon. The results of the three simulations are shown
alongside the observations, in Figures 2c and d.

   It is encouraging that the magnitude of the simulated
precipitation rates and totals are representative of those observed. However direct comparison between observations and model
results must at all times be made with care. In the atmosphere
local fluctuations in convective activity result in a wide range
of cloud sizes. At best the model is able to produce one such
realization which,within the confines of the various approximations, will be representative in so far as the upper air ascent
is typical of the real atmosphere.

   The bimodal distribution apparent in the model results
obtained using the midnight ascent resulted from successive cells
of a system which therefore continued to produce precipitation
for a longer period than others.

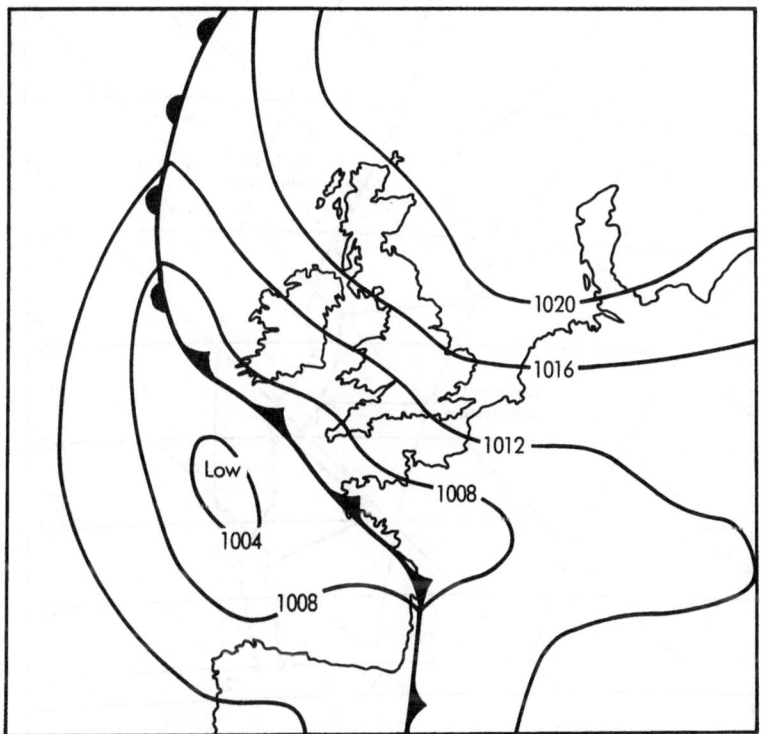

Figure 3. The synoptic situation at 0000GMT on 29th July 1980

The above case study was of winter-time convection. However, since extreme rainfall normally occurs when the air is warmer and more unstable, a similar comparison was made for a summer-time case: 29th July 1980.

3.   RESULTS OF THE CASE STUDY ON 29th JULY 1980

The synoptic situation is shown in Figure 3 and the feature of interest is the active cold front to the west of the United Kingdom. The thermodynamic data are shown in Figure 4 and the vertical wind profile (not shown) was almost unidirectional with a speed difference over the depth of the clouds of about 10 ms^{-1}, equivalent to a wind shear of 12.8 x 10^{-4} s^{-1}. Study of Figures 3 and 4 suggests that the convection was initiated by low level convergence probably associated with the cold front.

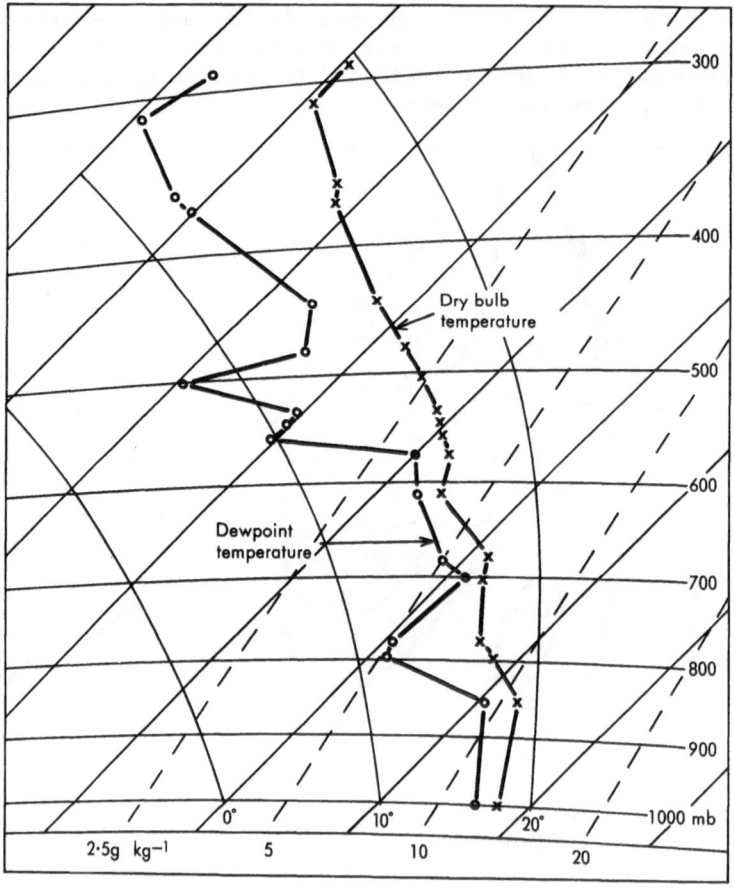

Figure 4.   The 0000GMT radiosonde ascent from Camborne.

    The frontal zone was active and a radar picture (Figure 5)
taken at 0023GMT shows widespread convective activity producing
light rain with rates of a few mm h^{-1} but with a few pockets of
very heavy rain, occasionally with rates of over 100 mm h^{-1}.

    Figure 6 shows the frequency distribution of the maximum
instantaneous surface rainfall rate to occur at any time during
a cloud's evolution. Only those clouds observed over a significant part of their lifecycle are included. Two peaks are
evident, the lower being associated with the smaller, more
isolated clouds and the higher with the larger cloud complexes.
The aim of this section is to understand this large variability
in the observed rainfall and to reproduce it in the model.

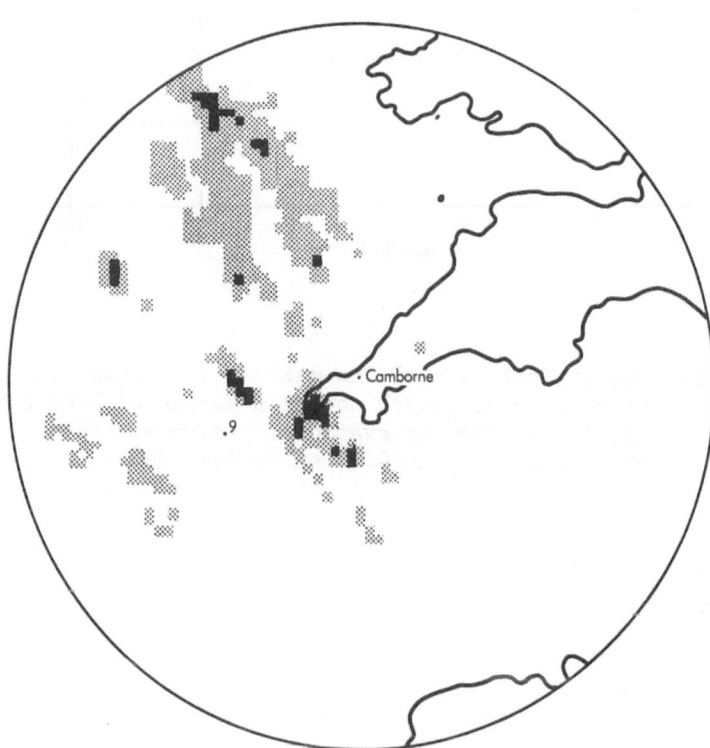

■ Heavy rain (≥ 50 mm h^{-1})

※ Light rain (< 10 mm h^{-1})

Figure 5.   Rainfall distribution as seen by radar at 0023
            on 29 July 1980.

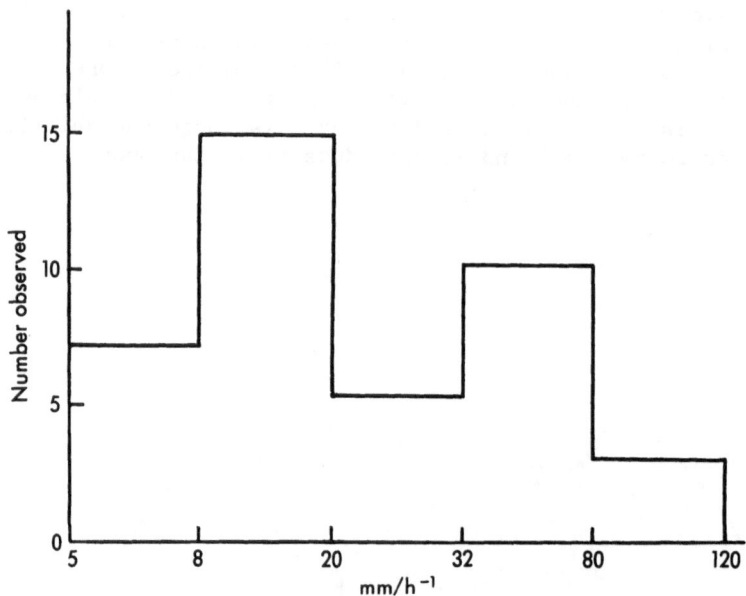

Figure 6.   The distribution of the maximum instantaneous
            rainfall rates which occurred during a cloud's
            lifetime.   The observational period was from
            2000 on 28 July to 0200 on 29 July 1980.

a.   Single Cell Clouds

     The thermodynamic and dynamic data obtained from the radio-
sonde ascent were used as input to the numerical model and
convection was initiated by an assumed form of low level convergence designed to reproduce the main features of the frontal
zone. Figure 7 shows a vertical section of the mature phase of
the simulated cloud. Even though there are quite high concentra-
tions of cloud water and precipitation within the centre of the
cloud, little rain actually reaches the ground. This is partly
because the freezing level is high allowing the hail to melt,
thus reducing the mean fall speed of the precipitation, and
partly because the ambient air is relatively dry inducing rapid
evaporation. Consequently surface rainfall is light, never
exceeding 8 mm h^{-1}.

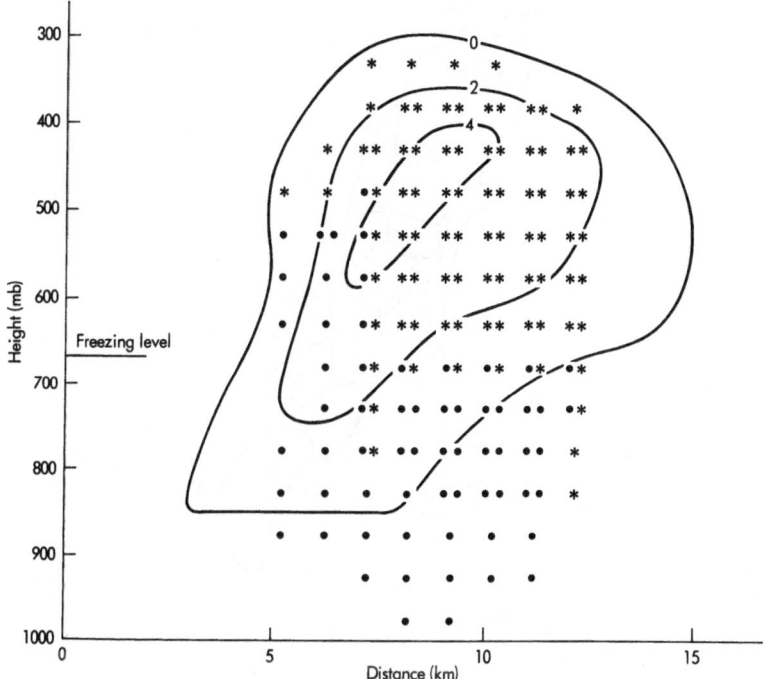

Figure 7.   The mature phase of the single cell cloud. Contours
            are of cloud water substance in gm kg^{-1} and (• , ✳)
            represent rain and hail respectively. The number of
            symbols indicates rainfall intensities of < 8, 8 - 44
            and > 44 mm h^{-1}.

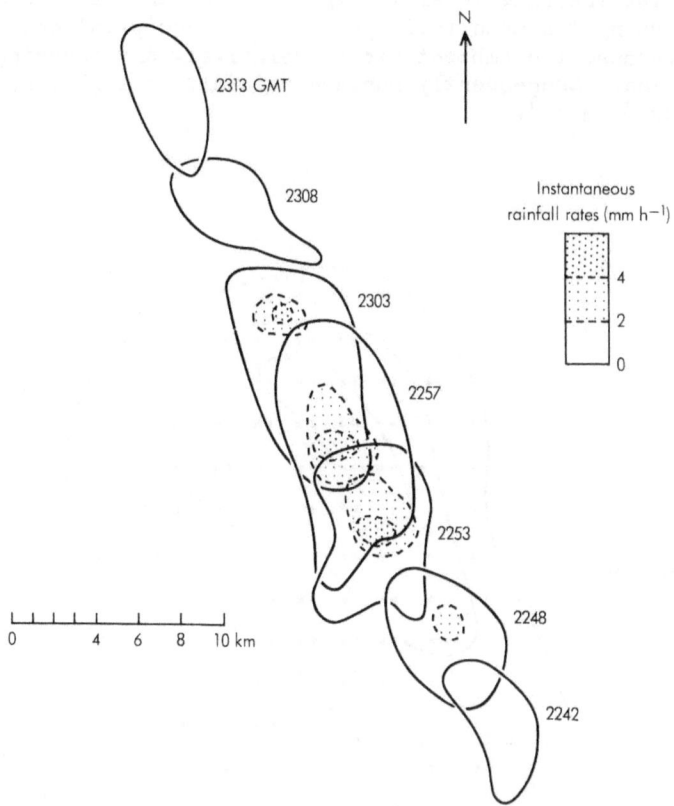

Figure 8.   The time evolution of a typical weak echo.

Such behaviour is characteristic of the observed weaker echoes shown in Figure 5. Figure 8 shows the time evolution of one typical weak echo and Figure 9 shows a comparison between the modelled rates at 875, 925 and 975 mb, meaned over 4 km^2, and observed radar derived values, also over 4 km^2, which were representative of the 950 mb level. (The elevation of the radar beam was $1\frac{1}{2}^{o}$ and the approximate range of the cloud was 20 km). For this particular cloud the agreement is good. Of course the atmosphere exhibits considerable variability as was discussed in section 2 and some assessment of this is revealed by Figure 6.

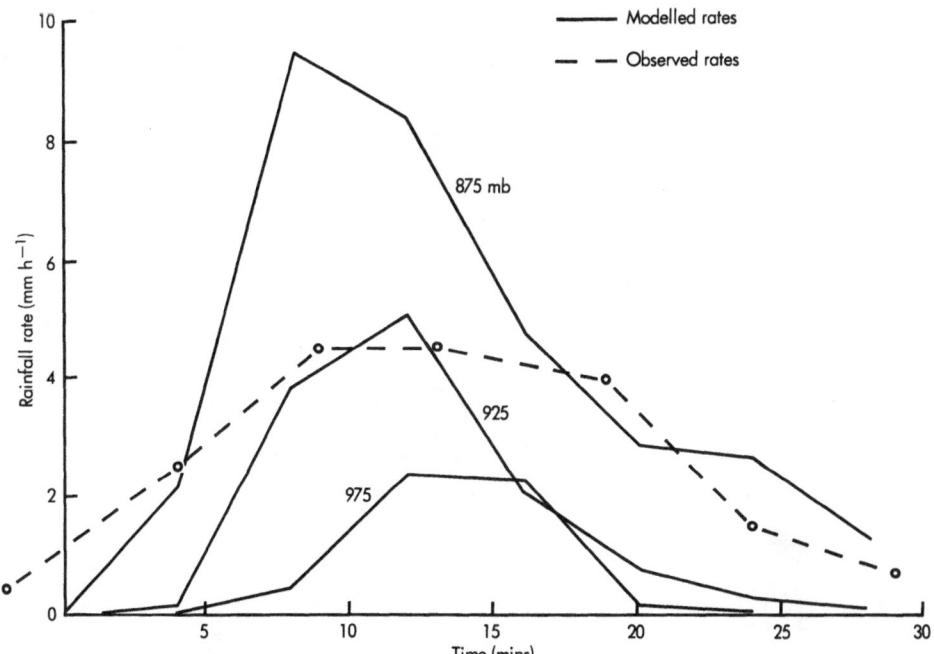

Figure 9. The instantaneous maximum rainfall rate during the lifecycle of one cloud. The three modelled rates are obtained from the lowest three grid levels and illustrate the rapid evaporation. The observed rates are from a height of 950 mb, the radar being at an elevation of $1\frac{1}{2}^{o}$ and the cloud being at a range of approximately 20 km.

b.   Underline Interacting Clouds

     Interest centres on the intense echoes, for the rainfall in
these far exceeds that expected from model predictions.
Inspection of Figure 6 shows that these echoes also form a
distinct sub-group of the observations. For reasons which will
become clear, the behaviour of the intense echoes was indistin-
guishable, in all respects except in intensity, from the weaker
echoes. However study of other occasions exhibiting similar,
but less well marked, characteristics revealed that this form of
behaviour could arise from the merging of two initially isolated
cells. Figure 10 shows one such example. The important features
were observed to be that

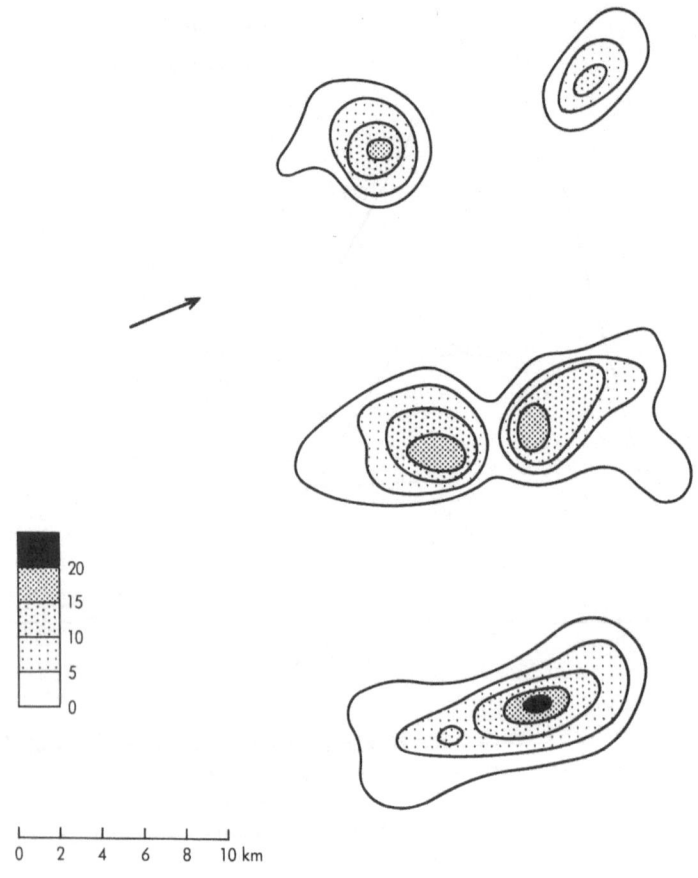

Figure 10. A merging pair of clouds observed by radar.

    i)   The cloud centres were aligned along the wind
   ii)   The clouds were at different stages of development
  iii)   The upshear cloud was the larger.

Such behaviour was simulated in the model by initiating two separate clouds, one a few minutes after the other. An early stage in the development is shown in Figure 11. Figure 12 shows a later stage (cf Figure 6). There is a copious concentration of rain and hail in the centre of the cloud and the maximum simulated instantaneous surface rainfall was 100 mm h^{-1} which agreed well with the maximum observed value of 104 mm h^{-1}. Merging can therefore explain the highest observed rainfall rates shown in Figure 5.

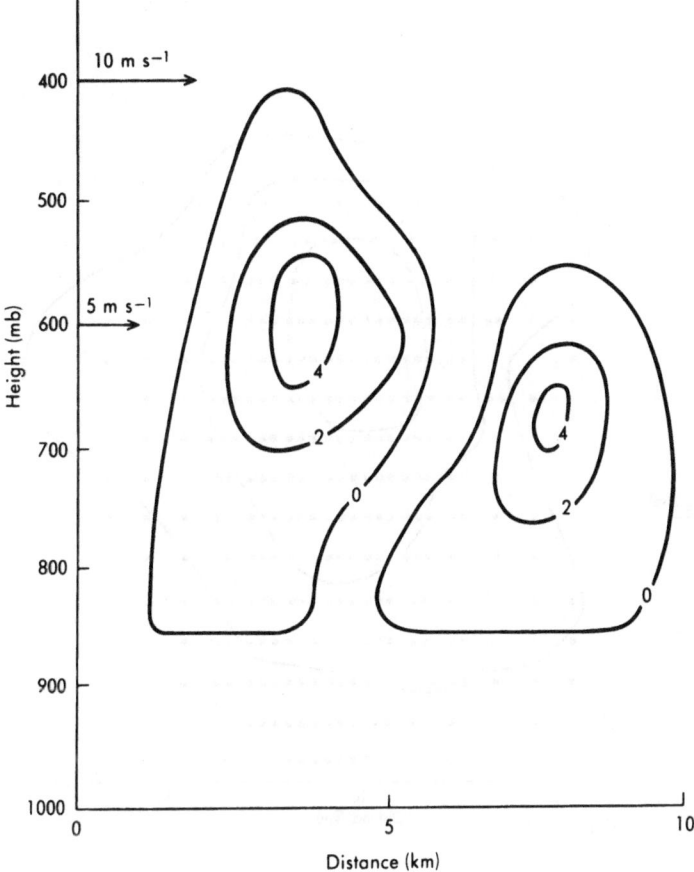

Figure 11.   An early stage of the simulated merging clouds. Contours are of cloud substance (gm kg^{-1}). The ambient shear relative to cloud base is indicated.

c.   Discussion

Comparison between the merged and two isolated single cell
clouds showed that the merged cloud produced approximately 7
times more total surface precipitation than two single cell
clouds growing in isolation and there was about a 10 fold increase in the instantaneous surface rainfall rate. This increase
was due to

(i)   the enhanced accretion as embryo precipitation from the
       larger cloud fell through the high liquid water content
       of the smaller and

(ii)   the reduced evaporation because the distance that the
       precipitation had to fall in clear air decreased.

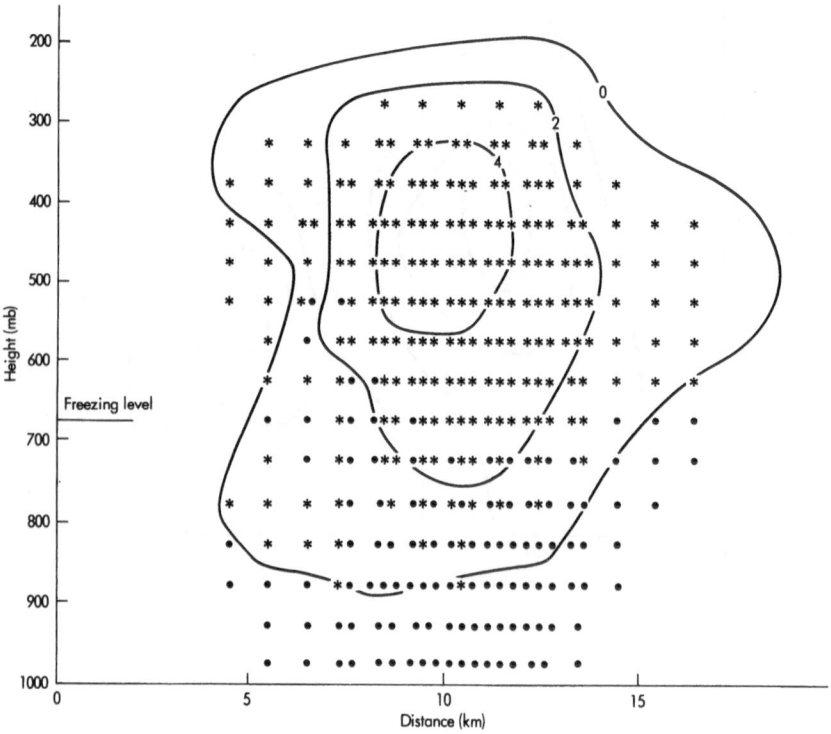

Figure 12.   A later stage in the development of the simulated
             cloud after merging had taken place. Contours are
             of cloud substance (gm kg^{-1}). Symbols indicating
             rain and hail are as in Figure 7.

Supplementary integrations showed that the most successful mergers, ie those that produced the greatest enhancement in rainfall, occurred when the majority of the embryo precipitation particles from the larger, upshear cell, fell through the younger cloud at the moment when its liquid water content was at a maximum. This occurred just before precipitation formed within it and consequently the most efficient mergers are extremely difficult to identify by radar as the smaller cloud is not visible before it becomes part of the merged cloud. However, clouds can interact and join in other less efficient ways, for example the enhancement in rainfall shown in Figure 10 is less than a factor of two. This contrasting behaviour illustrates the importance of the structure within the clouds and emphasises that enhancement is a consequence of the initial relative configuration of the clouds rather than of cloud interaction "per se".

This form of cloud merging has some similarity to that described by Orville et al (1980) in that the interaction takes place between clouds of different sizes and contrasts with the form described by Simpson et al (1980) where the two original clouds were of similar size and remained visually separate. Orville et al (1980) simulated many merging systems but restricted their study to clouds growing in the absence of vertical wind shear and found rainfall enhancements of less than a factor of two. The presence of vertical wind shear allows a much wider variety of cloud interactions, some of which are considerably more efficient at producing rain at the ground.

4. REFERENCES

Bennetts D A and Rawlins F	1981	Quart.J.R.Met.Soc.107, 477-502.
Browning K A	1979	Met.Mag. 108, 161-184.
Ludlam F H	1980	Clouds and storms. University Park (Pennsylvania State Univ press)
Miller M J and Pearce R P	1974	Quart.J.R.Met.Soc.100, 133-154.
Orville H D, Kuo Y H Farley R D and Hwang C S	1980	J.Rech.Atmos., 14 499-516
Simpson J, Westcott N E Clerman R J and Pielke R A	1980	Arch. Met. Geoph. Biokl. A29, 1-40.
WMO	1973	Operational Hydrology Report No.1. Manual for estimation of probable maximum precipitation. WMO No 332.

OBSERVED AND NUMERICALLY SIMULATED STRUCTURE OF A MATURE SUPERCELL THUNDERSTORM

Peter S. Ray

NOAA/ERL/National Severe Storms Laboratory
Norman, Oklahoma   73069   U.S.A.

Joseph B. Klemp

National Center for Atmospheric Research[1]
Boulder, Colorado   80307   U.S.A.

Robert B. Wilhelmson

University of Illinois
Urbana, Illinois   61801   U.S.A.

ABSTRACT. Sixteen tornadic storms occurred in the afternoon and evening hours of 20 May 1977 and the early morning hours of 21 May 1977. One of the tornadic storms was observed for two hours during its growth stage prior to becoming tornadic. This storm, named the Del City storm, exhibits certain important features also found in other tornadic storms observed in this outbreak. These features are essential to its longevity and its transition to a tornadic phase. Through the interactive use of Doppler-radar analysis and a three-dimensional storm simulation, these features are detailed and examined. The structural similarities that exist in the observed and simulated storm suggest that the large scale environment dominates the structuring of many storm features.

I.   INTRODUCTION

Time evolving storm structure has been revealed in unprecedented detail through the analysis of observations from two or more Doppler radars. Various analysis techniques for recon-

E. M. Agee and T. Asai (eds.), Cloud Dynamics, 379–393.
Copyright © 1982 by D. Reidel Publishing Company.

structing the storm's wind field have been employed, including
recently developed ones which involve application of physical
principles such as that described by Ray et al. [1]. Concurrent
advances in understanding storm structure and evolution have
been made with the aid of three-dimensional cloud model simulations. Both approaches have been used to study the severe and
tornadic storms that occurred in Oklahoma on 20 May 1977.

During the 1977 National Severe Storm Observing Season, 16
tornadoes occurred in western and central Oklahoma during the
afternoon hours of 20 May and early morning hours of 21 May.
The damage paths associated with these tornadoes are shown
schematically in Fig. 1 (numbered in order of their occurrence)
superimposed on the network of observing facilities. Our study
will concentrate on the storm associated with the tornado track
labeled 9. The observing network included four Doppler radars,

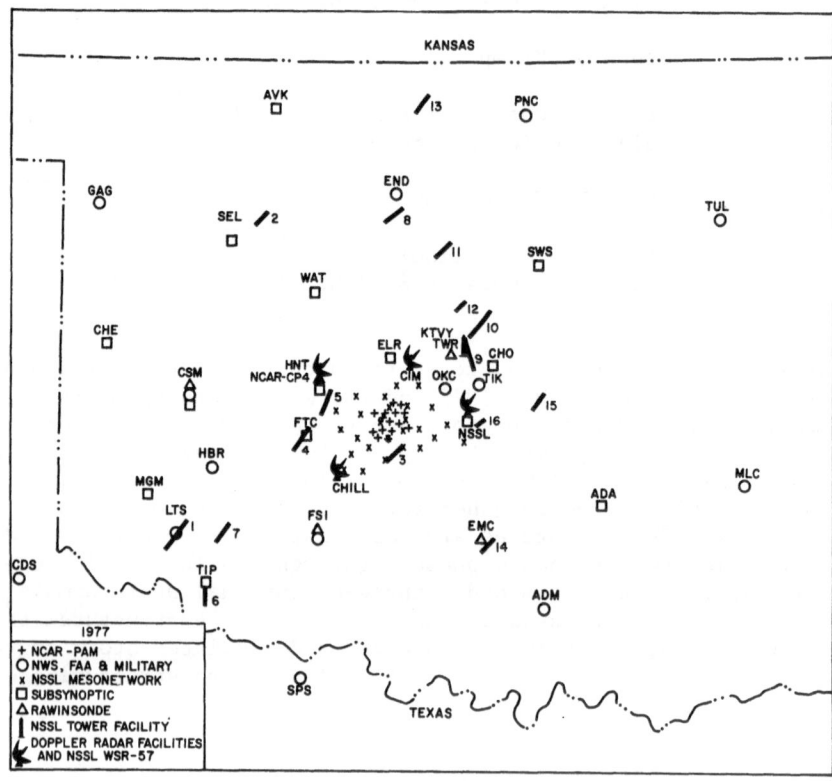

Figure 1. Schematic map of Oklahoma showing location of radars,
surface instrumentation, tower and rawinsonde sites for the 1977
spring program. Damage paths in order of their occurrence of
the 16 tornadoes that occurred on 20 May are indicated.

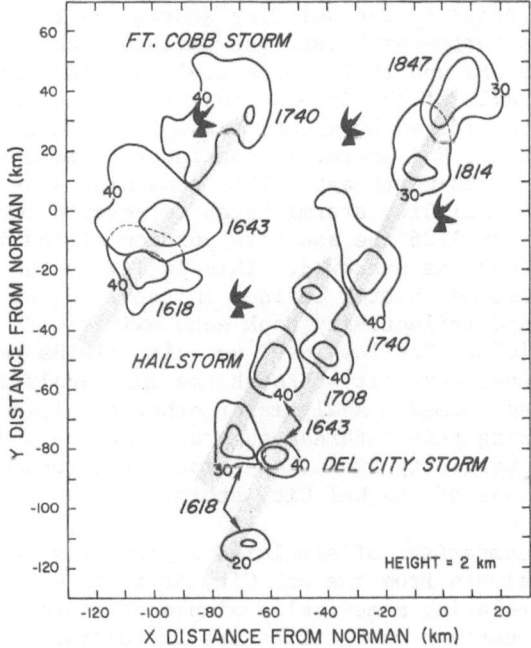

Figure 2. Time history of storm cores at a height of 2 km for
the Ft. Cobb, Hailstorm, and Del City storms. Inner core is
10 dBZ greater than outer contour. Location of the four radars
are indicated by the radar symbols and the grid origin is at the
Norman radar. Stippled arrows indicate storm track.

40 instrumented surface stations, 4 rawinsonde locations, a
440 m tower instrumented at seven levels, instrumented aircraft,
and satellite imagery. However, poor visibility hindered ground
intercept teams and aircraft operations throughout the day.
Further details about the network and the analysis procedures
are given in Ray et al. [2].

    Time evolution of storm cores at 2 km height are shown for
three storms in Fig. 2. The three storms shown have been designated the Ft. Cobb storm, the Hailstorm, and the Del City storm
which is analyzed here. All times referenced are in Central
Standard Time (CST).

    The Ft. Cobb storm was the first to intensify, reaching a
mature stage by 1643, producing a tornado at this time and
another 30 minutes later. The other two storms approached the
observing network together, with the Del City storm located
about 30 km south of the Hailstorm (1618-1643). The Hailstorm,
due to its greater size and intensity during this time, had

evidently formed prior to the Del City storm. As these storms
propagated to the north-northeast, the Hailstorm was overtaken
by the Del City storm and their reflectivity fields merged
around 1708. Details of the interactions that occurred during
this merger are considered by Klemp et al. [3]. As the Del City
storm merged with the Hailstorm, it continued to intensify and
propagate to the north-northeast. This remaining storm (still
referred to as the Del City storm) began to develop rotational
characteristics. By 1726 the southern portion of the 40 dBZ
reflectivity contour has narrowed. This is likely the embryonic
hook echo which becomes better defined in the 1740 analysis. By
1800 a well-defined reflectivity hook echo exists with substantial vorticity aloft. The low level velocity fields reflect
increasing low-level vorticity through the 1826 analysis. By
1847 a tornado had formed. Analysis of other tornadic storms
that occurred during this outbreak suggest that the updraft,
reflectivity and vorticity evolution through the tornadic phase
nearly parallel that of the Del City storm.

A detailed comparison of simulated supercell structure with
observational analysis from the Del City storm is made in Klemp
et al. [4]. In relating numerically simulated storm structures
to observations, certain ambiguities are unavoidable. The model
is initialized with a horizontally homogeneous sounding. Inhomogeneities in the environment, large scale forcing, the effects
of neighboring storms are not represented. In the analysis of
observations, questions arise due to sampling artifacts (poor
boundary conditions, side lobe effects, hail effects, etc.).
The agreement found is therefore quite encouraging and suggests
that simulation and Doppler radar data can interactively and
synergistically contribute to understanding the evolution and
structure of storms.

The coevolving three-dimensional wind and reflectivity
structure is given in Section 2 along with an overview of the
setting in which these storms evolved. In Section 3 the simulation is presented. Comparison of the simulated and observed
storms is made in Section 4 with concluding remarks in Section 5.

2.   THE OBSERVATIONS

The 0600 Oklahoma City sounding revealed a very unstable
air mass (lifted index of -6) with strong vertical wind shear
(4×10^{-3} s^{-1}) in central Oklahoma. Southeasterly low-level flow
continued to bring warm moist air into Oklahoma at northern
Texas. Adding further to conditions suggestive of a severe
storm outbreak was a strong shortwave evident at 500 mb
approaching from the southwest.

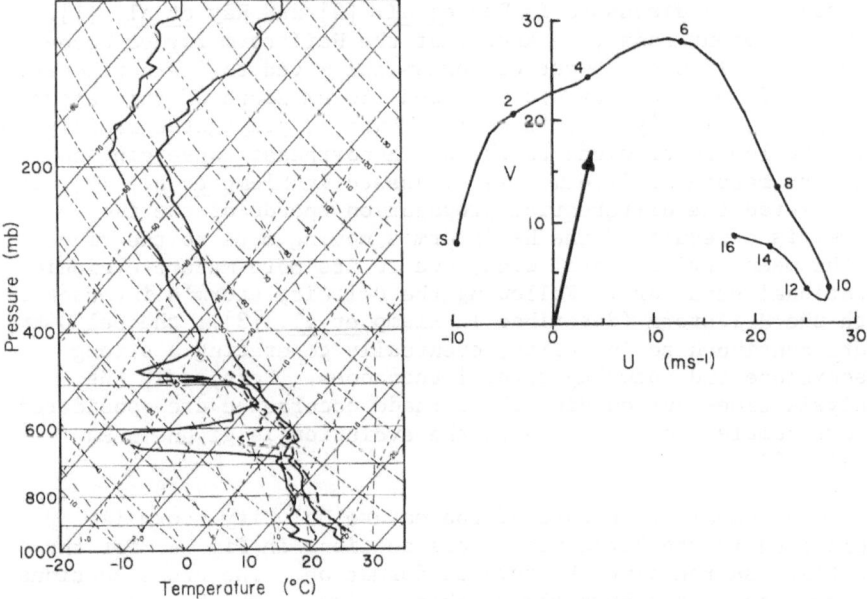

Figure 3. Temperature and dewpoint sounding from Elmore City,
Oklahoma at 1620 (heavy solid lines) and Ft. Sill, Oklahoma at
1500 (heavy dashed lines) on 20 May 1977. The hodograph is a
composite of environmental winds from these same soundings.
The heights (km) above sea level are labeled along the profile.
The observed propagation speed of the Del City storm during
its mature phase is indicated by the heavy arrow.

     The first storms formed near Lubbock, Texas. Storms built
slowly toward the northeast along a surface cold air boundary
formed by storms the previous night. By 1400, lifting associated with a shortwave, caused these storms to increase in
intensity. A representative sounding and hodograph is shown in
Fig. 3. These show strong low-level shear, which has been shown
in numerical simulations (Wilhelmson and Klemp [5], Schlesinger
[6]) to be important in producing long-lived storms. In addition,
the hodograph displays a clockwise turning of the wind shear
vector with height, a condition which enhances the development
of cyclonic storms which move to the right of the mean tropospheric wind (Klemp and Wilhelmson [7]). Because of its interaction with the Hailstorm to its north, the evolution of the Del
City storm is considerably more complex than storms previously
studied and simulated.

     Multiple Doppler radar analyses for these storms begins at
1618 and extends through 1847 at 13 separate analysis times.

Methodology is discussed in Ray et al. [2] and Ray et al. [1].
From incoherent radar, we know that the Hailstorm formed some-
time before 1530 southwest of Norman and moved to the northeast.
About 1600 the Del City storm formed 30 km south of this storm.
During the Del City storm's early growth it propagated rapidly
with the low level winds to the north-northeast converging on
the more mature Hailstorm. As discussed in Klemp et al. [3], it
is believed the differential propagation speeds of the two
storms is a result of the Hailstorm's moving more to the right
of the mean environmental wind, due to its more mature cyclonic
rotational structure. Following the Del City storm's interaction
with the Hailstorm (described in Klemp et al. [3]), the Del City
storm continued to intensify, eventually generating a strong
mesocyclone and spawning several tornadoes. Here, only the
analysis times surrounding the tornado occurrence are considered.
A more complete description of the evolution is given in Ray
et al. [2].

   The overall structure of the mature Del City storm is
summarized in the horizontal cross sections in Fig. 4. for the
two times surrounding the tornado formation. The cross sections
are at 1 and 4 km above the earth's surface. The horizontal
wind vectors are relative to the storm motion which is estimated
to be 17.5 m s^{-1} at 14° during this time. The updraft originates
at low levels within the strong convergence zone which extends
along the boundary between the moist inflow approaching the
storm from the east and southeast and the cold downdraft outflow.
The updraft at 1 km height seems to be elongating at the 1833
time in response to the gust front below. By this time vorticity
maxima exceed 1.4×10^{-2} s^{-1} at all levels below 10 km. At mid-
levels a single updraft with a diameter ~10 km exists with
maximum vertical velocities exceeding 30 m s^{-1}. The updraft
region at 4 km contains significant amounts of precipitation
with environmental air moving cyclonically around the southern
and eastern side of the updraft, sweeping rain from above around
to the northside of the updraft.

   The tornadic phase (at 1847) is illustrated in Fig. 4(c,d).
Although the overall structure is similar, significant small
scale structure has developed in association with tornadogenesis.
Within the updraft region there are now two strong vertical
vorticity maxima which have amplitudes at 2 km height of
~2.5×10^{-2} s^{-1} and ~1.5×10^{-2} s^{-1} with the western and eastern
branches, respectively. Between these two maxima is a vorticity
minimum of -1×10^{-2} s^{-1} associated with a downdraft region
between the updraft maxima. Strong low level cyclonic rotation
distorts the convergence zone at 1833, into a horseshoe-like
structure at 1847 (Fig. 4c). As the downdraft behind the convergence line progresses cyclonically around the mesocyclone
(where the tornado exists), the air supporting the updraft is

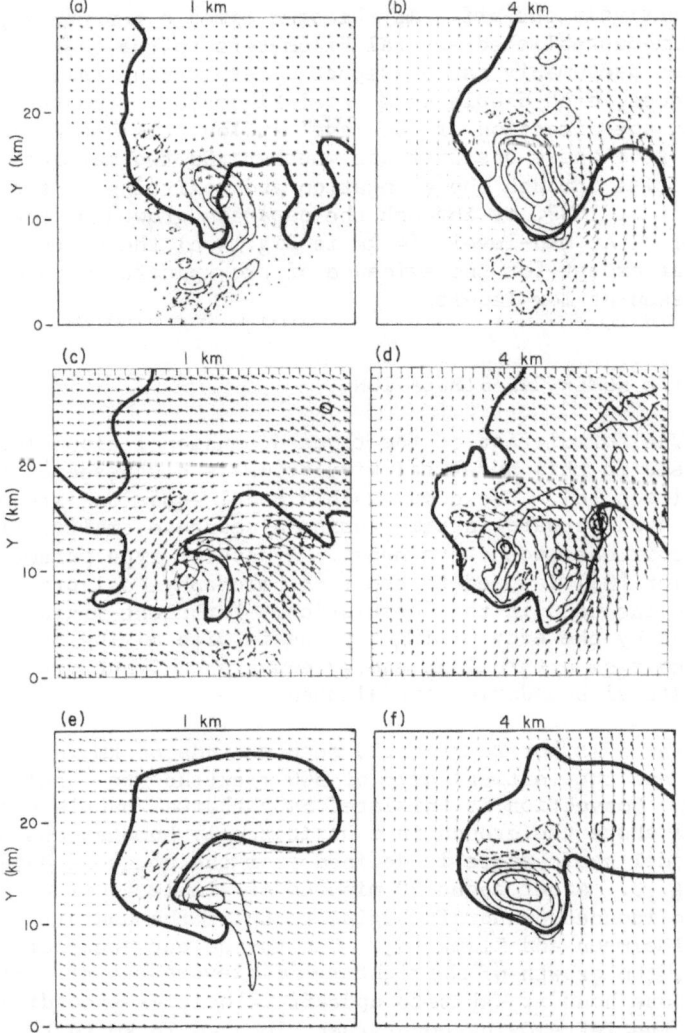

Figure 4. Horizontal cross sections of observed (a,b,c,d) and
modeled (e,f) storm at 1 (a,c,e), and 4 (b,c,f) km above the
ground. Observed times are (a,b) 1833 and (c,d) 1847 with the
heavy solid line corresponding to the 30 dBZ contour. Grid
origin for 1833 is 14 km west and 5 km north of Norman, Oklahoma
and 10 km west and 20 km north for the 1847 time. Modeled storm
is at 2 hr. of simulation. Wind vectors are scaled such that
one grid interval represents 20 m s^{-1}. Updraft velocities
(solid lines) and downdraft velocities (dashed lines) are contoured at 10 m s^{-1} for 1847 and at 5 m s^{-1} increments at 1833
and in the simulations. Shaded regions designate areas of
negative vertical velocity (w<-1 m s^{-1}). The heavy solid line
outlines the rainwater field enclosed by the 0.5 g kg^{-1} contour.
Wind vectors are scaled such that one grid interval represents
20 m s^{-1}.

progressively cut off. At the same time, a new convergence zone
is being established several kilometers to the east. The same
process presumably can be repeated, explaining families of
tornadoes, each displaced several kilometers from the previous
one in the direction of low-level inflow. It can not be determined from these observations whether the two main updrafts at
1847 arise through the elongation and splitting of the single
updraft at 1833, or through the separate growth of the second
updraft. At mid-levels (4 km in Fig. 4.c) the structure is
similar except for the evidence of two updraft cores whose roots
were seen at low-levels.

3.   THE NUMERICAL SIMULATION

The formulation of the three-dimensional cloud model which
was used in this numerical simulation is discussed by Klemp and
Wilhelmson [8]. The compressible model equations are numerically
integrated forward in time, using a separate smaller time step
to accommodate soundwave modes. A Kessler type parameterization
(Kessler [9]) is used to represent microphysical processes with
no ice included in the formulation. Turbulent mixing is approximated by the eddy mixing hypothesis with mixing coefficients
derived from a turbulent energy equation. Transports through
the lateral boundaries are allowed.

The 1 km horizontal grid mesh extends 48 km in both dimensions and the vertical grid of 500 m extends to a height of
16 km. Convection is initiated by a thermal bubble centered
1500 m above the ground in a horizontally homogeneous atmosphere.
The model basic state is defined from the sounding data presented
in Fig. 3. After about 30 minutes of simulated time, the precip-
itated induced downdraft begins to split the updraft as described
by Wilhelmson and Klemp [5]. Due to the curvature of the hodograph, the growth of the right mover is favored as discussed earlier.
After one hour it exhibits updraft velocities exceeding 30 m s^{-1}
while the left-moving counterpart located 18 km to the northwest
has a maximum updraft of only ~10 m s^{-1}. Frequently, these
left-moving counterparts produce little rainwater and are difficult to detect with radar.

A nearly steady updraft-downdraft structure is reached by
1.5 h. A gust front supporting the updraft has developed with
strong convergence (exceeding 10^{-2} s^{-1}). After 2 hours of
simulated time the storm structure is essentially constant.
Horizontal cross sections of the horizontal wind, up- and downdrafts, and precipitation field are shown for 1 and 4 km above
the ground in Fig. 4. This figure displays only a 29x29 km
window of the total simulation domain, the same area shown in
the accompanying radar derived analyses. The horizontal wind

vectors are relative to the moving model framework (14.3 m s^{-1} at 12°). This is nearly storm relative since the storm is drifting northward at only ~1.5 m s^{-1}. Strong cyclonic rotation is evident in the wind field and also in the low-level rainwater field.

The downdraft region rotates in a counterclockwise direction around the updraft as it descends. The downdraft is situated with respect to the environment winds at each level such that it is downstream from the updraft core. A similar relationship can be found in the Doppler radar derived winds.

The model fields are considerably smoother than those depicting the observations. The small-scale structure in the observations is in part due to the interpolation to a 1 km grid with little subsequent filtering. Additionally, inhomogeneities in the environment may be a source for some of the observed fine-structure. The possible effect of errors in data sampling and in the analysis are not well known. However, the model effectively filters smaller scale disturbances through parameterized turbulence. Klemp and Rotunno [10] have increased model resolution significantly and have captured many of the smaller scale features in these observations, but not illustrated in these simulations. More detail on these simulations can be found in Klemp et al. [4].

4.   COMPARISON OF MODELED AND OBSERVED STORMS

4.1 Trajectory Analysis of Updraft Structure

To further examine storm structure and compare observed and simulated storms, air parcel trajectories which illustrate significant storm features are considered. Each trajectory is derived by assuming the flow is steady state and by computing the line which is everywhere parallel to the local three-dimensional wind vector. Strictly speaking these are streamlines which represent trajectories when there is little flow field variation during the residence time of the parcel in the domain. This is more nearly true for the model flow than for the observed flow. A more accurate computation would change some of the details but not the fundamental course of any of the presented trajectories or the conclusions that are inferred from them. Selected precipitation trajectories using parameterized terminal fall velocities were also computed. Although clearly not a trajectory of individual particles, precipitation transport (through the storm relative to the air motion) is qualitatively represented. These trajectories for the model and observed fields are overlayed on the 1 km rainwater field in Fig. 5. For convenience, trajectories will be referred to by

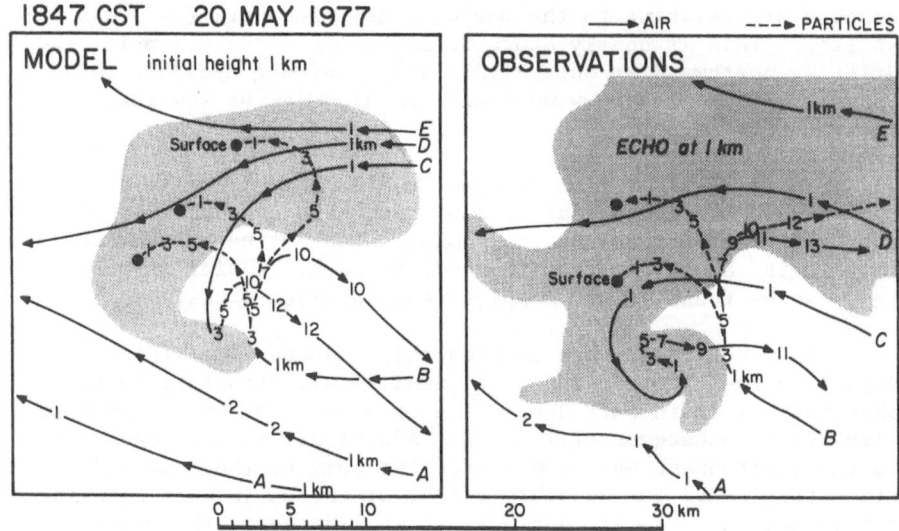

-Figure 5. Model and observed trajectories of air parcels (solid
lines for air originating in the vicinity of 1 km. Elevations
above the ground (km) are labeled among each trajectory. Precipitation trajectories (dashed lines) are included beginning at
selected locations along air parcel trajectories. The heavy
solid line outlines the rainwater or reflectivity field as in
Fig. 4.

the capital letter at their origin. A striking feature documented by these trajectories is that air rising in the updraft
turns clockwise or anticyclonically with height (B and C) even
though the updraft itself is imbedded in a cyclonic windfield
(see Fig. 4). For example, moist air approaching the storm from
the east (B and C) at an elevation of 1 km feeds the updraft
along and between the trajectory locations indicated by B and C.
In the southern portion of this easterly flow (B) the air is
lifted over the gust front and turns anticyclonically to the
north as it rises rapidly in the updraft, finally turning anticyclonically at high levels and existing toward the east.
Easterly inflow further north (D) flows through the rain area
north of the updraft as it turns cyclonically at the western
edge of the low-level updraft area, rising with a northerly
component of motion, finally turning anticyclonically in a
similar fashion to (C) as it leaves the updraft.

The model storm trajectories which pass through the western
and eastern portion of the central updraft behave similarly to
those through the two separate updraft cores in the 1847 analysis.
The updraft structure at 1833 more nearly resembles that of the

modeled storm and trajectories at this time behave very much
like those in the modeled storm. These trajectories support the
interpretation that the main updraft is decaying and the secondary
updraft (to the east) is intensifying and will soon dominate.
Trajectories at this level approaching from south of B or north
of D flow around the storm converging behind the storm to the
west.

The anticyclonic curvature exhibited by some of the air
parcels appears to be caused by the clockwise turning with
height of the storm relative environmental winds. The horizontal flow direction within the updrafts is biased by the
environmental wind flow direction at that level, due partly to
entrainment of environmental air into the updraft.

The precipitation trajectories in Fig. 4 illustrate rain
being transported away from the updraft, where the rain was
formed, into downdrafts on the north and west sides of the
storm. The rain trajectories in Fig. 4 start at heights of 3, 5
and 7 km along trajectory (B). Those that descend to the surface
follow a cyclonically curved trajectory. Thus, the updraft
remains relatively precipitation-free, while rain accumulates on
the back side of the storm, providing substantial negative
buoyancy through water loading and evaporation which drives the
downdraft circulation. Downdraft outflow near the ground sus-
tains convergence along the gust front which is important in
supplying moist inflow to the updraft. As emphasized by
Browning [11], this updraft-downdraft configuration seems to be
crucial in maintaining supercell storm longevity.

4.2 Vorticity Structure

A storm's rotational characteristics is strongly linked to
a structure which promotes storm persistence. Rotunno [12] and
Davies-Jones [13] discuss mechanisms by which vorticity evolves
within a supercell storm. Once rotation is established, the
accompanying flow structure can persist and maintain the storm's
intensity. The vorticity structure and evolution of the Del
City storm has been discussed by Brandes [14] and is similar to
structures observed in other tornadic storms.

The vertical distribution of maximum positive vorticity in
the simulation and as observed are nearly identical with a
maximum value ($\sim 2.5 \times 10^{-2}$ s^{-1}) near the earth's surface dropping
to a more nearly constant value with height ($\sim 1.5 \times 10^{-2}$ s^{-1})
above 1 km. The maximum positive vorticity as increased at all
levels (2×10^{-2} s^{-1}) and increasing below 3 km to a maximum
(3.5×10^{-2} s^{-1}) at the earth's surface. For all cases, the
maximum negative vorticity ($\sim -1. \times 10^{-2}$ s^{-1}) remain nearly the
same with height. These illustrate the low-level cyclonic

vorticity created by tilting of environmental shear and the
increases due to stretching associated with the gust-front-
induced convergence.

Horizontal distributions of vertical vorticity for the model
and observed storms are shown in Fig. 6 at heights of 1, 4, and
7 km. Updraft regions are hatched and downdraft regions are
stippled with updraft cores indicated with a plus mark. The
pretornadic structure of 1833 closely resembles the modeled
structure. Large low-level positive vorticity values are
expected through the tilting of horizontal vorticity in the
environment and the subsequent enhancement through stretching.
Low-level negative vorticity maxima are smaller due to negative
stretching from the divergent downdraft. As expected from this
vorticity generation model, low-level vorticity maxima is near
the updraft-downdraft interface but in within the updraft. At
mid-levels (4 km) the vorticity maxima are more nearly coincident
with the updraft maxima. Negative stretching and the increase
of negative vorticity occurs in the region of downdraft, negative
vorticity coincidence. Although similar, but more intense,
structures are found in the 1847 analysis, the presence of
extensive small scale features complicates simple analysis.
Klemp and Rotunno [10] used high resolution numerical simulations to reproduce some of the enhanced observed small-scale
structures present in the 1847 analysis.

5.   SUMMARY

Numerical model simulations and Doppler radar analysis have
been used to study storm structures on 20 May 1977. The complete and self consistent fields provided by the model aid in
the interpretation of the more complex and smaller scale observation. In turn, the higher resolution typically offered by
observations indicate areas of possible model deficiencies or
where small-scale environmental variability may be important.
When the two approaches are in agreement, the results can be
interpreted with greater confidence and when they differ, a
closer scrutiny can lead to improvements in either the model or
the Doppler radar analysis.

The Del City storm studied here evolved over a couple of
hours during which it strongly interacted with another major
storm; whereas the numerically simulated storm grew in complete
isolation in a horizontally uniform environment from an artificial thermal perturbation. The similarities in their mature
phase suggest that larger scale environmental conditions are
strongly linked to the structure of many important storm features. It is significant (and encouraging) that other storms
that occurred on the same day possessed many of the same features
as the Del City storm selected for closest examination.

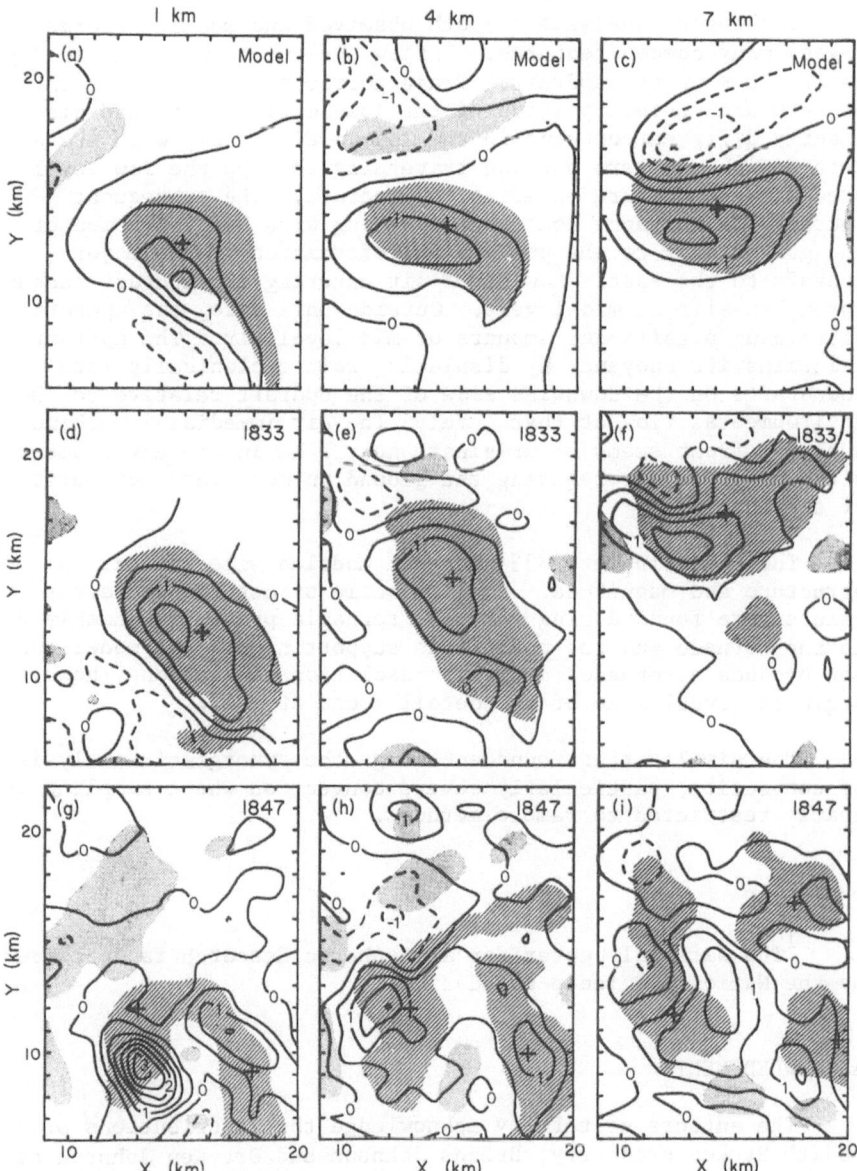

Figure 6. Horizontal distributions of vertical vorticity for
the model simulation and the observed Del City storm at 1833 and
1847 at elevations of 1, 4 and 7 km above the ground. Vorticity
is contoured in increments of 0.005 s^{-1} with labels in units of
s^{-1} 10^2. Updraft regions exceeding 5 m s^{-1} at 1 km and exceeding 10 m s^{-1} at 4 and 7 km are shaded with diagonal lines while
downdraft regions exceeding −5 m s^{-1} are stippled. The main
updraft centers are shown with a plus mark.

Trajectory analysis of both observed and modeled storms reveal many common features. Inflow air turns anticyclonically as it rises in the updraft. Trajectories passing through the eastern and western portion of the low-level updraft had distinctly different origins in the low-level inflow, with the air entering the western portion traversing through the low level precipitation before entering the updraft. The subsequent decrease in buoyancy contributes (along with the intrusion of the gust front) to the successive reformation of the major updraft to the east. Low-level air entering the updraft passes through a slit at mid-levels. Outside this core, the updraft is entraining significant amounts of mid-level air. The updraft maintains its buoyancy by displacing rain cyclonically into downdrafts on the downwind side of the updraft relative to the environmental flow at that level. The air immediately behind the gust front seems to originate near 2 km in the environment with mid-level air reaching the ground in more interior portions of the storm.

The vorticity at 1833 and that modeled were similar in structure and magnitude. Smaller scale structure and larger values were found during the 1847 tornadic phase, presumably due to the tornado and the conditions supporting it (the model could not produce a tornado). Higher resolution simulations, however, begin to reveal some of the detail found at 1847.

The similarities found encourage the synergistic analysis of convection, particularly severe convection where sampling is nearly restricted to remote methods.

FOOTNOTES

[1]The National Center for Atmospheric Research is sponsored by the National Science Foundation.

ACKNOWLEDGMENTS

The authors gratefully acknowledge the contributions of Judith Stokes Bradberry, Brenda Johnson and Dr. Ken Johnson at NSSL in analyzing the Doppler radar data and the assistance of Morris Weisman and Ron Krubeck at NCAR in providing programming support. Dr. Wilhelmson was supported through NSF Grants ATM-78-01010 and ATM-80-11984 to the University of Illinois and NOAA Grant PO-04-6022-44034. The acquisition and processing of the Doppler radar data was supported in part by the Experimental Meteorology Weather Modification Division of the Atmospheric Sciences, National Science Foundation under Grants ATM 77-04285 and ATM 78-27420 to Florida State University.

REFERENCES

1. Ray, P., Ziegler, C.L., Bumgarner, W., and Serafin, R.J.:
    1980, Mon. Wea. Rev. 108, pp. 129–147.
2. Ray, P.S., Johnson, B.C., Johnson, K.W., Bradberry, J.S.,
    Stephens, J.J., Wagner, K.K., Wilhelmson, R.B., and Klemp,
    J.B.: 1981, J. Atmos. Sci. 38, pp. 1643–1663.
3. Klemp, J.B., Ray, P.S., and Wilhelmson, R.B.: 1980,
    Preprints 19th Conf. Radar Meteorology, Miami Beach, Amer.
    Meteor. Soc., pp. 317–324.
4. Klemp, J.B., Wilhelmson, R.B., and Ray, P.S.: 1981,
    J. Atmos. Sci. 38, pp. 1558–1580.
5. Wilhelmson, R.B., and Klemp, J.B.: 1978, J. Atmos. Sci.
    35, pp. 1975–1986.
6. Schlesinger, R.E.: 1978, J. Atmos. Sci. 35, pp. 690–713.
7. Klemp, J.B., and Wilhelmson, R.B.: 1978, J. Atmos. Sci.
    35, pp. 1097–1110.
8. Klemp, J.B., and Wilhelmson, R.B.: 1978, J. Atmos. Sci.
    35, pp. 1070–1096.
9. Kessler, E.: 1969, Meteor. Monogr., No. 2, Amer. Meteor.
    Soc., 84 pp.
10. Klemp, J.B., and Rotunno, R.: 1982, Proceedings of the
    IUGG/ IUTAM Symposium on Intense Atmospheric Vortices,
    Reading, England.
11. Browning, K.A.: 1964, J. Atmos. Sci. 21, pp. 634–639.
12. Rotunno, R.: 1981, Mon. Wea. Rev. 108, 577–586.
13. Davies-Jones, R.P.: 1982, Proceedings, IUGG/IUTAM Symposium
    Symposium on Intense Atmospheric Vortices, Reading, England.
14. Brandes, E.A.: 1981, Mon. Wea. Rev. 109, pp. 635–647.

TURBULENCE PARAMETERIZATION IN A DEEP CONVECTION MODEL

J.L. Redelsperger and G. Sommeria

Laboratoire de Météorologie Dynamique CNRS Paris
24, rue Lhomond 75231 Paris Cedex 05

ABSTRACT

   A deep convection model derived from previous work by
J.W. Deardorff and G. Sommeria is presented, with particular
emphasis on the parameterization of sub-grid scale turbulent
processes. The model covers a 40x40x16 kms domain with a
grid size of 1x1x.4 km, and uses the anelastic approximation.
It has been run on several study cases, in order to test the
relative importance of various features in the turbulence
parameterization : use of adequate thermodynamic variables for the
expression of second-order moments, use of a prognostic turbulent
kinetic energy equation, choice of a parameterization method for
the water variables, water vapor, cloud water and rain water.
   Some of the results are discussed, in order to show how
the above features may affect the simulated development of one
cumulus cloud up to its raining stage.

1.- INTRODUCTION

   The purpose of this paper is to present preliminary
results from a three-dimensional convection model, developed
at the Laboratoire de Météorologie Dynamique as part of the
"C.O.P.T." (COnvection Profonde Tropicale) Research programme
on tropical convection.
   The presentation of the model will put the emphasis on
the subgrid turbulence and microphysics parameterization,
which have been tested by comparative simulations in one
case of precipitating convection.

E. M. Agee and T. Asai (eds.), Cloud Dynamics, 395–409.
Copyright © 1982 by D. Reidel Publishing Company.

2. PRESENTATION OF THE MODEL

The dynamical frame-work is taken from Deardorff (1972), Sommeria (1976) and Redelsperger and Sommeria (1981, 1982). The Deardorff's model has been first extended by Sommeria to include a water cycle for no-precipitating clouds and a new treatment of the subgrid-scale turbulence. Then in order to be adapted to deep convection, it has been modified in two steps. The first step was to improve the parameterization of subgrid turbulent processes (Redelsperger and Sommeria 1981) by the use of an evolution equation for subgrid scale turbulent kinetic energy and of quasi-conservative variables during the condensation process. The second step included the representation of precipitation and the correlative adaptation of the previous subgrid scale turbulent parameterization. After validation tests in cases of slightly disturbed boundary layers, it is used here for deep convection simulation.

(a) Basic dynamical frame-work

The fluid is supposed to consist of a mixture of two perfect gases, air and water vapor, containing only two classes of water droplets, the first in suspension, the second falling with a mean terminal velocity. It is defined locally by six scalar variables (density ρ, specific humidity q, specific content of cloud liquid water q_c, specific content of rain liquid water q_r, potential temperature Θ, pressure P) and the three components of velocity u_i.
The governing equations after application of grid volume averaging are :
For momentum —

$$\frac{\partial \overline{u_i}}{\partial t} = - \frac{1}{<\rho>} \frac{\partial}{\partial x_j} (<\rho>\overline{u_i}\,\overline{u_j}) - \frac{1}{<\rho>} \frac{\partial}{\partial x_j} (<\rho>\overline{u'_i u'_j})$$

$$- \frac{1}{<\rho>} \frac{\partial \overline{P}}{\partial x_i} - 2\varepsilon_{ijk}\Omega_j \overline{u}_k \qquad (1.1)$$

$$+ \delta_{i3} \left[\frac{g}{<\Theta_{v1}>} (\overline{\Theta}_{v1} - <\Theta_{v1}>) - g \right]$$

The Einstein summation convention for indices is used, ε_{ijk} is the alternating unit tensor, δ the Kronecker/delta, Ω_j the jth component of the earth's angular velocity and g the gravitational acceleration. In this equation, the acoustic modes are filtered with the anelastic approximation. The buoyancy uses the virtual potential temperature Θ_{v1}, taking in account the weight of both cloud and rain water :

$$\Theta_{v1} = \Theta(1+.61q-q_c-q_r)\Theta \qquad (1.2)$$

The overbars indicate Reynolds averaging over grid volumes and the primes are deviations from this quantities. This formulation has to be completed by expressions for the grid averaged "turbulent terms" which require a specific parameterization presented in the next section.

For the mass continuity :

$$\frac{\partial}{\partial x_j}(<\rho>\overline{u}_j) = <\rho>\frac{\partial \overline{u}_j}{\partial x_j} + \overline{W}\frac{\partial <\rho>}{\partial z} = 0 \tag{1.3}$$

For pressure :

$$\frac{\partial^2}{\partial x_j \partial x_j}\overline{P} = \frac{\partial S_i}{\partial x_i} \tag{1.4}$$

where S_i refers to the source terms of u_i.

The pressure equation is obtained by taking the divergence of (1.1) associated with the continuity equation (1.3).

For heat and the water variables :

$$\frac{\partial \overline{\Theta}}{\partial t} = -\frac{1}{<\rho>}\frac{\partial}{\partial x_j}(<\rho>\overline{u}_j\overline{\Theta}) - \frac{1}{<\rho>}\frac{\partial}{\partial x_j}(<\rho>\overline{u'_j\theta'})$$
$$+ \frac{L\Theta}{TC_p}\left[(\frac{\partial \overline{q}_c}{\partial t})_{CE} + (\frac{\partial \overline{q}_r}{\partial t})_E\right] \tag{1.5}$$

$$\frac{\partial \overline{q}}{\partial t} = -\frac{1}{<\rho>}\frac{\partial}{\partial x_j}(<\rho>\overline{q}\overline{u}_j) - \frac{1}{<\rho>}\frac{\partial}{\partial x_j}(<\rho>\overline{u'_j q'})$$
$$- \left[(\frac{\partial \overline{q}_c}{\partial t})_{CE} + (\frac{\partial \overline{q}_r}{\partial t})_E\right] \tag{1.6}$$

$$\frac{\partial \overline{q}_c}{\partial t} = -\frac{1}{<\rho>}\frac{\partial}{\partial x_j}(<\rho>\overline{q}_c\overline{u}_j) - \frac{1}{<\rho>}\frac{\partial}{\partial x_j}(<\rho>\overline{u'_j q'_c})$$
$$+ \left[(\frac{\partial \overline{q}_c}{\partial t})_{CE} - (\frac{\partial \overline{q}_r}{\partial t})_A - (\frac{\partial \overline{q}_r}{\partial t})_{A.C}\right] \tag{1.7}$$

$$\frac{\partial \overline{q}_r}{\partial t} = - \frac{1}{<\rho>} \frac{\partial}{\partial x_j} (<\rho> \overline{q}_r \overline{u}_j)$$

$$- \frac{1}{<\rho>} \frac{\partial}{\partial x_j} (<\rho> \overline{u'_j q'_r}) \tag{1.8}$$

$$+ \left[(\frac{\partial \overline{q}_r}{\partial t})_E + (\frac{\partial \overline{q}_r}{\partial t})_A + (\frac{\partial \overline{q}_r}{\partial t})_{A.C} \right]$$

$$+ \frac{1}{<\rho>} \frac{\partial}{\partial z} (<\rho> V_t \overline{q}_r)$$

where the CE and E subscripts refer respectively to the effects of evaporation-condensation of cloud water and the evaporation of rain-water. $(\frac{\partial \overline{q}_r}{\partial t})_{A.C}$ is the auto conversion rate from cloud water into rain water, $(\frac{\partial \overline{q}_r}{\partial t})_A$ the transfer rate from cloud water into rain water due to the collection of cloud droplets by rain drops. V_t is the average terminal velocity of rain drops within one grid volume. All these terms are detailed in Redelsperger and Sommeria (1982) and are closely similar to the parameterization presented by Kessler (1969).

(b)   Subgrid scale turbulence and condensation parameterization

        Until recently the parameterization of subgrid scale
turbulence proposed by Smagorinsky (1963) and Lilly (1967) was
the most commonly used in convection models. It assumes the
Reynolds fluxes to be proportional to the gradients of
corresponding quantities with a proportionality coefficient
(so-called eddy diffusion coefficient) proportional to the
deformation of the velocity field. The heat and water
subgrid-scale fluxes are parameterized the same way with an
eddy coefficient proportional to the eddy coefficient for
momentum. This method has been successfully used by
Deardorff (1972) for neutral and unstable boundary layers.
However, in stable layers, this formulation leads to unrealistically
high subgrid fluxes. It can be improved either by a better
parameterization of second order moments, which incorporates the
effect of thermal stratification (as in Sommeria, 1976), or by
using conjointly a prognostic equation for subgrid turbulent
kinetic energy (from which an eddy-coefficient is derived)
and a subgrid-scale mixing length depending on thermal stability
(Deardorff 1975, 1980).

Another-short-coming of these parameterizations when used in convection models has been pointed out by Cotton (1975) : all these approaches ignore the role of condensation and precipitation processes as turbulence generation mechanisms. This problem has been avoided by Redelsperger and Sommeria (1981, 1982) by using quasi-conserved variables during condensation and precipitation processes in a precipitating-cumulus model. This allows for example to obtain realistic subgrid fluxes in the cloud layer, which reinforces cloud convection by adding the subgrid contribution . These two types of considerations lead to the following parameterization, used in the present convection model. It is based on troncated expressions obtained from the evolution equations of second order moments, with the help of simplificatory assumptions as in Sommeria (1976), except for turbulent kinetic energy which is computed by using a prognostic equation.

The basic subgrid scale variables are now the total water content q_W and a liquid water potential temperature Θ_1 (introduced by Betts, 1973), which are fairly well conserved, even in the precipitating stage :

$$\begin{cases} \Theta_1 = \Theta - L\Theta/C_pT \ (q_c + q_r) \\ q_W = q + q_c + q_r \end{cases}$$

The expressions obtained for the second-order moments can be summarized the following way :

$$\overline{u_i'u_j'} - \frac{2}{3}\delta_{ij}\overline{E} = -\frac{4}{15}\frac{\Delta}{C_m}\overline{E}^{1/2}(\frac{\partial \overline{u}_i}{\partial x_j} + \frac{\partial \overline{u}_j}{\partial x_i}) \tag{2.2}$$

$$\overline{u_i'\Theta_1'} = -\frac{2}{3}\frac{\Delta}{C_s}\overline{E}^{1/2}\frac{\partial \overline{\Theta}_1}{\partial x_i}\Phi_i \tag{2.3}$$

$$\overline{u_i'q_w'} = -\frac{2}{3}\frac{\Delta}{C_h}\overline{E}^{1/2}\frac{\partial \overline{q}_w}{\partial x_i}\psi_i \tag{2.4}$$

$$\overline{\Theta_1'q_w'} = C_2\Delta^2\left(\frac{\partial \overline{\Theta}_1}{\partial x_i}\frac{\partial \overline{q}_w}{\partial x_i}\right)\left(\Phi_i + \Psi_i\right) \tag{2.5}$$

$$\overline{q_w'^2} = C_1\Delta^2\frac{\partial \overline{q}_w}{\partial x_i}\frac{\partial \overline{q}_w}{\partial x_i}\Psi_i \tag{2.6}$$

$$\overline{w'\theta'}_{vl} = -\frac{2}{3}\frac{\Delta}{C}\ \overline{E}^{1/2}\left[E_\theta\ \frac{\partial\overline{\theta}1}{\partial z}\ \Phi_3 + E_q\ \frac{\partial\overline{q}_w}{\partial z}\ \Psi_3\right] \qquad (2.7)$$

$$\overline{\theta'^2_1} = \ C_1\Delta^2\ \frac{\partial\overline{\theta}_1}{\partial x_i}\ \frac{\partial\overline{\theta}_1}{\partial x_i}\ \Phi_i \qquad\qquad (2.8)$$

$$\overline{u'_i q'_r} = -\frac{2}{3}\frac{\Delta}{C}\ \overline{E}^{1/2}\frac{\partial\overline{q}_r}{\partial x_i}\ x_i \qquad\qquad (2.9)$$

where C_m, C_s, C_h, C_2, C_1 and C are constants and Φ_i, Ψ_i, x_i are weight functions depending on thermal stability through dimensionless numbers similar to Richardson numbers. The main effect of these functions is to decrease the subgrid scale kinetic energy and the ratio of vertical to horizontal exchanges of heat and moisture when thermal stability increases. This formulation allows for example counter-gradient heat fluxes in slightly stable layers.

The above parameterization needs to be completed by specific hypothesis concerning the statistical distribution of condensed water within grid volumes. Generally cloud models contain the assumption that a given grid volume is either entirely saturated or entirely unsaturated. Even for small grid volumes such as those used in boundary layer models of the order of $(100m)^3$, this approximation appears to be rather crude. This is all the more the case for deep convection models with grid sizes of the order of $(1km)^3$. Mellor (1977), Manton and Cotton (1977), Sommeria and Deardorff (1977) have developed methods to solve this problem, by using assumptions for the distributions of variables within grid volumes.

Although part of the drawbacks mentioned above are solved by the use of quasi-conservative variables, hypothesis on the distribution of water substance within grid volumes are needed to determine for example the liquid water content of the grid volume, subgrid-scale buoyancy fluxes and longwave radiative fluxes. Among the results presented below, some of them show the importance of taking into account this "subgrid condensation". The parameterization used here, which is not detailed, follows Sommeria and Deardorff (1977) assuming gaussian distributions for quasi-conservative variables within grid volumes. The variances of these distributions are directly obtained from the second order moment equations presented above (equations (2.5), (2.6) and (2.8)).

3.- OTHER GENERAL FEATURES AND NUMERICAL METHODS

The domain of integration is 40 km in both horizontal directions and 16 km in the vertical, with three coordinate directions. Each grid volume has the following dimensions : $\Delta x = \Delta y = 1000$ m and $\Delta z = 400$ m. A staggered grid is used in which the wind velocities are defined at the sides of a grid cell and all the others variables at its center. The centered-difference scheme in space and the second-order leap-frog scheme in time are used to integrate all prognostic equations. The time step is determined from the linear stability criterion and in the present simulations takes values of 10 or 20 seconds. The pressure equation is solved by fast Fourier transform in the horizontal directions and by linear recursion in the vertical. At the top and bottom of the domain, the vertical gradients of all quantities and the vertical velocity are assumed to vanish and the lateral boundary conditions are periodic. These features have been chosen for simplicity in the present case, they should be revised, especially the bottom boundary condition, in the simulation of a real case.

4.- PRESENTATION OF ONE STUDY CASE

(a) Initial conditions

The simulation presented here is designed to test the various characteristics of the convection model, especially those related to the subgrid turbulence parameterization. For this reason a rather idealized case has been chosen, with a clear distinction between the various layers, and which allows a sufficient spatial resolution of the physical features of one convective cell. The initial temperature and moisture profiles are depicted on figure 1. A mixed layer homogenous in potential temperature and relative humidity is assumed to extend from the ground up to a height of 2000 m and to be surmounted by a conditionally unstable layer up to 10000 m. To initiate convection, a bubble-shaped temperature perturbation $\Delta\Theta$ is introduced in the center of the domain with the following characteristics :

$$\Delta\Theta = \begin{cases} 1,5 \cos^2\left(\dfrac{\pi r}{2}\right) & \text{, for } r<1 \\ 0 & \text{, elsewhere} \end{cases} \qquad (3.1)$$

$$\text{where } r = \left[\left(\frac{x-x_c}{x_r}\right)^2 + \left(\frac{y-y_c}{y_r}\right)^2 + \left(\frac{z-z_c}{z_r}\right)^2\right]^{1/2}$$

with $x_c = y_c = 20$ km ; $z_c = z_r = 2$ km ; $y_r = x_r = 17$ km.

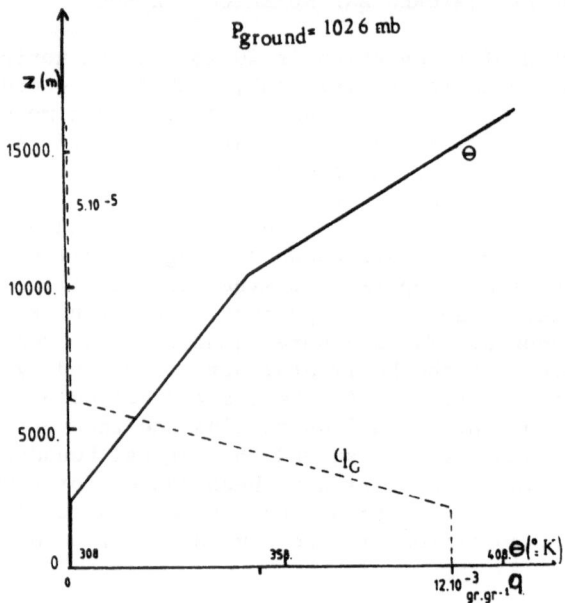

Figure 1.-   Initial profiles for potential temperature and
              specific humidity.

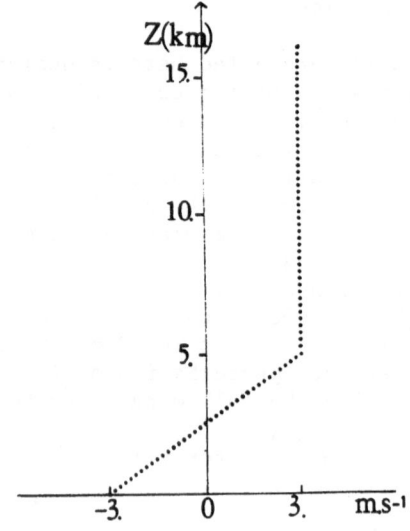

Figure 2.-   Initial profile of the wind component along O_x.

In all simulations presented here, the initial wind
profile along the direction O_x (Figure 2) presents
a weak horizontal shear of 6 m.s^{-1} accross the first 4000 m
of the domain.

There is no initial wind in the y-direction.

(b) Results of the basic simulation

Initial conditions presented above allow a rather fast
development of one convective cell which can be adequately
resolved by the model network. An ascending current starts at
the location of the temperature disturbance and a cloud
develops after 7 minutes. At 36 min model time, the cloud
reaches a mature stage with a vertical extension of 4800 meters
and ascending velocities up to 15 m.s^{-1}. This is also the
time for which rain reaches the ground. A maximum value of
25 m.s^{-1} for the ascending velocity is reached at 41 min.
in the upper part of the cloud as convective activity is
already weaker near cloud base (Figure 3).

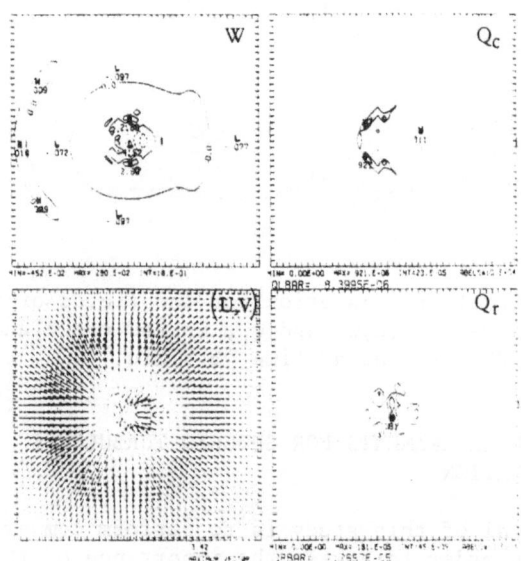

Figure 3.- Horizontal cross-section for the fields of w, q_c,
      horizontal wind and q_r, at level 2.6 km and
      time 41.6 min. Solid and dotted lines for w
      represent respectively positive and negative values.
      Intervals between isopleths are denoted INT in
      MKS units.

Accumulation of rain water in the lower part of the
cloud progressively turns part of the ascending current into
a descending flow. It however does not suppress the low level
moisture inflow because of the effect of the wind shear. The
ascending current is then divided into two parts (Figure 3) with
the occurrence of two maxima for ascending velocity and cloud
water around the precipitating area. At time 48 min., a vertical
cross-section of the various fields (Figure 4) clearly shows the
formation of two cells as previously observed by Klemp and
Wilhelmson (1978) in a shear situation.

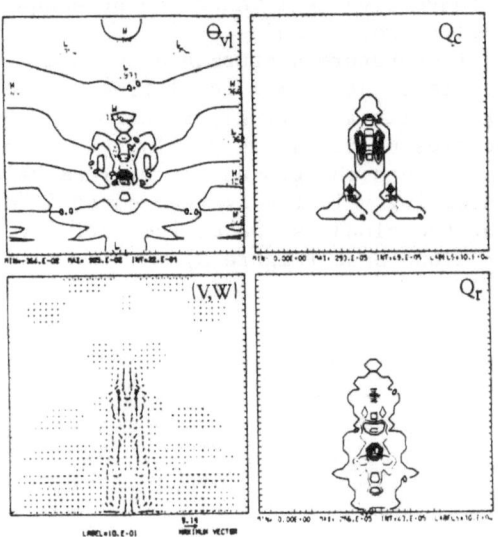

Figure 4.- Vertical cross-section for the fields of Θ_{vl}, q_c,
           wind vector (u,w) and q_r, in the y-z plane for
           x = 23,5 km and at time 48,5 min.

5.- COMPARATIVE EXPERIMENTS FOR SUBGRID TURBULENCE
PARAMETERIZATION

The goal of this study is to perform comparative
simulations, in order to assess the importance of the various
features of the subgrid parameterization scheme. These simulations are also designed to allow comparisons with previous
studies by Klemp and Wilhelmson (1978) and Clark (1979).

The run previously described, later designated as
Run A, is considered as the basic run, containing the whole
subgrid parameterization scheme. Two other numerical experiments

were performed with the same set of initial conditions but
with simplified features for subgrid turbulence : in Run B the
subgrid condensation scheme has been eliminated, in Run C both
subgrid condensation and the weight functions ψ and ϕ for the
turbulent fluxes of q_w and Θ_1 have been eliminated (replaced by 1).

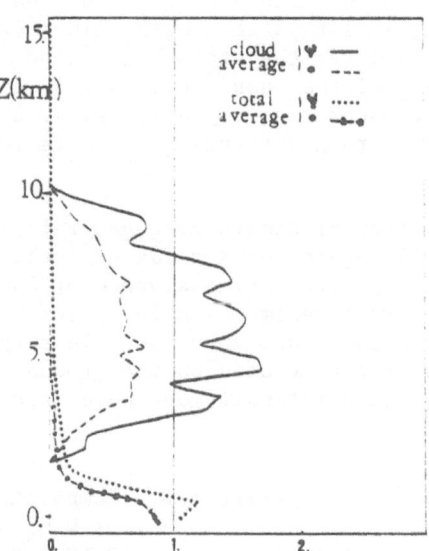

Figure 5.- Weight functions ψ and ϕ. (5 min average from
           42. to 47. min).

        Differences between runs B and C are illustrated on
Figure 5 where weight functions ψ and ϕ for Run B averaged over
the model area and averaged over the cloudy domain are presented.
One notices the decrease with height of the domain-averaged
weight functions from about one in the mixed layer to small
values of the order of 1. In Run C, weight functions are assumed
to keep a uniform value of 1.

        For the set of initial conditions presented above, all
runs show qualitatively similar general features ; however major
differences occur in the magnitude of the various parameters
describing the storm intensity.

        The temporal evolution of the maximum and minimum values
of vertical velocity throughout the whole domain (Figure 6) is
a good indicator of the intensity of convection. In the basic
Run A, W_{max} presents two peaks at time 2450 s (25.6 m.s^{-1}) and

2750 s (21.3 m.s^{-1}) during the mature stage of the storm. It
then decreases to values of the order of 8 m.s^{-1} during the
decaying stage. The maxima in runs B and C occur slightly
earlier (at around 2000 s model time) and reach values of the
order of 10 m.s^{-1} only. Convection seems then to be weakened
in runs B and C, although the mature stage occurs earlier.
W_{min} presents a more irregular pattern with values of the order
of −10 m.s^{-1} in all runs after the storm has reached the mature
stage. At the beginning of the runs the location of W_{min} in the
domain corresponds to compensating dry downdrafts, as later
on it occurs within the strong descending currents induced
by precipitations.

Temporal evolution of domain-averaged rain intensity
at the ground (Figure 7), again shows a faster evolution in
runs B and C compared to A with a time advance of about 200 s
in the occurence of the rain maximum. Rain intensity reaches
similar values in runs A and B and about 30 % less in run C.
The total amount of water accumulated on the ground is however
largest in run A, where rain intensity decreases slower after
its maximum.

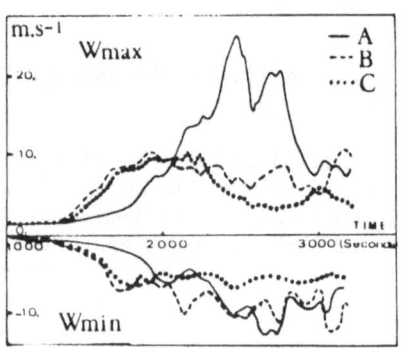

Figure 6.- Comparative evolution
of W_{max} and W_{min} for
RUNS A,B and C.

Figure 7.- Comparative evolution
of mean rain intensity
at the ground for
RUNS A,B and C.

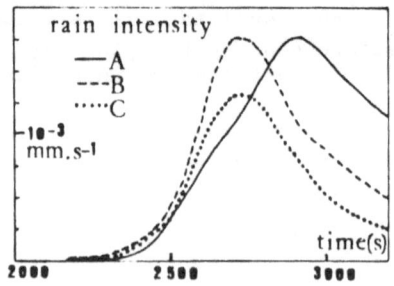

Similar remarks are valid for the comparison of the temporal evolution of total cloud volumes (Figure 8). The cloud development starts earlier in Run A and its volume is always larger than in Runs B and C. It is about 50 % larger at the initiation of the raining stage, this value increasing with time due to the much slower decay of the storm in Run A.

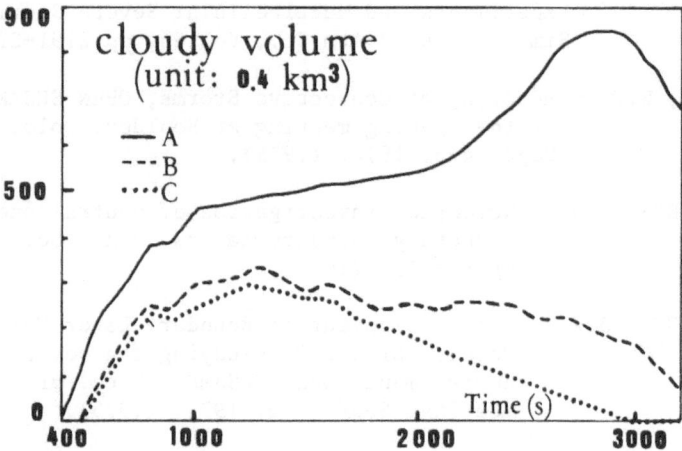

Figure 8.- Comparative evolution of total cloud volume for RUNS A, B and C.

Some of these results could already be expected from the conclusions of the study of subgrid condensation in a non-precipitating boundary layer by Sommeria and Deardorff (1977), which used a smaller grid size $(50m)^3$ and a smoother initialization (a weak random temperature perturbation at the first level above ground). The present comparison confirms that taking into account fractional cloud cover within grid volumes in a convection model tends to increase all parameters related to cloud activity. This increase is all the more noticeable as the grid size is larger, as shown by the present simulation with a $(1km)^2$ horizontal resolution. Concerning still the magnitude of the convection parameters, the difference between runs B and C is not as large as between runs A and the others ; this tends to show that the model is not as sensitive to the value of the weight functions used in the expression of subgrid fluxes as it is to subgrid condensation. The slower evolution of the convection cycle noticed in Run A compared to B and C may be related to the larger value of cloud edge entrainment in Run A. With the present "bubble type" initialization, the cloud is narrower when subgrid condensation is not included and does not mix as much with the environment ; this allows a faster development and evolution cycle.

REFERENCES

BETTS, A.K. - Non precipitating cumulus convection and its
              parameterization. Quart. J.Roy.Met.Soc.,
              Vol.99, pp.178-196 (1973).

CLARK, T.L. - Numerical simulations with a three-dimensional
              cloud model : Lateral Boundary Condition
              Experiments and Multicellular Severe Storm
              Simulations. J.Atm.Sc., Vol.36, pp.2191-2215 (1979).

COTTON, W.R. - Modeling of Convective Storms, OPEN SESAME : Proc.
               of the opening meeting at Boulder, Colo.,
               Sept. 4-6, 1974. (1975).

DEARDORFF, J.W. - Numerical investigation of neutral and unstable
                  planetary boundary layers. J.Atm.Sc., Vol.29,
                  pp.91-115. (1972).

DEARDORFF, J.W. - The development of Boundary Layer Turbulence
                  Models for use in studying the severe storm
                  environment, OPEN SESAME : Proc. of the opening
                  meeting, Sept. 4-6, 1974. (1975).

DEARDORFF, J.W. - "Strato-cumulus capped mixed layers derived from
                  a three-dimensional model". Boundary-Layer
                  Met., 18, pp.495-527. (1980)

KESSLER, E. - On the distribution and continuity of water sub-
              stance in atmospheric circulation. Meteor.Monogr.,
              10, n°32,p.84. (1969)

KLEMP, J.B. and R.B. WILHELMSON - The simulation of three-
              dimensional convective storm dynamics. J.Atm.Sc.,
              15, pp.1070-1096. (1978)

LILLY, D.K. - The representation of small-scale turbulence in
              numerical simulation experiments. Proc.I.B.M.
              Computing Symposium on Environmental Sciences
              IBM for N° 320, 1951. (1967).

MANTON, M.J. and W.R. COTTON - A consistent set of approximate
              equations for moist deep convection. Part II.
              Submitted to J.Atmos.Sci. (1977).

MELLOR, G.L. - The gaussian cloud model relations. J. Atmos.Sci.,
               34, pp.356-358. (1977).

REDELSPERGER,J.L. and G. SOMMERIA - Méthode de représentation
                de la turbulence d'échelle inférieure à la
                maille pour un modèle tri-dimensionnel de
                convection nuageuse. Boundary-Layer Met., 21,
                pp.509-530. (1981).

REDELSPERGER,J.L. and G. SOMMERIA - Méthode de représentation
                de la turbulence associée aux précipitations
                dans un modèle tri-dimensionnel de convection
                nuageuse. Boundary-Layer Met. (to appear) (1982).

SMAGORINSKY, J. - General Circulation experiments with the
                primitive equations : I - the basic experiment.
                Mon. Wea. Rev. 91, pp. 99-164. (1963).

SOMMERIA, G. - Three-dimensional simulation of turbulent process
                in an undisturbed trade wind boundary layer,
                J.Atm.Sc. Vol 33, pp. 216-241. (1976).

SOMMERIA, G. and J.W. DEARDORFF - Subgrid scale condensation in
                models of non precipitating clouds. J.Atm.Sc.,
                Vol.34, pp.344-355. (1977).

Dr. M. T. Abshaev
High-Mountain Geophysical Institute
Nalchik-2, VGI, USSR

Professor Ernest M. Agee
Department of Geosciences
Purdue University
West Lafayette, Indiana 47907 USA

Professor Tomio Asai
Ocean Research Institute
University of Tokyo
1-15-1, Minamidai, Nakano-Ku
Tokyo 164 JAPAN

Dr. R. Auria
Universite de Clermont
Centre de Recherches Atmospheriques
Campistrous - Cidex B 47, 65300
Lannemezan, FRANCE

Dr. M. J. Badar
Meteorological Office
Bracknell, UNITED KINGDOM

Dr. S. Bakan
Max-Planck-Institut fur Meteorologie
Bundesstrasse 55
D-2000 Hamburg 13 WEST GERMANY

Professor Louis J. Battan
Institute of Atmospheric Physics
The University of Arizona
Tucson, Arizona 85721 USA

Dr. D. A. Bennetts
Meteorological Office
Bracknell, UNITED KINGDOM

Dr. Alan K. Betts
West Pawlet, Vermont 05775 USA

Dr. N. Sh. Bibilashvili
High-Mountain Geophysical Institute
Nalchik-2, VGI, USSR

Professor R. R. Braham, Jr.
Dept. of Geophysical Sciences
University of Chicago
5734 S. Ellis Avenue
Chicago, IL 60637 USA

Dr. V. N. Bringi
Atmospheric Sciences Program
The Ohio State University
Columbus, Ohio 43210 USA

Dr. B. Campistron
Universite de Clermont
Centre de Recherches Atmospheriques
Campistrous - Cidex B 47, 65300
Lannemezan, FRANCE

Dr. M. T. Chahine, Manager
Earth and Space Sciences Division
Jet Propulsion Laboratory
California Institute of Technology
4800 Oak Grove Drive
Pasadena, California 91103 USA

Dr. Charles F. Chappell
Office of Weather and Research
  and Modification
U.S. Department of Commerce - NOAA
Environmental Research Laboratories
Boulder, Colorado USA

Dr. Christopher R. Church
Department of Aeronautics
Miami University
Oxford, Ohio 45056 USA

Professor Mladjen Curic
Institute of Meteorology
University of Belgrade
Dorbacina 16, P.O.B. 550
11000 Beograd, YUGOSLAVIA

Professor Jean Dessens
Universite de Clermont-Ferrand II
Centre de Recherches Atmospheriques
Campistrous - Cidex B 47, 65300
Lannemezan, FRANCE

Professor Kerry A. Emanuel
Department of Meteorology
Massachusetts Institute of Technology
Cambridge, Massachusetts 02139 USA

Dr. Y. Fouquart
Laboratorie de Meteorologie
  Dynamique de CNRS
24, rue Lhomond 75231
Paris CEDEX 05 FRANCE

Dr. G. Held
Atmospheric Sciences Division
Council for Scientific and
  Industrial Research
National Physical Research Laboratory
P.O. Box 395
Pretoria 0001 SOUTH AFRICA

Dr. L. Ray Hoxit
USDC - NOAA
Environmental Data and Information Service
National Climatic Center
Federal Building
Asheville, North Carolina 28801 USA

Professor R. J. Hung
School of Science and Engineering
University of Alabama in Huntsville
P.O. Box 1247
Huntsville, Alabama 35807 USA

Dr. L. G. Kachurin
Leningrad Hidrometeorological Institute
Malookhtinsky Pr., 98
Leningrad, I95196 USSR

Dr. R. D. Kelly
Dept. of Geophysical Sciences
University of Chicago
5734 S. Ellis Avenue
Chicago, IL 60637 USA

Dr. Joe Klemp
NCAR
P.O. Box 3000
Boulder, Colorado 80307 USA

Professor Yeong-Jer Lin
Department of Earth and Atmospheric
  Sciences
Saint Louis University
3507 Laclede Avenue
Saint Louis, Missouri 63103 USA

Dr. R. Pasken
Department of Earth and Atmospheric
  Sciences
Saint Louis University
3507 Laclede Avenue
Saint Louis, Missouri 63103 USA

Dr. J. R. Peterson
Atmospheric Sciences Program
The Ohio State University
Columbus, Ohio 43210 USA

Dr. P. S. Ray, Chief
Meteorological Research Group
National Severe Storms Laboratory
1313 Halley Circle
Norman, Oklahoma 73069

Dr. J. L. Redelsperger
Laboratorie de Meteorologie
  Dynamique du CNRS
24, rue Lhomond 75231
Paris CEDEX 05 FRANCE

Dr. Henri Sauvageot
Universite de Clermont
Centre de Recherches Atmospheriques
Campistorus - Cidex B 47, 65300
Lannemezan, FRANCE

Professor Thomas A. Seliga, Director
Atmospheric Sciences Program
The Ohio State University
Columbus, Ohio 43210 USA

Dr. Hampton N. Shirer
Department of Meteorology
The Pennsylvania State University
University Park, Pennsylvania 16802

Dr. N. S. Shishkin
Voeikov Main Geophysical Observatory
Leningrad, USSR

Dr. R. E. Smith
School of Science and Engineering
University of Alabama in Huntsville
P.O. Box 1247
Huntsville, Alabama 35807 USA

Dr. G. Sommeria
Laboratorie de Meteorologie
  Dynamique du CNRS
24, rue Lhomond 75231
Paris CEDEX 05 FRANCE

Dr. Wen-Yih Sun
Department of Geosciences
Purdue University
West Lafayette, Indiana 47907 USA

Professor Owen E. Thompson
Department of Meteorology
University of Maryland
College Park, Maryland 20742 USA

Dr. Ph. Veyre
Laboratorie de Meteorologie
  Dynamique du CNRS
24, rue Lhomond 75231
Paris CEDEX 05 FRANCE

Professor K. J. Weston
Department of Meteorology
The University of Edinburgh
James Clerk Maxwell Building
King's Buildings
Mayfield Road
Edinburgh EH9 3JZ UNITED KINGDOM

Dr. Bob Wilhelmson
Meteorological Research Group
National Severe Storms Laboratory
1313 Halley Circle
Norman, Oklahoma 73069